AI Meets Strategy

A Product Manager's Guide to Leading Innovation

Anshuman Srivastava
Abhinav Garg
Anshuman Mishra

Apress®

AI Meets Strategy: A Product Manager's Guide to Leading Innovation

Anshuman Srivastava
Noida, Uttar Pradesh, India

Abhinav Garg
Hyderabad, Telangana, India

Anshuman Mishra
Gurgaon, Haryana, India

ISBN-13 (pbk): 979-8-8688-2138-7   ISBN-13 (electronic): 979-8-8688-2139-4
https://doi.org/10.1007/979-8-8688-2139-4

Copyright © 2026 by Anshuman Srivastava, Abhinav Garg, Anshuman Mishra

This work is subject to copyright. All rights are reserved by the Publisher, whether the whole or part of the material is concerned, specifically the rights of translation, reprinting, reuse of illustrations, recitation, broadcasting, reproduction on microfilms or in any other physical way, and transmission or information storage and retrieval, electronic adaptation, computer software, or by similar or dissimilar methodology now known or hereafter developed.

Trademarked names, logos, and images may appear in this book. Rather than use a trademark symbol with every occurrence of a trademarked name, logo, or image we use the names, logos, and images only in an editorial fashion and to the benefit of the trademark owner, with no intention of infringement of the trademark.

The use in this publication of trade names, trademarks, service marks, and similar terms, even if they are not identified as such, is not to be taken as an expression of opinion as to whether or not they are subject to proprietary rights.

While the advice and information in this book are believed to be true and accurate at the date of publication, neither the authors nor the editors nor the publisher can accept any legal responsibility for any errors or omissions that may be made. The publisher makes no warranty, express or implied, with respect to the material contained herein.

> Managing Director, Apress Media LLC: Welmoed Spahr
> Acquisitions Editor: Shivangi Ramachandran
> Development Editor: James Markham
> Project Manager: Jessica Vakili

Distributed to the book trade worldwide by Springer Science+Business Media New York, 1 New York Plaza, New York, NY 10004. Phone 1-800-SPRINGER, fax (201) 348-4505, e-mail orders-ny@springer-sbm.com, or visit www.springeronline.com. Apress Media, LLC is a Delaware LLC and the sole member (owner) is Springer Science + Business Media Finance Inc (SSBM Finance Inc). SSBM Finance Inc is a **Delaware** corporation.

For information on translations, please e-mail booktranslations@springernature.com; for reprint, paperback, or audio rights, please e-mail bookpermissions@springernature.com.

Apress titles may be purchased in bulk for academic, corporate, or promotional use. eBook versions and licenses are also available for most titles. For more information, reference our Print and eBook Bulk Sales web page at http://www.apress.com/bulk-sales.

If disposing of this product, please recycle the paper

Table of Contents

About the Authors xiii

Part I: AI: Foundations and Strategy 1

Chapter 1: The AI Makeover: How Product Management Is Evolving 3
   From Intuition to Intelligence: The Rise of AI-Driven Product Management 4
   Showcasing the Role of AI/ML in Creating Outstanding Products 6
      Uber 7
      Grammarly 13
      Meta 19
      Amazon Marketplace 23
   All Future Product Managers Will Be AI Product Managers 26
      The Future of Product Management in an AI-Driven World 28
      Evolving Expectations from Product Leadership Roles in an AI-Driven Era ... 29
      Building AI Fluency: Key Skills for Future-Ready Product Managers 32
   Summary 35

Chapter 2: Foundations of AI: Understanding Key Concepts and Terminology 37
   From Cradle to Cognition: The Evolution of Human Intelligence 38
      Decoding Human Intelligence: Foundations for Building Artificial Minds 42
   Thinking Machines: A Philosophical and Scientific Introduction to AI 46
      The Necessity of Intelligence: What Drove AI's Creation 48
      The Father of AI: Turing's Legacy of Thinking Machines 49

TABLE OF CONTENTS

    Human Needs, Intelligent Systems: What Every Product Leader Should Know 50

    Foundational Paradigms of AI: Symbolic, Connectionist, and Emergent Approaches 51

    Bridging Minds and Machines: Cognitive and Metacognitive Insights for AI.....57

  Data, Information, Knowledge, Wisdom (DIKW): Foundations for AI Thinking 61

    The DIKW Framework: A Conceptual Model of Intelligence...........................61

    DIKW as a Model of Human Reasoning 63

    Turning Data into Action: Architecting Scalable AI Using DIKW 65

  Summary...68

Chapter 3: Building a Truly AI-Powered Organization: Strategy, Integration, and Transformation 69

  From Hype to Reality: Navigating AI's Adoption Challenges................................70

    Why AI Initiatives Underperform 71

    The Path Forward: What Product Leaders Must Do 76

  Why AI Products Are Different..80

    Uncertainty Is Built into the Product..80

    Behavior: Static vs. Evolving..80

    Continuous Learning Replaces Static Releases 81

    The Business Case Is Less Predictable 81

    Value Creation Shifts from "Only" Code to "Data and Code" 82

    User Trust: Predictability vs. Explainability 82

    Dealing with Deterministic vs. Opaque Failures...82

    UX, Brand, and Trust Are Intertwined..83

    Ethics and Bias: Optional vs. Essential...83

    AI Demands a Product Culture Shift 83

    Product Management Best Practices for AI Products....................................84

TABLE OF CONTENTS

From Pilots to Platforms: Navigating AI Maturity 86
    Why AI Maturity Assessment Matters 87
    AI Maturity Scoring and Evaluation Framework 91

Building an Enterprise AI Strategy Framework 101
    What Is a Good AI Strategy? 102
    What Is a Bad AI Strategy? 103
    Strategic, Actionable AI Strategy Framework: A Six-Layer AI Strategy Framework 104

Structured Approach to AI Use Case Selection 121
    What Is a Bad AI Use Case? 121
    Mental Models That Undermine AI Success 124
    Techniques for Spotting High-Impact, Feasible AI Opportunities 127
    AI Use Case Feasibility Study Framework 130

Defining Actionable AI Problem Statements 135

From Strategy to ROI: Financial Evaluation of AI Initiatives 138
    What Is Financial Viability in the AI Context? 138
    Why Financial Viability Matters Strategically 139
    Economic Fundamentals Behind AI Use Case Viability 141
    Assessing Financial Viability of AI Projects Through Key Financial Metrics 142

Operationalizing AI: Managing Change for Enterprise Adoption 147
    Before You Adopt: What AI Will Change in Your Organization 147
    Actionable Framework: AI Change Management Roadmap 149

Summary 152

TABLE OF CONTENTS

Part II: Building the Data Platform for AI Success 153

Chapter 4: From Strategy to Execution: Building a Data Operating Model for AI-Ready Growth 155

What Is a Data Operating Model? 156

The Relationship Between a Data Operating Model and a Data Platform 156

Introducing the EDGE Data Operating Model: End-to-End Data for Growth and Enablement 158

    EDGE Implementation Stack: Structure, Flow, Governance 161

    Business Context Layer 161

    Analysis Planning Layer 163

    Data Collection Layer 166

Data Centralization layer 173

    Ingestion Layer 173

    Curated Data Layer 175

    Data Activation Layer 185

    Data-Driven Growth Layer 185

Summary 186

Chapter 5: Mastering Data Quality: Techniques and Best Practices for Product Managers 187

Why Product Leaders Must Care About Data Quality 188

Introduction to Data Lineage and Quality in Modern Data Architectures 191

    What Is Data Lineage? 191

    What Is Data Lineage (and Why Does It Matter to PMs)? 192

    How Data Lineage Enhances Data Quality, Transparency, and Governance 193

    Key Challenges in Managing Data Lineage and Quality at Scale 194

Key Dimensions of Data Quality 196

Data Quality: Definition and Key Dimensions 196

    What Is Data Quality? 197

TABLE OF CONTENTS

Key Dimensions of Data Quality 197
Data Quality Management Framework: The PROMT Framework 210
    The PROMT Framework Components Explained 211
    Designing SLAs for Data Quality 224
Measuring Data Quality: Practical Approaches 226
Profiling and Quality Checks in Large Datasets 227
Data Quality Challenges in Unstructured and Semi-Structured Data 228
Data Quality for AI/ML Workflows: Ensuring Data Quality for
Model Training and Testing 233
Data Quality for AI/ML Workflows Implementing Quality Gates in
AI/ML Model Development Pipelines 235
    Schema Validation Gate 236
    Data Distribution Gate 236
    Label Quality Gate 237
    Bias and Fairness Gate 237
    Model Readiness Gate 238
    Mapping Data Quality Gates to ML Lifecycle Stages 238
Summary 244

Part III: Navigating the ML Landscape: Essential Knowledge for AI Product Leaders 247

Chapter 6: Introduction to Machine Learning for Teaching Systems 249

Challenges That Detail AI Implementation 251
Designing Successful AI/ML Systems 255
    From Pitfalls to Best Practices: Smarter AI Implementation Planning 256
    Designing AI Implementation Plan That Succeed: Leason Learned 263
Types of Machine Learning Algorithms 274
    Classical Machine Learning Algorithms 274

vii

TABLE OF CONTENTS

    Handling Real-World Data Situations 289

    Deep Learning 293

    Decision Framework: When to Use Deep Learning vs. Classical ML 306

    GenAI 306

Summary...310

Chapter 7: From Prototype to Production: Scaling ML at Speed and Scale 311

    From Prototype to Production: Why Product Leaders Need MLOps for AI Product Success 312

    MLOps Framework..314

        Model Deployment (De) 315

        Model Monitoring and Maintenance (Mo)...334

        MLOps Governance (G) 354

        Model Testing (T) 356

        Model Deployment Infrastructure...371

        Summary...378

Chapter 8: The Ethical Framework: Building Trust in AI Through Human Values 381

    How Humans Trust 384

    Applying Human Trust Factors to AI Systems..387

    Framework for Trust in AI..388

    Responsible AI..389

        AI Ethics Principle: Fairness 391

        AI Ethics Principle: Inclusiveness 409

        AI Ethics Principle: Safety 417

        AI Ethics Principle: Accountability 426

AI Governance 435
    Why It Is Relevant for Product Leaders 436
    AI Governance Model 437
AI Risk Management 439
    Types of AI Risks 442
    Risk Assessment and Mitigation in AI 446
Robust AI 452
    Design Principles for Robust AI 453
Summary 456

Part IV: AI in the Real World 457

Chapter 9: Designing AI-Powered Products 459

Unlocking AI's Value in Product Design: Benefits and Designer Strategies 460
Why Product Managers Should Care About AI System Design 461
Core Techniques and Patterns for Designing AI Products 463
    AI Interaction Patterns: Personalization and Recommendation Systems 465
    Balancing Personalization with Usability 476
Navigating AI Interaction Design: Common Pitfalls and Best Practices 477
Practical Framework: Balancing AI Personalization and Privacy 482
Designing Accessible AI-Powered Personalization 487
Building for Trust and Ethics 490
    Ensuring Transparency in AI-Driven Recommendations and Decisions 490
    Communicating AI Behavior to Users: Clarity and Trust-Building 493
    Ensuring Ethical AI in Personalized Experiences 497
Summary 502

TABLE OF CONTENTS

Chapter 10: Putting AI to Work: Case Studies in Applied Intelligence 503

List of Case Studies 504

Case Study Format and Team Roles 505

Case Study 1: AI Model for Human Exercise Recognition Using Computer Vision 506

    Business Use Case 506

    Proposed Solution 506

    Solution Approach 506

    Model Development 508

    Model Performance 509

    Deployment 510

    Challenges and Limitations 511

    Business Impact 512

Case Study 2: Optimizing Healthcare Triage Through a GenAI Chat Bot 512

    Executive Summary 512

    Solution Approach 513

    Model Development 513

    Model Performance 514

    Deployment 515

    Monitoring 516

    Metrics 516

    Challenges and Limitations 517

    Business Impact 517

Case Study 3: Dynamic Pricing Based on Demand Prediction by the AI Model 518

    Executive Summary 518

    Solution Approach 518

TABLE OF CONTENTS

Model Development 520
Model Performance 520
Deployment 521
Monitoring 522
Metrics 524
Challenges and Limitations 525
Business Impact 526

Case Study 4: Optimizing Ads Recommendation on Search Results Page for E-commerce Platform 526
Executive Summary 526
Solution Approach 527
Model Development 529
Model Performance 530
Deployment 530
Monitoring 532
Metrics 535
Challenges and Limitations 536
Business Outcome 536

Summary 537

Chapter 11: Executive Perspectives: Navigating AI, Strategy, and the Future 539

Interview 1: Chief Digital Officer: Harmeen Mehta 541
  Strategizing AI at Scale: A Conversation with Harmeen 542
Interview 2: Managing Director, Consumer Digital: Harry Singh 553
  Harry on AI: Strategy, Risk, and Reinvention 553
Interview 3: Head of Innovation: Abhishek Singh 567
  Abhishek's Take: How AI Is Redefining Innovation Playbooks 568
Interview 4: EVP of Product Management: Nimish Kulshrestha 574

TABLE OF CONTENTS

  AI-Driven Product Thinking: Insights from Nimish 575
 Interview 5: Big 4 Consulting Firm Partner AI Practice: Vishal Agarwal 595
 Interview 6: Engineering Leader: Himaanshu Gupta 605
  Building AI-Ready Infrastructure: Himaanshu Gupta's Perspective 606
 Interview 7: Data Product Manager: Peeyush Panthari 611
  Building Data-Driven Products: Inside the Role of a Data Product Manager 611
 Authors' Perspective on Architecting Agentic AI 621
  Building Blocks of Agentic AI Systems 623
  Ethics and Agentic AI 626
 Summary 628

References **629**

Index **635**

About the Authors

Anshuman Srivastava is a product and AI leader with over 12 years of experience driving enterprise scale platforms at the intersection of business strategy and technology execution. Having spent most of his career in AI across both strategic and hands on implementation roles, he specializes in building and scaling enterprise-grade, AI-driven platforms that solve real-world business problems across core digital systems underpinning revenue, scale, and customer value. He is recognized for translating AI strategy into measurable outcomes, helping enterprises scale innovation with production-grade governance, reliability, and performance.

Abhinav Garg brings 15 years of expertise in AI and data-driven solutions, spanning research, startups, and global e-commerce like Tokopedia. Combining technical depth with real-world impact, he turns advanced machine learning and computer vision innovations into practical, scalable solutions that solve complex business challenges.

ABOUT THE AUTHORS

Anshuman Mishra is an entrepreneurial product and digital leader with 20 years of experience driving large-scale transformations and building enterprise B2C and B2B/SaaS products. An early adopter of product management in India, he has collaborated with global product leaders and data experts to scale organizations, unlock growth, and deliver exceptional experiences to millions of users.

PART I

AI: Foundations and Strategy

CHAPTER 1

The AI Makeover: How Product Management Is Evolving

> *"In any given moment we have two options to step forward into growth or step back into safety."*
> —*Abraham Maslow*

Artificial intelligence (AI) is no longer just hype; it has become the new operating system on which businesses will build their moats, define their key differentiations, and shape their long-term competitiveness. As AI's adoption increases, most organizations have realized a successful AI implementation isn't just about sophisticated algorithms; it is equally about how well AI initiatives are aligned with the organization's strategy. As product leaders are often considered the thought leaders in how technology can be used to best suit the business interests, they are now at the forefront of this transformation.

This chapter will help product leaders understand how AI is fundamentally changing product management as a trade. We aim to make product leaders realize that today the question is not *if* AI will affect their product strategy but how ready they are to lead it. And leading AI doesn't mean simply using AI for innovation; it means embedding it thoughtfully into the very core of your organization's strategy and product offerings.

CHAPTER 1   THE AI MAKEOVER: HOW PRODUCT MANAGEMENT IS EVOLVING

This requires identifying the right user pain points where AI will make an impact, evaluating the implementation from strategic alignment perspective, ROI perspective, and organization's readiness point of view, and then supporting each AI initiative by providing right technical and organizational support required.

When AI is treated as a side hustle or a late-stage add-on or a tech-led initiative, it rarely delivers meaningful outcomes. This is because AI on its own does not create any value. It delivers value only when supported by the right structures, context, and deep understanding of customer problems. Just like salt enhances the flavor of a dish without becoming the dish itself, AI must be applied with intention. Too much of it might overcomplicate or misguide, and too little may lead to missed opportunities.

This chapter will highlight how product leaders and managers must evolve to gauge where, when, and how much AI to add to enhance a product's flavor without losing its soul.

From Intuition to Intelligence: The Rise of AI-Driven Product Management

If you ask us what is most successful products have in common, you would hear it is the ability to build a technology-based solution to an unfulfilled customer need. From the invention of the telegram and telephones to modern-day instant messaging apps, these successful innovations address a human need to make communication faster and easier. These solutions have become successful because cutting-edge technology was used to meet a customer's need. Currently we are in middle of a technological revolution that is unlocking opportunities for lasting, impactful solutions. This revolution is the rise of AI. Much like previous industrial revolutions, AI is set to change the way the world functions, each revolution building upon the success of the last.

CHAPTER 1   THE AI MAKEOVER: HOW PRODUCT MANAGEMENT IS EVOLVING

The current revolution also called the Industrial Revolution 4 and will revolutionize the way decisions are made and executed. Decisions will now be more fact based as this revolution will help generate more accurate data by enabling the capture of subtransactional data (IOT and other forms of digitization). AI will make decision-making easier. In some cases, agentic AI can also help act based on the recommendation made. This era is reshaping industries and changing how products are conceptualized, developed, and delivered.

To thrive in this new world powered by data, product managers must embrace new ways of thinking and operating. Product management as a discipline is about creating a balance between creativity and pragmatism. As product leaders, we must put into use our understanding of technology to solve and understand user needs and build products accordingly while acting as the innovation beacon for the organization. In other words, it is our responsibility to help organizations understand the changing technology landscape and to ensure that organizations stay ahead of the competition by putting the technology to use in driving innovation.

Product managers are now required to be strategic thinking leaders who cannot just ship great products but also play a significant role in defining an organization's strategy and technology adoption roadmap. They are also expected to be capable of enough of analyzing vast amounts of data and leveraging cutting-edge technologies to make informed product decisions while designing products powered by AI. In a nutshell, data has become the backbone of product strategy, and product managers must be comfortable analyzing real-time data, predicting future trends, and anticipating customer needs

Early in our careers we navigated the same terrain that today many peers find themselves in, relying on intuition, using minimal data in market research, and using good old-fashioned gut instinct based upon limited data points to shape products. And like many, we began to feel the weight of this approach as the pace of change accelerated. Data points were piling up faster than we could process them. User behavior was evolving at lightning

CHAPTER 1   THE AI MAKEOVER: HOW PRODUCT MANAGEMENT IS EVOLVING

speed, and no amount of brainstorming sessions or focus groups seemed to offer clarity. It was then our data background helped because we could put data science methods to use in designing and developing our products. We soon started observing that our products were more data-driven than others. Our comprehensive understanding of the data ecosystem enhanced our ability to process a large of variety of data at every step of product design and execution. It also enabled us to integrate AI into our product features, allowing us to achieve our product goals more efficiently.

As we reflect on our individual journeys today, we see AI has completely transformed the role of the product manager, not just in terms of what we do but in how we do it. In this book, we want to take you through that transformation. This book covers the key concepts and professional advice required for meeting the changing needs of the industry and transforming you into a truly AI-powered product manager.

In next section, we will explore how great products use AI in solving problems faced by users in their industry. The goal is to showcase how AI is overcoming previously unsolvable problems and utilizing its potential to develop truly outstanding products.

Showcasing the Role of AI/ML in Creating Outstanding Products

In this section we will be exploring four industry leading products and understand how these products are using AI to best serve customer needs. We will be exploring the following:

- Uber
- Grammarly
- Meta (Facebook)
- Amazon Marketplace

One might wonder why we chose these organizations over others. Well, the answer to this question is all these four organizations belong to very diverse industries: Uber is in the transportation industry, Facebook is social media, Amazon Marketplace is a classical ecommerce platform, and Grammarly is a prominent player in AI-powered writing assistance. These companies operate in different industries, and their business models are also different. What is common in these companies is the way data and AI are used extensively in their daily operations. All of these companies are AI-first companies. These organizations perfectly illustrate that, regardless of the industry, product leaders must integrate data and AI into all aspects of their offerings. And this also proves a point: the future belongs to those leaders who not only understand AI well but are also comfortable creating products that are AI powered.

Uber

Mission statement: "We reimagine the way the world moves for the better."[1]

Uber provides a convenient platform for on-demand transportation and ride services. It allows passengers to book rides and drivers to accept ride and get paid via a smartphone app. These drivers are freelance drivers with their own cars who use this app to use their car as taxi. Uber addresses the following use cases for its customers:

1. Uber allows users to get a confirmed and safe ride with a predefined fare within minutes, solving the problem of waiting for a taxi and thus removing the uncertainty associated with traditional car services especially in areas where taxis are difficult to get especially during peak times.

2. Uber shows the price up front while booking the app, which allows riders to make an informed choice and helps in reducing the confrontation with driver. Uber's dynamic pricing model changes the price according to the demand in that area. The model reduces the price during nonpeak seasons and increases the price during the peak season. This gives both riders and drivers the option to choose when they want to commute.

3. Drivers working for Uber are freelancers. Uber offers them the flexibility to offer rides at their own chosen times, making it attractive for people who are looking for supplemental incomes or people who prefer working part-time. On the platform level, this gives Uber the ability to have infinite supply of taxis ready to cater its riders.

Overall, Uber is solving the problem of inconvenient, expensive, and inefficient traditional transportation options by creating an accessible, reliable, and flexible transportation system. To achieve this, the Uber technology platform has to perform following primary tasks:

1. Match a ride with a willing driver.

2. Compute and recommend a price that is acceptable to both the rider and the driver while considering traffic and demand situations.

3. Ensure there is an ample supply of taxis during peak hours, recommending high-demand areas to drivers.

4. Recommend best and shortest available routes for rides.

CHAPTER 1   THE AI MAKEOVER: HOW PRODUCT MANAGEMENT IS EVOLVING

AI and data are integral to Uber's operations, with nearly every product or workflow incorporating some form of AI. Let's take a broader look at how AI is applied across key workflows at Uber.

1. **Expected time of arrival (ETA) prediction:**[2]

    One of the most crucial features of Uber is its expected time of arrival prediction for riders. It helps them plan their journey better. But predicting ETA in real time is a very daunting task that requires considering various forms of data in real time and using them for real-time prediction. This feature would lose its sheen if the ETA conveyed by Uber was not accurate. This makes the task of calculating ETA even more difficult. Uber relies heavily on sophisticated machine learning algorithms to predict this. These models analyze various factors such as time taken during historical rides at the same time and day, current traffic conditions and road conditions, and networks.

    Uber improves ETA predictions by using routing information such as map data and real-time traffic, road conditions, etc., and first predicts a base estimate. The system that predicts the base ETA is called the *routing engine*. Once a base estimate is calculated, then a deep learning (an advance field of algorithms which are primarily neural networks based) model adjusts it using real-world patterns. This model is called DeepETA, and it analyzes factors like location, time, traffic, and request type to refine accuracy. DeepETA uses encoder-decoder self-attention architecture (a deep learning algorithm) for this purpose.

2. **Marketplace forecasting:**[3]

Marketplace forecasting is a very important for Uber as it helps match demand (i.e., users waiting for a ride) with supply (drivers waiting). Forecasting marketplace demand helps in directing Uber drivers to high-demand areas before the actual demand surges, thereby helping both riders in getting a ride easily and drivers in increasing their trip count and earnings. The company has split the solution into three steps to solve the problem.

1. Estimates supply (i.e., number of cabs required)
2. Estimate demand (i.e., number of riders waiting for cab)
3. Estimate queue consumption rate

To ensure optimized utilization of cabs, Uber has to ensure adequate availability of cabs in high-demand areas such as airport/railway stations, etc. The way Uber tackles this is by forecasting available cabs and then using this to optimize its allocation. It is achieved by estimating expected time to request (ETR) by using a machine learning model for the cabs in queue. ETR is the expected time a driver must wait at that location before they receive a trip request. This helps drivers identify areas where demand is high and the supply of cabs is low and vice versa. To make this complex information simpler for consumption by drivers, Uber shows a heatmap indicating areas of high demand. Additionally, a price multiplier is showcased that informs drivers about how much more they will earn if they pick a ride from a high-demand area. Here a tree-based algorithm is used to predict the position of the last car in the queue every minute.

Uber uses past data to identify how many cabs are required at a location on a given date and time. Various deep learning models are used to predict rider demand in real time. Rider demand is very dynamic in nature as it depends on exogenous factors that can change any moment. To counter this dynamic nature, Uber employees use a tree-based approach (machine learning algorithm) to estimate rider demand.

In the last stage of the solution, the supply queue's last position and predicted riders demand is fed to another algorithm to estimate the final ETR, which helps drivers identify areas where demand is high. The model continuously evolves and recalculates the ETR as demand is met.

3. **Fraud detection:**[4]

   Uber is a global platform on which numerous rides are booked every minute. Uber follows a demand-supply match operating model, also called a *marketplace model*, and takes all the payment-associated risk. Hence, fraud detection is critical for maintaining trust and safety. Uber relies heavily on machine learning for fraud detection.

   Uber has created an in-house AI fraud detection and mitigation system with a human in the loop called RADAR.

RADAR is designed to detect early warning signs of fraud, and alerts are raised for a human fraud analyst to analyze the case and make a decision accordingly. Since RADAR is a machine learning model, it can analyze huge amounts of data and detect fraud patterns at a much faster rate. Human analyst are involved to investigate only in complicated situations.

4. **Customer Support:**[5]

    For a company like Uber, customer support is a crucial role in providing a good experience for their customers. Ensuring customer satisfaction is essential for maintaining good demand on platform. To facilitate the best customer service, Uber relies heavily on machine learning systems, utilizing an AI platform that supports customer agent communication channels. This AI platform is called COTA, and it uses NLP algorithms to help agents deliver better customer support. This how COTA functions:

    1. As soon as a customer creates a new ticket, all associated information associated with ticket is collect by using a back-end service, fed to a machine learning model.

    2. The model predicts scores for all probable solutions for this ticket. Each ticket agent receives a top resolution for each ticket. An agent can pick the best-suited resolution of these three.

COTA has delivered promising results by reducing ticket resolution time by 10% while delivering similar or higher customer satisfaction.

Grammarly

Mission statement: "To improve lives by improving communication."

To fulfill its mission of improving lives, by improving communication Grammarly empowers users to write effectively and clearly. It thus boosts the confidence of its users, while expressing their thoughts. But how does Grammarly achieve this? Grammarly is an advanced writing assistant that leverages AI for providing feedback on a user's writing. Internally Grammarly uses large language models (LLMs) to recommend suggestions. Launched in 2009, Grammarly is widely used by professionals, students, academics, and content creators to enhance the quality of writing across emails, documents, and online platforms. Grammarly provides the following services to its users:

- **Grammar and spelling corrections:** Grammarly detects and corrects grammar, spelling, and punctuation mistakes in real time.

- **Tone and clarity suggestions:** Grammarly analyzes tone and suggests improvements to make writing more formal, friendly, or professional.

- **Rewriting and paraphrasing:** Grammarly provides alternative sentence structures to improve readability and flow.

- **Plagiarism detection:** Grammarly compares text against a vast database to identify potential plagiarism.

- **Personalized writing insights:** Grammarly tracks writing patterns and gives personalized feedback on areas for improvement.

- **Autocorrect and predictive text:** Grammarly predicts words and phrases to speed up writing.

- **Multilingual writing assistance:** Grammarly helps non-native speakers improve their English writing.

- **Context-aware suggestions:** Grammarly understands the meaning behind sentences and offers relevant edits.

Grammarly serves as a classic example of how AI can drive both innovation and successful business. With its entire product offering built on AI, it is more than just a writing assistances tool. Grammarly demonstrates how AI can be the foundation of a thriving business model. Today Grammarly enjoys a unicorn status.

By leveraging AI for solving a common but latent problem faced by writers, Grammarly's technology-driven innovation has unlocked tremendous value not just for its creators but also for its millions of users around the world. Grammarly confirms the hypothesis that AI can unlock opportunities beyond automation, and this is precisely why we included Grammarly's journey as a case study.

For product leaders, Grammarly serves as a powerful reminder that AI is not just a supporting tool; it can be the core driver of a scalable, high-value business.

Let's take a broader look at how AI is applied across key workflows at Grammarly

1. **How Grammarly achieved superior accuracy in grammatical error correction:**[6]

    Writers trust in Grammarly's ability to improve their writing is the fundamental reason why they use this solution for verifying their most important work. It delivers consistent and reliable grammar, spelling, and other writing enhancement suggestions. Let's evaluate how Grammarly handles grammar error correction (GEC).

CHAPTER 1   THE AI MAKEOVER: HOW PRODUCT MANAGEMENT IS EVOLVING

1. To understand the grammar rules of a language, one must have a deep understanding of language usage and its rules. For this, Grammarly employs several linguistic experts who work together to create rules of the language. From a machine learning point of view, these linguistic experts help create domain knowledge for their machine learning engineering colleagues who use this knowledge to write complex machine learning algorithms.

2. Next a series of three AI models is created that helps generate recommendations.

    a. First a "sequence-to-sequence rewriting" model is developed. This is a large sequence-to-sequence machine translation model based on a deep learning architecture called Transformer. This model takes text input written by a user and produces error-free text. This model is augmented by a second model in the series.

    b. The second model in the series is called sequence tagging. This model tags specific errors in a sentence and then corrects each error separately. This method of error correction is different than the one used in step 1. The reason why two different approaches of generating error correction recommendations are combined is that the recommendations become very comprehensive as the approach offers the best of the both approaches. Grammarly gets the context of an entirely rewritten sentence while being able to identify localized issues one by one.

c. Last is a rule-based method based on syntax patterns of the language. These rules are generated by linguistic experts. This engine is built on the outcomes of the previous models and further fine-tunes the recommendations.

2. **How Grammarly makes user experience better by reducing input lag:**[7]

Imagine how frustrating it would be if your favorite text editor slowed down your computer, making it harder for you to write efficiently. To prevent this, Grammarly employs advanced engineering in addition to advanced machine learning models to deliver real-time and accurate writing recommendations without affecting the system performance. This section will help you understand that for an AI product to succeed, managing user experience is as important as its ability to deliver accurate results.

1. Defining the metric correctly such that it captures the real problem

When the Grammarly team started to investigate the issue, one of the most critical findings was that users who experienced lag either were using slower CPU or were working on long documents while using the browser-based Grammarly extension. This finding could have easily led to issue closure if product leaders would have taken it at face value. The leaders at Grammarly decided to first identify a metric that would clearly capture the essence of the problem. The metric that mattered in this case was keyboard input latency overhead introduced by the Grammarly browser extension. They also set the definition right of which text would be classified as long text and which as medium.

2. Designing a system for real-world data collection

   To effectively gauge the lag users were facing and identify the actual situations in which latency was experienced, the Grammarly team decided to establish a baseline for the metric. A separate environment was set up on Playwright, and Chrome performance profiling and metric values were captured by performing common user actions in a browser with and without the Grammarly extension. This exercise proved that the latency existed. As the next step, the Grammarly team decided to develop a system to serve the dual responsibility of capturing real-world data and helping compare the proposed solution's performance. This system employed advanced sampling techniques to capture event-level data in sample sessions such that all user types and session types were equally represented.

3. Solution designing approach

   With data captured from events and the Chrome performance analyzer, key areas of improvement were identified and a technical solution was developed. However, these technical solutions led to a change in the overall user experience of the browser extension product. The new UX, unlike the existing one, delayed the overlap check until the user stopped typing.

4. Capture impact by mapping technical metric (keystroke lag to a business metric)

   This optimization helped Grammarly improve input latency by up to 50%, depending on the type of editor the Grammarly browser extension was integrated into. By improving the input latency, the number of users disabling the extension decreased by 9%.

The problem that users faced was outside the Grammarly environment and did not affect product functionality (i.e., accuracy of suggestions) but still affected the user experience, which indirectly affected a business metric (users disabling the extension). The solution did not employ any machine learning solution but employed classical software engineering optimization techniques well supported by strong product management techniques. This story about Grammarly reinforces the importance of considering user experience not only within your platform but also in the broader ecosystem it impacts. For AI products, ensuring a seamless user experience is trickier as the model that makes recommendations often interacts with other engineering and design components before it serves the user. The limitation of these components can influence the

CHAPTER 1   THE AI MAKEOVER: HOW PRODUCT MANAGEMENT IS EVOLVING

overall product experience and performance. Leaders should adopt a very strategic approach as described earlier to affectively address these challenges.

How does Grammarly protect its users from harmful text?[8]

One of ethical responsibilities of Grammarly is to protect its users from harmful text, a duty it takes seriously. They have defined concept of "delicate" text. Delicate text is defined as any text that is sensitive or distressing and can potentially trigger negative emotions or has the potential to result in harm. This text may or may not be explicitly offensive or vulgar but might carry a risk for users or AI models that are exposed to it.

To tackle this problem, Grammarly worked with a team of linguistic experts to prepare a dataset of delicate text. They have made this dataset, along with the annotation instructions and baseline model trained on this dataset, public.

This initiative not only upholds its ethical responsibility but also helps in advancing the AI industry's efforts in harmful speech detection. By making its dataset and guidelines public, Grammarly has enabled millions of researchers to contribute to the broader mission of fostering a safer, hate-free digital world.

While this initiative might not yield any tangible results currently, this move has reinforced its position as a leader in the AI writing assistant industry, gaining valuable intangible benefits such as trust, credibility, and influence.

Meta

Mission statement: "Build the future of human connection and the technology that makes it possible."

How Meta Uses Data and AI for Generating Ad Revenue

Meta generated $48.3 billion in revenue in the last quarter of 2024[10], with approximately 96% ($46.7 billion) coming from advertising. This ad revenue is primarily driven by the platform's huge volume of daily active users and high user engagement across meta family apps.

19

CHAPTER 1   THE AI MAKEOVER: HOW PRODUCT MANAGEMENT IS EVOLVING

All applications under Meta (Facebook, Instagram, and WhatsApp) are targeted to solve the needs of connecting people socially and hence compete for users' attention. As users spend more time on Meta applications, they are exposed to more relevant ads and content contributing directly to revenue. In Q4 2024, Meta earned an average of $14.5 per user.[10] With user attention of such importance, Meta employs various strategies such as suggesting friends ("People you may know") and recommending personalized content that aligns with user preferences to boost user engagement.

Let's understand how Facebook utilizes data and AI for recommending engaging content to its user using data.

Facebook employs a multistep approach to personalize content and ad recommendations. In step 1, user behavior and preferences are analyzed including user interactions (clicks, views, purchases), demographics, and engagement patterns on various Meta platforms.[11] To this data sequence learning algorithms[14] are applied. This provides an understanding of the order in which user actions were made and then predicts the future sequence of user actions based on past action sequences. The content type and content topic with which the user interacted are also identified. Post user analysis, various machine learning algorithms are used to identify ads that a user might find most relevant.

To maintain an engaging advertisement ecosystem, Meta leverages advanced AI on the audience side but also provides AI-powered tools for advertisers. Meta has introduced AI Sandbox and Advantage + suites[12, 13] which focuses on automation and campaign optimization. These platforms help advertisers maximize ROI on their campaign by streamlining ad creation, improving targeting, and optimizing performance efficiently.

While engaging content on Facebook will keep users hooked, the real reason behind users visiting the family of apps is they want to feel connected with their friends and family and share life moments with them. On WhatsApp, it is a straightforward method; anybody can search for you using your phone number, and by default all the numbers saved in your phone's contact book can be seen in your WhatsApp list.

But on Instagram and Facebook, it isn't that simple. Meta uses complex data algorithms including Graph DB to identify and recommend friends to you. This step is so crucial for Facebook that in earlier days Facebook used a metric "7 friends in 10 days" as critical metric for Facebook discovery program. The reason behind this was people who were not able to find friends on Facebook early in their journey were more likely not to be retained. While Facebook has not officially disclosed the exact algorithms it uses for powering the "People you may know" feature. This feature recommends a list of people a user might know and would want to connect with on Facebook. Meta uses a combination of graph-based algorithms and machine learning models to recommend a list of probable people a user may want to connect to.

Graph-based storage and algorithms help analyze complex relationships between users by analyzing interactions and connections between them (LinkedIn's second/third connects, and in Facebook users see mutual friends). Here users are considered as *nodes*, and connections between two users, such as number of mutual friends, reaction to same type of content, commonality in current city, home town, school, organization user works in, places they have visited, or common hobbies, are called *edges*. The stronger the connection between the two nodes, the higher the probability that these two nodes are known to each other. The following are some most prominent graph-based algorithms that are generally employed for analyzing a social network:

> Link prediction algorithms: Predicts the probability of two users becoming friends
>
> Common neighbor analysis: Calculates the number of mutual friends between two individuals

CHAPTER 1   THE AI MAKEOVER: HOW PRODUCT MANAGEMENT IS EVOLVING

Generally, once graph analysis is performed, various supervised and unsupervised algorithms are employed to predict the likelihood of two users being friends

How Meta Keeps Its Users Safe from Harmful Content[9]

With 3.35 billion daily active users,[10] Meta is a major social media platform. Given its vast user base and high engagement rate, Facebook can easily become a place where fake and destabilizing news and content can flourish. While the spread of misinformation poses ethical and moral questions for Facebook, it can seriously impact the revenue of Facebook by affecting the trust of its users. The rise of artificial intelligence has further simplified creation and spread of such content while making detection increasingly challenging.

Facebook takes this responsibility very seriously and has created a comprehensive ecosystem to combat harmful content. This includes policy formulation, development of technology for misinformation detection, and organization structure to counter fake and harming content. Facebook has developed Simsearchnet and Simsearchnet++, self-supervised learning models to identify and label fake content. This model can also identify modified versions of fake content and functions as an end-to-end image indexing and matching system on images uploaded to Facebook and Instagram.

Misinformation can appear in multiple formats, including text, images, embedded text, etc., and Facebook has built a solution to identify and mark all of these. For example, for images with text, the AI solution can group similar content with high accuracy using optical character recognition (OCR).

Another key challenge in detecting misinformation at scale is that false claims can appear in infinite variations over time. To address these issues, Facebook has developed a set of systems capable of recognizing content with same context despite having superficial differences. Alongside technological solutions, Facebook also relies on independent human fact checkers, whose input further strengthens detection models.

CHAPTER 1   THE AI MAKEOVER: HOW PRODUCT MANAGEMENT IS EVOLVING

Facebook has also built a proprietary computer vision solution called ObjectDNA and a cross-language sentence-level embedding solution called LASER to detect misinformation to enhance misinformation detection.

For AI-driven initiatives to succeed at scale, having a well-defined organizational structure with clear roles and responsibilities is very important. Cross-functional teams facilitate better execution. Senior leadership support ensures smoother adoption and inclusion in company working manuals. Another point of learning for product leaders is to understand that AI systems will continuously evolve and one must always look at it as a work in progress with scope of improvement. Additionally, a human-in-the-loop approach remains essential, ensuring AI-driven content moderation is accurate, reliable, and responsible.

Amazon Marketplace

Mission statement: "Amazon strives to be Earth's most customer-centric company, Earth's best employer, and Earth's safest place to work."[9]

Amazon is one of the largest marketplaces in the world with revenue estimated in the tens of billions. To achieve this scale, Amazon must maintain high customer satisfaction to encourage more frequent purchases. Amazon has integrated AI across all stages of the customer journey to drive high customer satisfaction. Let's explore how these AI initiatives are applied at each stage.

Amazon makes product discovery easy in its marketplace by allowing users to search by different modalities. For example, a user can search by typing the name or product category, a user can perform a search by uploading an image of the product, and a user can also perform search by voice. By providing multiple product discovery options, Amazon has made its platform more inclusive to users with different needs, as well as, provided itself with an opportunity to make more money.

CHAPTER 1   THE AI MAKEOVER: HOW PRODUCT MANAGEMENT IS EVOLVING

While text search (i.e., the ability to perform a search by writing text) requires employing advance machine learning and deep learning algorithms. Enabling search by image and audio further adds layers of complexity to it.

More recently, Amazon has launched a Gen AI shopping assistant that can help users identify their shopping needs as per the event they are shopping for. This brings the online shopping experience one step further toward brick-and-mortar shopping where a human shopping assistant would help you discover what you need, answer questions about product you are buying, and help you make a purchase decision

Amazon makes product discovery easy by recommending products to its users. It also bundles the products together that can be used together. For example, if you are buying an air conditioner, then Amazon might recommend extended warranty, mechanical fixtures to fix your air conditioner's outer unit, etc.

Amazon also helps its users review the feedback of the product they are planning to buy easily while providing them with an AI-generated summary of all the reviews provided by other users who have purchased this product.

Amazon also leverages AI to help sellers create accurate and clear product descriptions by analyzing product features, customer reviews, customer search trends, etc. This not only saves seller's time but also enhances product visibility and relevance of product listing. As a result, customers are able to discover products that best match their needs, leading to smother product purchase experience.

Amazon's relentless focus on making product search simple and seamless is the cornerstone of its customer-first strategy. In the modern world where a user's attention has decreased a lot and they are overwhelmed with choices, helping them quickly discover the product of their choice can be the deciding factor of the shopping experience. When customers can effortlessly discover relevant products without getting stuck in identifying how to make a search, they are far more likely to complete a

CHAPTER 1   THE AI MAKEOVER: HOW PRODUCT MANAGEMENT IS EVOLVING

purchase. By doing so, you also open your system to customers who would otherwise shy away from using it because of language barriers or their inability to spell the product name.

This ease of discovery is made possible through Amazon's deep understanding of user behavior. The platform doesn't just focus on what users are searching for; it also considers how users shop, what influences their decision, and what are the barriers that stop users from coming on the platform and making a transaction. All the previously mentioned features are examples of this. Amazon has wonderfully put AI to use in identifying and solving for these challenges.

Such a comprehensive and integrative approach toward blending AI in solving challenges encountered by users reflects Amazon's mature product management strategy. The company isn't just using AI for the sake of innovation; it's using it to solve real-world customer problems at scale. This is a classic example of how a company can align its technological investments with its organizational strategy. The synergy between Amazon's business goals and its AI initiatives is what makes these efforts so effective. To attain this level of synergy, it demands creation of shared vision across the organization and robust data infrastructure, cross-functional collaboration, and an ongoing commitment to learning and iteration.

For product leaders, Amazon offers a powerful blueprint. By only focusing on identifying hidden customer needs and inhibitions and using the right technologies to address them, the actual business impact can be achieved.

It's not just about layering AI onto a product as an afterthought; it's about embedding intelligence into the very fabric of product experience and design. The product leader must ask themselves, what does the customer need? What is stopping customers from using our product, and how can technology help you create value faster and at scale?

The final lesson that Amazon's implementation of AI teaches us is when strategic clarity meets technological capabilities, the result is not just better products but also stronger business.

All Future Product Managers Will Be AI Product Managers

In the previous section, we explored how four distinct organizations across different industries have leveraged AI to gain a strategic edge. The advancements in AI have not only helped organizations become more efficient and innovative but also paved way for the creation of new business models built around solving problems previously unthought of. Grammarly is just one of the many notable names.

AI is truly unlocking new possibilities for both businesses and consumers. As product leaders, we are at a pivotal moment to embrace this transformative wave of AI. However, care should be taken in not just adopting AI at a superficial level but integrating it deeply and mindfully into every stage of product development from identifying problem to building solutions. To support this transformation, organizations must focus on building strong synergies across cross-functional teams, aligning them around a shared vision and strategically directing both organizational and financial resources toward AI-driven initiatives.

The momentum behind the AI adoption in business and product development is not just growing; it is accelerating at an unprecedented pace. Recent studies have shown that AI is deeply embedded across various industries. According to Gartner (2023 study),[15] close to 80% of organizations are already using or planning to use Gen AI in their business by 2026. This clearly indicates that AI is no longer a future investment but a current strategic imperative. In another survey by New Vantage

CHAPTER 1   THE AI MAKEOVER: HOW PRODUCT MANAGEMENT IS EVOLVING

Partners (2022),[16] 91% of the participants are actively investing in AI. In the same survey, 92% of participants have observed tangible outcomes. This statistics has seen significant improvement from 48% in 2017.

The financial stakes made on AI are massive. In 2022 PWC claimed that AI-powered offerings will contribute $15.7 trillion to the global economy. With this tremendous value being created by AI, it is reshaping industries and redefining value creation across industries and markets. This massive economic and leadership shift has already started influencing the product strategy.

However, not everything is smooth sailing; many organizations continue to face significant challenges in implementing AI. These issues range from ethical concerns, such as ensuring that AI systems are free from bias, to difficulty in explaining AI-driven recommendations. To designing of underlying data and organisational systems which are critical to AI success. While there can be multiple factors contributing to these challenges, a key reason is the shortage of AI skilled professionals, especially at the senior level.

These seismic shifts in the business and product landscape are also influencing the job market and the role of the product manager itself. AI fluency is becoming a fast differentiator today, and in the near future it will become a requirement, owing not just to the extreme demands of skilled resources both on the technical and product sides but also from challenges that organizations are facing. Even AI expert Andrew Ng has acknowledged the growing demand for AI product management roles. Forbes has referred to AI product management as a future-forward career.[18] A quick Google search will reveal just how high the demand for these positions currently is.

The writing is on the wall: AI is not just another technology; it is a revolution that will redefine the very nature of how industries operate, which in turn will change the very nature of product management. Those who embrace it early will lead and shape the future.

27

CHAPTER 1   THE AI MAKEOVER: HOW PRODUCT MANAGEMENT IS EVOLVING

Now that we have seen how the industry is responding to AI resolution, with significant investment and strong support from senior leadership, this trend is only expected to grow. Now let's explore how the role of product management professionals is set to evolve.

The Future of Product Management in an AI-Driven World

A product manager is responsible for overseeing a product from its concept to its launch and beyond. Their primary duties include aligning product with organizational strategy while addressing user needs and defining product vision. They build and prioritize features, create and maintain product roadmaps, and work very closely with cross-functional teams such as design, marketing, and engineering in realizing the product map. They also ensure that products are built on time and within budget and that they meet business objectives.

With the rise of AI, the traditional responsibilities of product manager will also evolve significantly. As AI is being used in products extensively, PM will have to be fluent in AI. A PM should be able to understand machine learning models, data pipelines, and AI capabilities. A core aspect of this fluency is also understanding the data and AI lifecycles.

If a PM lacks this understanding, they might risk launching suboptimized products or, even worse, biased products. Either of these outcomes could negatively affect user trust and the brand's reputation.

AI-driven products require a different approach. Rather than managing static features, the PM has to oversee the dynamic and continuously evolving systems that can adapt and improve through user interactions and data. Understanding how AI models learn and change over time is essential to effectively managing product iteration cycles. A product manager who is not fluent in AI will find it difficult to recognize the limitations of a machine learning model or the nuances of AI project lifecycles, including the importance of sampling and continuous training.

In addition to transforming traditional products, AI introduces new business models that rely on data as its core offering. Grammarly is a classic example; its product leverages AI to aid writers in enhancing the quality of their writing. As AI enables products like, this data is an asset that often drives new value streams and business opportunities.

Without AI fluency, a product manager would struggle to recognize how to identify these opportunities and how to create these products. They may also miss opportunities to innovate by leveraging user data in ways that drives organizational growth.

Another key aspect of the PM role is to monitor and measure the process of their product, and if a PM does not understand AI well, it would become difficult to monitor and measure the success of AI products. Another area where the PM might see difficulty is the ability to set realistic expectations on product performance and timelines.

To summarize, AI is reshaping product management by revolutionizing traditional approaches. It is enabling the creation of new business and shifting static product features into personalized and dynamic ones that can adapt to individual customer needs. Data is becoming the central asset in this evolution. As AI continues to evolve, PMs must be equipped not only with strategic vision but also with technical knowledge to navigate the challenges and opportunities AI introduces.

Evolving Expectations from Product Leadership Roles in an AI-Driven Era

As AI continues to evolve and embed itself into the DNA of product development, it is not just tools and techniques that are changing; roles and responsibilities across product management will also evolve. From frontline product managers to senior product leaders, everyone will need to adapt to new skill sets, mindsets, and expectations. Let's take a look how each of these roles will evolve in the future.

CHAPTER 1   THE AI MAKEOVER: HOW PRODUCT MANAGEMENT IS EVOLVING

Product Manager: From Generalist to AI-Fluent Strategist

The PM role traditionally is grounded around customer empathy, user need discovery, solutioning, stakeholder management, and delivery timelines. But now it will demand a baseline technical fluency in AI. While PMs will not be expected to develop a model, they will be expected to have the following skills:

1. Understand AI workflows:

    a. Including basics of model training, data inputs/outputs, and evaluation metrics.

    b. Ability to translate product requirement into data and machine learning requirements. This may include defining training data, feedback loops, and ethical boundary requirements.

    c. Ability to communicate with data engineers and machine learning engineers with clarity and confidence.

    d. Stay up-to-date on the latest in AI regulations and responsible AI design, as bias, transparency, and fairness have become core product concerns. Also, with almost every major geography coming up with their own AI laws, any regional violation can bring severe financial and reputational loss.

    e. Ability to use AI-powered tools to enhance their own productivity.

Product Director: From Managing Roadmaps to Managing AI Readiness

Product directors typically oversee multiple teams and drive cross-functional execution. They are also responsible for managing and delivering product roadmaps. Their role will also evolve from roadmap

CHAPTER 1   THE AI MAKEOVER: HOW PRODUCT MANAGEMENT IS EVOLVING

management to capability orchestration. Their focus will expand to the following:

1. Building and scaling AI capabilities across product portfolios.
2. Leading teams through data maturity assessments, model lifecycle management, and AI-first experiment processes.
3. Developing shared infrastructure and design patterns for AI applications to reduce duplication and ensure consistency.
4. Championing a culture of data-driven decision making and ethical responsibility.
5. Directors will also be responsible for reskilling their teams and ensuring traditional PMs are AI fluent.

Chief Product Officer: From Visionary to AI Transformation Leader

At the top of a product organization, the CPO role is undergoing a fundamental transformation. The future CPO will no longer only set product vision; they will also be responsible for driving AI strategy at the organization level. This will include the following:

1. Defining and owning the AI innovation agenda by identifying areas where AI can unlock new business models or improve customer experience or operational efficiencies
2. Making key decisions about AI investments that may include partnerships and build versus buy strategies
3. Working toward building synergies cross-functional alignment
4. Owning ethical and reputational impact of AI products by setting clear standards for fairness, transparency, and accountability

CHAPTER 1   THE AI MAKEOVER: HOW PRODUCT MANAGEMENT IS EVOLVING

5. Enabling product organizations to scale responsibly with AI, by diverting investments in platforms, training, and governance structures

In essence, the future CPO is not just a product visionary but a transformation architect, driving AI adoption across the business while ensuring customer trust and long-term value.

Building AI Fluency: Key Skills for Future-Ready Product Managers

Up until this point in the chapter, we have explored how AI is driving rapid evolution of technological landscape and how the role of product management will be redefined. As we now move forward, it's these critical capabilities that product management professionals must develop to stay ahead of the curve and lead effectively in this fast-changing environment. This section outlines the core skill areas that will be essential for product management professionals.

1. **AI Fluency and Technical Foundations**
    - Product management professionals should be familiar with basic AI/ML principles, including supervised/unsupervised/Gen AI/deep learning algorithms, and how these algorithms are trained and evaluated. They should also understand the complete machine learning project management lifecycle. They should also be aware about the ethical considerations in AI development and design.
    - While not expected to be data champions, PMs must understand basic data lifecycle including

how data is getting generated, stored, and accessed in various systems. They should also understand concepts of data pipelines.

- The should be able to hold meaningful conversations with machine learning and data engineers. They should be able to interpret results, question assumptions, and make data-informed decisions.

- They should recognize the importance of clean, accurate, and reliable data is critical for successful AI implementation. PMs should be able to assess data readiness and work to improve data quality where necessary.

2. **AI Governance and Compliance**

As AI systems start becoming all-encompassing, an ethical implementation is becoming increasingly important.

- PMs should be able to understand AI ethical principles and translate them into product and machine learning requirements.

- PMs should be able to proactively identify and address biasness in data and models. They should be able to integrate ethical principles into product and model development.

- PMs should understand applicable laws around AI and design products that are compliant.

3. **AI-First Product Strategy**

   Creating AI-powered products requires a shift in strategic thinking. PMs should be able to perform the following activities:

   - PMs must evaluate whether an AI implementation is financial justifiable, considering development cost, time to market, and potential ROI.

   - PMs should be able to align AI outcomes to business goals.

   - PMs should be able to contribute to the design of robust data platforms and pipelines that enable continuous innovation.

   - PMs should understand how an AI solution fits into existing platforms, services, and infrastructure and how to drive change management for successful adoption and integration.

4. **Cross-Functional Collaboration**

   AI products are inherently multidisciplinary, requiring strong collaboration skills.

   - They should be able to work with engineers and data scientists by bridging the gap between business requirements and machine learning and data requirements. They should be able to clearly communicate what is expected from the machine learning model. They should also be able to define the evaluation criteria and translate it back into the business metric.

- Human-centric design remains critical, especially in AI where transparency and user trust are key. PMs should facilitate collaboration that brings up product requirements, technical feasibility, user needs, and ethical design.

Summary

This chapter highlighted the pivotal role of data and AI in transforming the business and product management landscape. Various successful implementations done by top organizations illustrated the importance of organizational strategies and structures needed to harness AI effectively. These examples also showcased how AI is being used to gain competitive advantage across industries.

This chapter also explored the importance of data platforms as a foundation before implementing AI solutions. It also showcased how with clever planning AI can be seamlessly integrated into the customer journey to solve real-world problems and drive value for both businesses and customers.

Once the need for AI was clearly established, the chapter delved into how the role of product managers will evolve in an AI-driven future by highlighting key skills and competencies needed by future PMs to stay competitive.

CHAPTER 2

Foundations of AI: Understanding Key Concepts and Terminology

In the current fast-paced technological age, artificial intelligence has become a central force, shaping and influencing various products and services we rely on and simplifying our daily lives. Therefore, gaining the understanding of the interplay between artificial intelligence (AI) and human intelligence (HI) has become increasingly important. For product leaders who are at the forefront of innovation and operate at the intersection of technology, business, and user needs, this understanding is particularly critical both for career progression and for meeting customer needs. While AI brings speed, efficiency, and unprecedented scalability in decision-making and problem-solving, human intelligence brings creativity, empathy, ethics, morality, and inquisitiveness to the table without which no problem can be solved effectively. Understanding the difference between the two will equip product managers to build products the right way.

CHAPTER 2   FOUNDATIONS OF AI: UNDERSTANDING KEY CONCEPTS AND TERMINOLOGY

This chapter is designed to help readers understand the deep and subtle concepts of how humans learn and develop wisdom, laying the foundation of understanding intelligence. We will examine how intelligence is developed, sustained, and evolved both in humans and machines. We will draw a contrast between how a human evolves from a dependent infant to a fully functional adult to a framework DIKW (data, information, knowledge, wisdom) to establish the importance of developing the right ecosystem in which artificial intelligence systems can thrive and contribute. We are including these fundamentals to enable readers develop a better understanding of the capabilities and limitations of AI. This chapter will also cover the brief history of AI, including the Turing test, key milestones, and essential terminology that every future ready product manager should be familiar with.

This chapter is more than just a technical overview about intelligence and how it is developed; it also offers a philosophical perspective about intelligence and the need to develop intelligent systems. You should realize that product managers are not just executors; they are thinkers and decision-makers. A closer understanding of these fundamentals will enable them to make smarter and ethical products that are more closely aligned with human values.

From Cradle to Cognition: The Evolution of Human Intelligence

All humans at birth are completely dependent upon their primary caregivers for all their needs such as feeding, cleaning, keeping them warm, and comforting them to sleep. They lack complete awareness about their surroundings, reasoning, and ability to function independently or be socially responsible. The journey from a helpless, knowledge-less infant to a socially contributing responsible adult is an incredible journey that involves grit, conviction, and multiple forms of learning. This

CHAPTER 2 FOUNDATIONS OF AI: UNDERSTANDING KEY CONCEPTS AND TERMINOLOGY

process unfolds gradually through observation, imitation, instructions, experiences, and self-reflection.

Have you ever observed how a newborn learns about their surroundings? From the very beginning, babies learn by watching the people and environments around them. For example, when a baby is hungry, they start associating feeding with a particular person, often their mother or any other primary caregiver. Over the time they begin recognizing voices, sounds, and facial expressions simply by observing. These observations turn into associations that grow into imitations. In fact, the first few words that a baby speaks are often the result of imitation.

This type of learning is called *observational learning* or *social learning*. Albert Bandura, a renowned psychologist, has presented a theory on it that emphasizes that people, especially early-age children, learn new behaviors by observing others. At an early stage, no direct instructions are needed; observation and imitation are powerful tools through which human learning begins.

As kids grow up a bit their learning becomes more intentional and organized as they start learning by following instructions and repeating tasks multiple times. Tasks can be as simple as reciting poems, counting, or identifying basic body parts. Here in addition to following a particular type of learning, the neurological development in brain enables them to learn, remember, and practice the new learning. This type of learning is called *explicit learning* or *direct instruction learning*. At this point in time, a child gains knowledge both from observational and explicit learning. The explicit learning helps a child gain academic and basic language skills.

As children grow and their brains continue to develop, they begin understanding the consequences of their action. They start associating the abstract concepts like emotions with their actions, which leads them connect actions with outcomes like rewards and penalties. A job well done is appreciated by society, which makes them feel happy, and this happiness acts as a psychological reward for them. In some cases, the reward can be tangible such as gift or recognition, reinforcing that positive emotional state. For instance, performing well on a test can evoke pride and increase social validation, which can further contribute to a sense of accomplishment again acting as a reward for the brain. Conversely, when a task is poorly executed, criticism or disapproval can evoke negative emotions such as sadness, which internally functions as a penalty. Over time, the brain learns to associate specific actions and behavior with reward or penalty. This helps children adjust their behavior to seek rewards and avoid penalties.

When faced with similar situations in the future, the brain recalls its past observations and experiences and favors actions that historically led to positive results. This type of learning through consequences where behavior is shaped by reward or penalty is called *operant learning*.

With children maturing further, their mental and physical abilities expand, and they expose themselves to more complex situations. This takes their learning beyond imitation, instructional, and operant. While these learnings still help a child to learn new things, at this stage, they develop the ability to put into practice multiple simpler concepts to solve a complex problem. For example, a child may use their understanding of shapes, color, and numbers to solve a puzzle. They also learn to transfer learning made from solving one type of problem to another simply by identifying patterns and drawing associations.

This learning type is more self-driven, intellectually at a deeper level, than the learning mentioned earlier and requires more analytical rigor. This is called *cognitive learning*. It involves understanding the problem statement, breaking it down into logical chunks, reasoning, and applying previously learned knowledge. This type of learning requires more application.

As learning matures further, children develop the ability to understand the importance of context in the application of learning. They learn that the same concept can be applied differently in different situations depending on the context. This ability to apply their learning based on situations makes them ready for real-world functioning. This ability is

CHAPTER 2   FOUNDATIONS OF AI: UNDERSTANDING KEY CONCEPTS AND TERMINOLOGY

called a *transfer of learning* and can be simply defined as the ability to apply the same knowledge differently in different situations depending on the context. A simple example could be a child learns to speak politely and softly while conversing with other, but they realize that raising their voice slightly to be heard during a group discussion is appropriate. By doing so, they have learned the art of moderating their voice depending on context and situation in which they are speaking. We will understand more about context in the next section.

All the learnings we have discussed so far primarily enhance the brain's ability to memorize, process, and apply information. These are also called *cognitive abilities* of the brain. However, supporting these abilities is a more advanced and silent function called *metacognition*. At its core, metacognition means thinking about thinking. It's awareness may sometimes be led by our subconscious. It is the prism through which one views the world, forms beliefs, make judgments, and builds habits. Metacognition helps learners observe their own thinking pattern and is the awareness of one's own thoughts.

In early childhood, metacognition helps a child by making them realize simpler things like how they learn faster. Is it by writing or reading loudly or by other means? As they mature, this ability deepens, and they start questioning their assumptions and beliefs that inherently changes their reaction to the world. Societal and environmental factors also influence this prism, as they prompt users to reflect upon how a task should be done or a thought should be formed. These reflections can reframe an individual's beliefs and behaviors essentially updating this prism through which they view this world.

Metacognition is also deeply influenced by the subconscious mind. Most of the reflections we do and questions we ask are influenced by our subconscious beliefs, emotions, and experiences formed over the years, often without us being actively aware of them. The subconscious mind stores patterns, memories, implicit biasness, narratives, and cultural norms that we have experienced and internalized from early life experiences, family, and social environments. These hidden mental models quietly shape how we perceive the world and react to the new information even before we become aware of it.

To summarize, metacognition transforms learning from a passive intake of information into an active and adaptive process. This capability is foundational to critical thinking, emotional intelligence, resilience, and other humanitarian aspects of learning.

Recognizing these learning types and how they influence the human brain to evolve will help product leaders understand the true capabilities of artificial intelligence and use this technology to design human-centric products.

Decoding Human Intelligence: Foundations for Building Artificial Minds

Understanding the word *intelligence* is central to understanding human development and learning as well as central to the idea of making machines learn and perform human-like tasks. The Oxford dictionary defines the word *intelligence* as a noun and describes it as the "ability to learn, understand, and think in a logical way about things and to do effectively." In simpler words, *intelligence* is not just the ability to learn but also the ability to apply that learning effectively in a real-life context and to grow from that experience. Intelligence is a complex subject, and it not only refers to learning and applying but also captures the ability of a person to understand the "why" behind the achieved outcome and use that answer to further augment one's knowledge base. It is because of its complexity that both philosophers and scientists have developed deep interest in this subject.

To cover the definition of the word *intelligence* holistically, let us now go through the philosophical and scientific definitions. From a philosophy standpoint, *intelligence* is often linked to reason, wisdom, ethics, and purposeful actions. Eastern philosophers view intelligence more holistically and associate intelligence with moral insights, compassion, and self-growth. On the contrary, Western philosophers associate intelligence with the ability to discover universal truths and apply them wisely. Both types of philosophers extend the concept of intelligence from just practical applications of learning to moral and ethical reasoning and its application in real-world situations.

Modern scientists view intelligence not just as a single entity but as multiple entities such as linguistic intelligence, analytical intelligence, interpersonal intelligence, etc. All these types of intelligence signify the ability of a person to practically apply their specific skill in real-world situations. The fundamental scientific view of intelligence is practical application of learning in real-world situations.

The Learning: Intelligence Nexus

By both philosophical and scientific definitions, intelligence is not merely the accumulation of knowledge; it is the ability to apply that knowledge effectively in practice in a way that brings about change. When an individual uses their knowledge to make a decision or to solve a problem or take any action, they are exercising their individual intelligence. This is the reason if two person are assigned same task, they may not only perform the task differently but also achieve totally different outcomes. This practical application of learning is more than just solving problems; it also feeds back the outcome into the learning process. If the outcome is successful, then it sends positive reinforcement, encouraging similar behavior in similar future situations. If the outcome is not successful, then it acts as negative reinforcement prompting reflection and adjustments in learning. In this way, intelligence is a constant cycle between learning, application and experience and refinement. Intelligence is never static; it evolves with each application in the real world.

CHAPTER 2 FOUNDATIONS OF AI: UNDERSTANDING KEY CONCEPTS AND TERMINOLOGY

Learning is the most essential exercise one must engage in to develop intelligence. In humans, it is facilitated by complex neurological structures that enable the brain to acquire, store, and apply knowledge, whereas in machines, learning is achieved through writing set of instructions in the form of programming. The programming instructions are supported by advanced semiconductor technologies like memory, storage, and processing prowess. Learning is a foundational and theoretical concept, and to apply this concept for a better outcome, it must be applied in the right context. Without contextual relevance, even the most sophisticated learning can fail to produce meaningful outcomes. In the next section, we will explore the concept of context and examine its role in human intelligence.

The Intelligence Multiplier: Role of Context in Human Intelligence

Context plays a foundational role in shaping human intelligence. It helps in shaping how we perceive, reason, decide, and behave. It is that one theme whose contribution in defining intelligence is agreed upon by both philosophical and scientific thinkers. The term *context* refers to the surrounding circumstances including historical, cultural, situational, physical, environmental, or any other factor that can affect or influence our understanding of a situation.

Humans rarely process information in complete isolation, and we continuously draw clues to assign meaning or adapt our response to a situation. These clues are often provided by adding context to a problem statement that we are trying to solve. Human actions, thoughts, point of view, and decisions are all influenced and acted upon by understanding the application context.

CHAPTER 2   FOUNDATIONS OF AI: UNDERSTANDING KEY CONCEPTS AND TERMINOLOGY

Ultimately, context makes human intelligence sensitive, which means a decision that is considered smart in one setting might not be appropriate in another. This ability to adjust behavior based on context and environmental cues is central for practical and emotional intelligence or in other words human intelligence.

The Inner Mirror: How Cognition and Metacognition Define Human Intelligence

Cognition and metacognition are core to the sophisticated human intelligence. While cognition refers to the core mental processes that enable individuals to perceive, attend to, store, retrieve, and reason and comprehend information. These processes together form the building blocks of intelligent behavior.

Metacognition, on the other hand, is the passive observer of the cognitive processes. It enables self-awareness, monitors performance of the cognitive task, evaluates the outcome achieved, and forms the foundation of human wisdom. Together cognition and metacognition not only allow humans to process information but also reflect on how effectively they are performing and adjust their approach accordingly. Metacognition also allows humans to transfer their learning from one problem space to another. Metacognition and cognition together form a dynamic feedback system where actions are informed by knowledge and refined by wisdom, making human intelligence system uniquely flexible and self-improving.

This ability of human intelligence of not just solving problems but also learning how to solve problems in a better way over the course of time is the defining feature of human intelligence and evolution. This defining feature of human intelligence distinguishes it from current artificial intelligence systems that often lag robust metacognition capabilities.

45

CHAPTER 2   FOUNDATIONS OF AI: UNDERSTANDING KEY CONCEPTS AND TERMINOLOGY

Thinking Machines: A Philosophical and Scientific Introduction to AI

John McCarthy, one of the founding members of the AI, defines artificial intelligence as "AI is the science and engineering of making intelligent machines, especially intelligent computer programs." In simpler words, AI is the technology that enables a machine to perform tasks that would otherwise require human intelligence such as understanding of language, recognition of image or patterns, ability to recommend a decision, and learning from experience, in this case, data.

This term was first coined in 1956 at the Dartmouth conference where AI pioneers like John McCarthy, Marvin Minsky, Claude Shannon, and Nathaniel Rochester first proposed that human intelligence can be simulated by machines. It was this bold hypothesis that revolutionized the world by directing research and efforts toward advancements in AI.

Academically artificial intelligence is a highly interdisciplinary field. It draws various fields such as computer science for designing algorithms and hardware, mathematics for developing logics and solutions that will help in developing algorithms, psychology for understanding how humans think and learn, neuroscience for understanding human brain structure, philosophy and ethics for understating the implication of AI's action and deciding limits until which AI can be used, and finally linguistics for understanding nitty-gritty details of the language.

AI development happened in waves, with the first wave hitting in 1950–1970 where efforts were focused on translating intelligence as symbols or logical rules. However, these systems struggled with ambiguity and real-world complexity, leading to slowed down progress. This period is also referred to as AI winter (the 1980s). This era saw disillusionment due to lack of research leading to the overpromised potential of early AI systems. The stagnation faced in technological advancements in this field led researchers to focus on fundamental issues like ethics and interdisciplinary issues.

CHAPTER 2   FOUNDATIONS OF AI: UNDERSTANDING KEY CONCEPTS AND TERMINOLOGY

The early 1990s saw a shift from rule-based artificial intelligence to data-based learning algorithms, also referred to as *machine learning algorithms*. This shift changed the way early rule-based AI systems were designed to more dynamic data learned rules, which would change with changes in the underlying data. These machine learning algorithms relied on patterns learned from the data using mathematical methods. Key innovations like support vector machine, random forest, and improved neural networks (LSTM) empowered AI to handle complex tasks like language processing and finer pattern recognition. These advances were complemented by the rise of technologies that enabled handling, processing, and storing large and diverse amount. This combination further enabled AI to process and learn from data and generate better recommendations. A famous example of this era would be IBM Deep Blue defeating world chess champion in a game of chess, clearly showcasing how data and computing can mimic human intelligence.

The present phase of AI development is also referred to as the era of generative and foundational models. In this era, AI is reaching new heights of progress by achieving unprecedented levels of versatility and accessibility. This era is driven by large-scale language models and advancements in architectural capabilities that are enabled by generations of human-like text, programming code, or other creations, which were until now reserved for humans. While the current wave of artificial intelligence has bought massive improvements in the technological advancements of AI, it is at the same time raising a lot of questions on the ethical social and regulatory challenges posed by the increasingly autonomous and influential AI systems.

CHAPTER 2 FOUNDATIONS OF AI: UNDERSTANDING KEY CONCEPTS AND TERMINOLOGY

The Necessity of Intelligence: What Drove AI's Creation

Long before the first computer was built, ancient philosophers were exploring questions about the nature of mind, thoughts, rationality, and whether cognition can be separated from the human body. Philosophers like Rene Descartes proposed the theory of dualism. The central idea was that the mind and body are two distinct entities, with the mind being immaterial in nature and the body being material. These two distinct entities interact to make a human functional. These questions and theories became the central part of the field of AI. These questions laid the conceptual groundwork for the idea that intelligence might not be exclusively biological and that rationality could be formalized in the form of set of rules. Thus, the desire to develop intelligent machines is more of a quest to answer these age-old questions with modern technology.

The following are the key philosophical needs that are driving AI progress:

- **The need to understand ourselves:** Humans have a long-standing desire to understand their own mind. Creating AI is just an attempt to understand and imitate how the human mind works.

- **The desire to transcend human limits:** Another deep desire that humans have is to overcome our natural cognitive and physical limitations. A human's ability to achieve is limited by time, energy, memory, and attention span. AI by contrast offers the possibility of extending human capabilities by processing a vast amount of data, working almost 24*7 without rest and maintaining consistency. This intention of extending our limits is contributing to technological advancements in AI.

- **The drive for creation and control:** Humans have a powerful impulse to create and to play the role of an inventor. Building intelligent machines satisfies this urge to shape the world and create life through reason and design.

While these philosophical motivations inspire humans to develop AI, the real-world practical benefits that AI promises are also pushing its rapid advancements. Because intelligent systems can function 24/7 objectively without getting tired, today AI is increasingly viewed as an ideal solution for automating repetitive or dangerous tasks and significantly enhancing overall performance and efficiency.

The Father of AI: Turing's Legacy of Thinking Machines

Alan Turing was one of the leading computer scientists who connected philosophy to computer science. He posed the question in 1950, "Can machines think?" in this groundbreaking paper "Computing machinery and Intelligence." This became the foundation for most work in the AI domain. He proposed a set of rules also known as the Turing test for gauging a machine's intelligence. He suggested that if a machine could engage in a conversation that is not distinguishable from that of a human, then a machine could be deemed intelligent. Turing's pragmatic approach of defining a machine as intelligent from a metaphysical definition of consciousness to observable, testable behavior led to the birth of AI as a formal field of study.

He further proposed that machines initially shouldn't be programmed to an adult level of intelligence; instead, just like humans, machines should be built to have a child-like mind, which is capable of learning and improving over the time. His notion of how machines should learn became the foundation of machine learning and associated learning models. Large

CHAPTER 2   FOUNDATIONS OF AI: UNDERSTANDING KEY CONCEPTS AND TERMINOLOGY

language models like BERT, GPT, and others support Turing's idea. These systems start with minimal knowledge (i.e., randomly initialized weights) and learn complex behavior through training and feedback.

Turing's learnings and ideas also shaped the ethical and philosophical debates that still surround AI. Turing played a crucial role in influencing not just the technical roadmap but also the intellectual framework required to guide AI development and policy today.

Human Needs, Intelligent Systems: What Every Product Leader Should Know

For product leaders working with or planning to work with AI, understanding the philosophical motivations behind its development is not just for intellectual gains but is also strategically essential. AI was not created for automating tasks but to replicate and explore the nature of human intelligence. This extends to the ability of reasoning, adapting response to a situation, and evolving through experience. By understanding the deeper motivations behind this, product leaders are better placed to design and develop AI tools that align better with human behavior and values. By understanding that intelligence is always evolving, deeply contextual, emotionally informed, and continuously shaped by feedback, leaders can design AI products that can learn, adapt, and improve just as humans do.

This perspective also encourages a more ethical and responsible approach toward AI development, enabling creation of AI, which supports humans achieving their goals rather than diminishing them. This understanding will help leaders make informed decisions required for developing trusted, relatable, and truly intelligent AI systems.

CHAPTER 2   FOUNDATIONS OF AI: UNDERSTANDING KEY CONCEPTS AND TERMINOLOGY

Foundational Paradigms of AI: Symbolic, Connectionist, and Emergent Approaches

Artificial intelligence has deep philosophical roots, and over time multiple conceptual frameworks have contributed to its evolution. Each has a diverse perspective of how machines can imitate human intelligence. Among these, three frameworks have emerged as the cornerstone of AI research and development:

1. Symbolic approach
2. Connectionist approach
3. Emergent approach

These paradigms differ not only in knowledge representation but also in their philosophical assumptions about intelligence itself. It is these differences that have contributed to progress in AI. These patterns play an essential role in enabling the understanding of these diverse methods and theories that contribute to the field today.

Symbolic Approach

The symbolic approach, also known as Good Old-Fashioned AI (GOFAI), talks about building artificial intelligence by defining how knowledge can be explicitly represented using symbols and formal rules. This approach is also referred as a rule-based or logic-based approach.

The symbolic approach involves transferring human knowledge into machines using logical statements, facts, and inference rules, allowing AI to reason, draw consultations, and solve problems by manipulating these symbols.

Simply put, symbolic AI is like teaching a computer by giving a set of clear rules. These rules are given to computer with the help of programming languages. In this approach, the knowledge that you want

CHAPTER 2   FOUNDATIONS OF AI: UNDERSTANDING KEY CONCEPTS AND TERMINOLOGY

computer to gain is written in the form of symbols, logic, and rules. For example, if you want your computer to learn that birds can fly and animals cannot, then all you must do is to write rules like "animals cannot fly," "birds can fly," and then create another set of rules that maps the names of animal and birds to a respective category. If you map dog to animal and sparrow to bird, then the computer can infer that a dog cannot fly, but a sparrow can.

From a technical perspective, this approach relies on putting programming instructions and formal logic into machines that further helps machines to make human-like decisions. This approach is effective in domains where rules and relationships between the entities are well established and can be explicitly stated. Simple examples could be an expert system in healthcare helping doctors with medical diagnoses or an expert system in banking handling petty loan applications. Programming languages like LISP and Prolog were developed to support symbolic computation.

Symbolic AI offers a high degree of explainability and interpretability, making it attractive for use cases where the need for transparency and auditability is paramount. However, this approach has a downside of being too rigid in adapting to new or ambiguous situations. Every new scenario or rule must be programmed to enable a symbolic system to adjust to changing business situations. The following implementations are the classical examples of this approach:

- MYCIN: A medical expert system built in 1970 to help doctors diagnose blood infections
- Knowledge graphs and ontologies used in Google Search or biomedical research (e.g., SNOMED CT)

CHAPTER 2 FOUNDATIONS OF AI: UNDERSTANDING KEY CONCEPTS AND TERMINOLOGY

Connectionist Approach

Unlike symbolic AI, which builds intelligence through explicit definition of rules, connectionist AI follows a more flexible approach inspired by the human brain learning approach. It is a data-driven learning approach, and systems developed using this type of learning approach learn from using examples rather than relying on hard-coded logical rules.

In simpler terms, connectionist AI is based on the idea of teaching computers the way our brains learn through experience. Instead of learning through a list of rules, we expose our brain to various examples and let it figure out the commonality in the examples and form its own rules. Human learning is facilitated by a special type of cell called a *neuron*. Neurons form a network that helps us learn.

Let's take an example of a child learning to recognize cats. We don't give that child a list of rules that would help them recognize a cat; instead, we show them lot of cat images and over time the child learns to recognize a cat by forming their own rules. This is how connectionist AI works. You expose the system to lot of data and then programmatically make the system identify the pattern in the data and use pattern-based rules to recognize an event.

Because it is a pattern recognition–based learning methodology, it works well in domains where rules are not very explicit, i.e., like recognizing faces, understanding speech, translating languages, and even playing video games or any study where rules to recognize an event are not very clear.

This is the type of AI that all of us are using in current time. The following are examples:

- **GPT-4 (OpenAI):** A large language model trained on vast textual data

- **AlphaGo (DeepMind):** Uses deep reinforcement learning to master Go

- **DALL·E/Midjourney:** Text-to-image generation using deep learning
- **Convolutional Neural Networks (CNNs):** Used in image classification (e.g., ResNet, YOLO)
- **Recurrent neural networks (RNNs) and transformers:** For sequence modeling like speech or translation
- Machine learning algorithms like support vector machines, random forest, etc.

Learning Type:

- Supervised learning (e.g., GPT, CNNs)
- Unsupervised learning (e.g., autoencoders, generative models)
- Reinforcement learning (e.g., AlphaGo, robotics)

Emergent AI: Intelligence Through Interaction

Emergence behavior refers to properties that arise from the interaction of simpler components or from the self-organizing process of these simpler components inside a system. In reference to artificial intelligence, emergence AI refers to systems where intelligence arises from the interplay of various individual algorithms, data, and computational processes. These simpler systems are not explicitly programmed for delivering intelligence but exhibit it through evolution, adaption, and coordination. However, this intelligent behavior is not evident when each of these elements is considered individually.

CHAPTER 2   FOUNDATIONS OF AI: UNDERSTANDING KEY CONCEPTS AND TERMINOLOGY

Emergence properties in AI are typically the result of several underlying mechanisms, including the following:

1. Interactivity between various subsystems like artificial neural network layers, learning algorithms, data points help in building emerging intelligent behavior

2. AI systems designed to build emergent intelligence adapt and learn from the data to which they are exposed, which leads to the development of intelligence

3. Aspiring emergent AI systems have the ability of self-organization to form complex structures and patterns without being explicitly programmed

4. Any aspiring emergent AI system has built in feedback loops complemented with reinforcement systems, which help them amplify or diminish certain behavior

While emergent AI showcases an enhancement in current AI capabilities, it is because of its emergence that nature poses many questions on the ethical, risk, and societal front that should be first answered before developing these systems.

Emergent systems are often hybrid; they combine symbolic reasoning, connectionist learning, and environmental feedback. Examples are as follows:

- GPT-4's in-context learning: Emergent capabilities not explicitly trained (e.g., arithmetic reasoning).

- Swarm robotics: Drones or robots collaborate with no centralized control.

- OpenAI's multi-agent hide-and-seek simulations

CHAPTER 2   FOUNDATIONS OF AI: UNDERSTANDING KEY CONCEPTS AND TERMINOLOGY

Learning Type:

- Evolutionary learning (e.g., genetic algorithms)
- Multi-agent reinforcement learning
- Self-organization and adaptive behavior

Conclusion

The history and future of AI can be best understood as a convergence of these AI paradigms

- Symbolic AI's foundational work of formalizing of knowledge through explicit rules
- Connectionist AI's ability to develop systems that could learn patterns from data, enabling rules formulation through experience
- Emergent AI's capability to learn and evolve organically through interactions within a complex system

This convergence of paradigms holds the promise of building more robust, interpretable, and general-purpose intelligence. One notable step in this direction is the ongoing research in the field of neuro-symbolic AI, which aims to combine symbolic reasoning with deep learning, paving the way for more transparent and high-performing AI systems.

Understanding these paradigms helps leaders choose the right approach for each context, whether it is ensuring transparency for legal compliance using symbolic AI, deploying scalable solutions using connectionist AI, or exploring adaptive intelligence using emergent AI.

CHAPTER 2   FOUNDATIONS OF AI: UNDERSTANDING KEY CONCEPTS AND TERMINOLOGY

Bridging Minds and Machines: Cognitive and Metacognitive Insights for AI

Human intelligence continues to inspire researchers working on the evolution of artificial intelligence. The way cognition and metacognition interplay give humans the ability to adapt, learn, evaluate, and apply wisdom to make intelligent action acts the blueprint for developing general-purpose AI.

While AI has made remarkable progress in cognitive tasks like image recognition, language processing, and pattern recognition, these capabilities are still context insensitive, narrow, and reactive in nature. Just to highlight one of the tests that conducted an image processing model Facenet developed by Google outperformed humans in face recognition test. However, these models are more suspectable to unusual or new poses in which an image is presented. When the same experiment was repeated with the noise being added to the same image dataset, humans outperformed the system. At its core, cognition involves the processes of acquiring, processing, and applying knowledge. Many contemporary AI systems like Gemmi, GPT, and Grok already mirror these basic cognitive abilities. These capabilities collectively allow AI to act with a degree of autonomy in a controlled environment; however, they often remain narrowly focused, and rigid, i.e., highly effective in specific tasks but easily disrupted with a change of context. This task-specific design leaves AI lacking the generalization and adaptability that are hallmarks of human intelligence. Humans get this ability by reflecting upon the outcome that they have received, and this ability is called *metacognition*.

Metacognition is the ability to monitor, evaluate, and adapt and is largely underdeveloped in current AI systems. This is a critical layer for achieving robust general intelligence. To develop AI with the interplay capabilities of cognition and metacognition, the work should be focused upon developing a system with the following capabilities:

- Learning of the system should not be limited to static datasets but also from its actions and real-world interaction.
- A system should be able to adapt itself and update its learning based on new information and feedbacks without requiring explicit reprogramming.
- A system should be able to reflect its own performance and revise its approach accordingly.

Human intelligence is shaped by the feedback loop of cognitive and metacognitive capabilities. Recreation of this loop in AI holds the key in creating systems that are not only reactive but proactive and self-improving just like humans.

While some foundational work towards developing metacognition capabilities in AI has already been started so lets us have a glance at those:

- Advances have been made enabling models to assign confidence estimates for each of their predictions. Techniques like Bayesian and entropy-based uncertainty metrics help identify outputs with high and low confidence.
- Advancements like meta learning or learning about learning as implemented in models like Model Agnostics Meta Learning allows systems to adapt quickly to new assignments by internalizing general learning systems. This capability mimics human' ability to transfer learning from an old experience to a new problem at hand.

CHAPTER 2   FOUNDATIONS OF AI: UNDERSTANDING KEY CONCEPTS AND TERMINOLOGY

- Explainable AI frameworks help to expose the reasoning behind each prediction that AI systems make. This helps in understanding how each prediction is made just like humans know the reasoning behind their decisions.

- Reinforcement learning also called closed-loop feedback helps machines to self-correct their behavior based on the real-time feedback that the machine receives. This helps machines take feedback at runtime just as humans can do, evaluate if the feedback is affirmative or negative, and then course correct their actions accordingly.

From Models to Minds: Emerging AI Architectures

In AI, various new paradigms have emerged recently. Let's understand how these align with different aspects of cognition and metacognition.

1. **Generative AI: Mimicking Perception and Expression**

    Generative AI systems like GPT, DALL, and Gemini excel in language generation, image synthesis, and text summarization by detecting patterns in massive datasets. These models have powerful cognitive abilities like pattern recognition, memory, and associative learning. However, they lack abilities like context awareness and self-awareness. While they generate text, they don't understand why a particular text was generated and also lack the ability to assess the quality of their output. These models are cognitively rich but metacognitively shallow.

CHAPTER 2   FOUNDATIONS OF AI: UNDERSTANDING KEY CONCEPTS AND TERMINOLOGY

2. **Agentic AI: Toward Self-Directed Systems**

   Agentic AI is a step further from passive responses to active decision-making systems. Systems like Auto-GPT were developed to accomplish goals and make plans required for achieving the goals. These systems also take and modify actions based on the feedback received. They are a step ahead of the Gen AI system and can perform metacognitive functions like progress monitoring, strategy adjustments at an elementary level, and a move toward adaptable goal-oriented intelligence.

3. **Reason-Based AI: Building Rational, Reflective Machines**

   This type of AI integrates models with logic, causal reasoning and symbolic processing to aid structured and explainable decision-making. These architectures combine learning power of neural networks with clarity of rule-based systems. This type of AI is also referred to as neuro symbolic AI.

   These models mimic a high level of human reasoning including deductive thinking, conscious problem solving, and ethical judgment. They allow a system to be able to explain their decisions and assess the degree to uncertainties. Reason-based AI offers a path toward rational transparent and ethically aware machine intelligence.

CHAPTER 2   FOUNDATIONS OF AI: UNDERSTANDING KEY CONCEPTS AND TERMINOLOGY

Data, Information, Knowledge, Wisdom (DIKW): Foundations for AI Thinking

DIKW collectively refers to the conceptual framework that explains how wisdom is created through a progressive transformation of data (Figure 2-1).

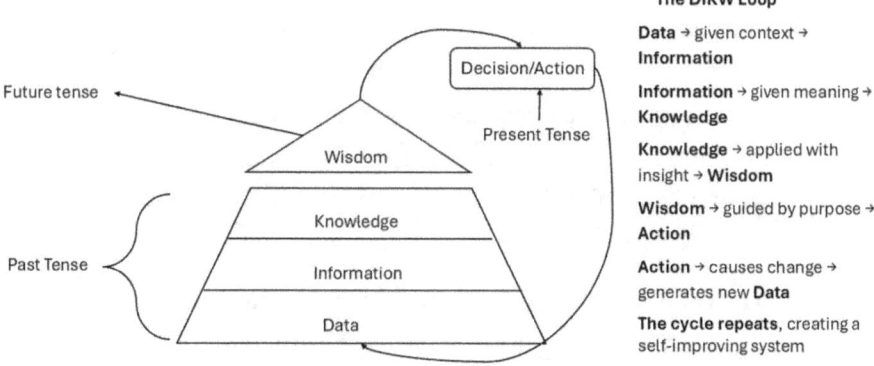

Figure 2-1.   DIKW loop in action

The DIKW Framework: A Conceptual Model of Intelligence

DIKW is a cyclic framework where data creates wisdom. Wisdom guided by purpose achieves an outcome to define real-world action, which in turn generates some data, which fine-tunes the wisdom. This framework provides a structured view of how data-driven wisdom is created and refined. It is because of its simplistic approach that this framework is widely referred to in knowledge management and information science domains. Since AI is also data dependent, it is gaining popularity here also. Let us understand each layer of this framework in detail.

CHAPTER 2   FOUNDATIONS OF AI: UNDERSTANDING KEY CONCEPTS AND TERMINOLOGY

- **Data:** Represents raw, unprocessed facts and signals with no interpretation attached it. The lowest level in the DIKW hierarchy. In digital systems, data may appear as numbers, text, sensor outputs, audio, or video. Example: 40, 32, 35.

- **Information:** Emerges from data when some context is associated with data. In simpler terms, information emerges from data when data is annotated, i.e., assigned a category name and given a structure. Information is basically used for answering fundamental questions like what, who, when, and where. Example: The temperature in New Delhi rose from 32° C to 40 ° C between 8 a.m. and 2 p.m.

- **Knowledge:** When information is interpreted and synthesized by recognizing patterns, understanding relationships and implications between various factors affecting the information, knowledge is created. Knowledge usually answers the "how" question. For example, an increase in morning temperature like this typically signals an onset of heat wave during summer months. The "how" question it answers is "How does this relate to the weather of the city?"

- **Wisdom:** Represents the highest value extraction from data; it also represents the highest level of the framework. Transformation of data into wisdom is a very sequential process. Data first translates into information and then information is synthesized into knowledge and then judgment, ethics, and insights are applied on top of the knowledge to generate wisdom. Wisdom is the actionable insight that can influence a decision. In summary, wisdom is the ability to apply knowledge in a way that is both contextually appropriate and aligned with the goal that must be achieved. It answers the "why" questions and guides in complex decision-making processes. Extending our temperature example, given the rising temperature, it is advisable to avoid going outdoors in the afternoon. In this example, an actionable insight is generated based on rising temperatures and what the outcome was when someone went outdoors.

- **Action:** Not directly associated with DIKW framework, but this is outcome of the decision made by applying the wisdom generated using this framework. It is the execution stage where wisdom is applied to create impact. Every action generates new signals such as user responses, metrics, or environment that become fresh data. This closes the feedback loop enabling continuous learning and refinement by enabling the DIKW cycle to repeat. In summary, action turns potential into progress and ensures that data-driven intelligence remains dynamic and evolving.

The DIKW framework is a generic concept that depicts how intelligence is created. To help readers understand better how this framework can be used to create evolving and dynamic AI systems, let's first understand how it helps create human intelligence.

DIKW as a Model of Human Reasoning

The DIKW framework is more than just a model for information systems. It also demonstrates how human intelligence evolves by perceiving the world, interpreting it, learning from experience, and making informed decisions

CHAPTER 2   FOUNDATIONS OF AI: UNDERSTANDING KEY CONCEPTS AND TERMINOLOGY

Understanding DILW from human intelligence perspective provides valuable insights for designing AI system, cognitive architecture, and product strategies that mimic ever-evolving human intelligence.

- **Data:** The data layer in humans is formed by capturing vast amounts of raw data through our five senses, i.e., sight, hear, smell, taste, and touch. For example, early in the morning you see dark clouds and feel a drop in temperature and smell an earthy scent.

- **Information:** These sensory inputs are processed by the brain to form information structures by adding context and prior experience. For example, you remember these conditions are generally followed by rain.

- **Knowledge:** Through learning and experience humans develop mental models that help them recognize patterns in the information gathered and the expected outcomes. Extending the cloudy weather example, based on your experience, you know that when the sky looks like this, it usually starts raining within an hour or two. Basically, now you have recognized the pattern that when it's cloudy accompanied with a drop in temperature, based on the historical trend, you estimate that it could rain.

- **Wisdom:** The highest form of human intellect is the ability to reason, consider future implications, and make decisions that integrate personal and social values. Wisdom also incorporates learnings from past experiences. Wisdom further incorporates real-world learning with knowledge appropriate to a context or application. In our example, if the context is you are about to leave for the office, the wisdom would say carry an umbrella, as you might get drenched.

- **Human action:** Once wisdom is applied to make a decision, humans take actions often associated with a goal to be achieved. These actions generate outcomes and consequences that are observed and re-sensed, creating new data. This closes the loop and enables continuous refinement of understanding and wisdom.

Turning Data into Action: Architecting Scalable AI Using DIKW

In the evolving landscape of enterprise technology, AI has moved from research concepts to a practical enabler of business value including innovation, competitiveness, and operational efficiency. However, its effectiveness does not solely depend on algorithms, but on the quality, coherence, and usability of the underlying data ecosystem that supports it.

Most organizations suffer from fragmented data systems, inconsistent data definitions, and silos, which limits AI's potential to deliver. A need of a cohesive framework that aligns all this is very profound. This is where DIKW becomes valuable. It provides a structured strategic model for building a cohesive and intelligent data ecosystem in which AI systems not only can function but also can thrive in alignment with organizational goals.

Data layer: The first layer in DIKW framework is the data layer; it is where raw facts, signals, and data are collected from diverse transactional systems and external systems. This layer is generally the most technically intensive requiring robust infrastructure capabilities for ingesting, storing, and securing data at scale. While technical prowess is important but not sufficient to develop this layer, an equal emphasis should be placed on governance issues like on making data a shared organizational asset with clearly defined ownership, standardized data formats, and strong governance policies to ensure privacy, quality, and compliance. This layer should instill trust in data, as the layers above cannot function reliably without trusting the data generated here. This layer should establish traceability, lineage uniformity in definition, and stewardship roles for ensuring data flows cleanly throughout the organization and serves as a reliable foundation for higher-order use.

Example: For a retail company with both online and offline stores, customer transaction data and purchase channel data need to be captured in a unified and traceable manner to enable any downstream analysis.

Information layer: Above the data layer sits the information layer, where data is structured, cleaned, and contextualized to give it meaning. In this layer, the objective is to translate raw data into understandable representations that can be used for basic decision support. This layer requires a mix of technical implementation required for data transformation and enrichment, strategic alignment on business logic, metric definition to make raw data meaningful. To extract maximum benefit from this layer, product leaders should ensure that definitions of these metrics and the transformation logic used for calculating these metric remains consistent across the entire organization. Promoting self-serve analytics and increasing data literacy across departments are the other initiatives that can help extract maximum value.

Example: Continuing our retail example, customer records from various different systems are mapped using a primary key (customer ID) and consolidated to create a single view of customer behavior.

Knowledge layer: This layer transforms structured information into insights by applying advanced concepts like analytics, machine learning, and experimentations. The knowledge created in this layer is not simply what algorithms learn and predict but also how an organization consumes and acts on these recommendations. To be effective, this knowledge layer must be continuously improved by implementing feedback loops, human judgment, and documentation. Models developed here must not only be accurate but also be interpretable, reliable, and auditable. Product leaders must ensure that various stakeholders like machine learning engineers, software engineers, domain experts, and business leaders are collaborative as the best knowledge is often created at the intersection of quantitative analysis and human expertise. Finally, all learnings made during the process of knowledge extraction, both failures and successes, should be retained.

Example: In our retail chain example, the development of a machine learning model for predicting customer churn and its integration in customer support workflows for proactive outreach is an example of the knowledge layer.

Wisdom layer: This is the highest layer in the DIKW framework. Wisdom refers to the ability to apply knowledge with judgment, foresight, and reasoning. In an organizational context, wisdom is seen as making decisions that go beyond optimization to consider the long-term impact, stakeholder well-being, and corporate values. These considerations help determine the ethical implications of AI and how to keep a human in the loop in high-stake AI implementations. Product leaders must create the organization's decision architecture such that AI augments human capabilities and does not replace human wisdom. AI development is driven by ethical principles like fairness, transparency, and responsibility.

Example: In our retail example, retraining data teams to use any variable that can cause biasness from being used in any data application is an example of organizational wisdom. Another example is to allow users to delete their personal information records from the system.

Action layer: Though not exactly the part of the DIKW framework, it is this layer that adds meaning to the entire DIKW framework by putting knowledge and wisdom generated through this framework into action. By capturing outcomes and user responses through feedback loops, this layer begins a new data cycle. This layer enables continuous learning and improvement by enriching the data synthesizing which knowledge and wisdom is created. Care should be taken that every action is measurable and tracked to validate assumptions and guide future decisions. This layer emphasizes learning from post-deployment data.

In conclusion, building a DIKW-aligned data ecosystem is not a simple linear technical project but an organizational-level transformation. It requires developing alignment across infrastructure, governance, talent, and leadership. For us product leaders, it offers a strategic lens to ensure the value of AI is fully utilized.

Summary

This chapter laid the foundational knowledge required to understand and design artificial intelligence by first exploring the nature of human intelligence. It explores how humans learn and evolve from a toddler depending on others to active, context-aware, and socially responsible adults.

The chapter delved further into the philosophical and scientific perspectives on intelligence, by emphasizing on the roles learning, context, cognition, and metacognition in shaping both human and artificial intelligence. It introduced the DIKW model as a conceptual outline for understanding how humans learn and how this model can guide the development of AI systems.

The chapter further explored the philosophical motivations behind the development of AI and highlighted the legacy of Alan Turing, the intellectual father of AI.

Finally, this chapter set the stage required for the rest of the book. The subsequent chapters will further explore key strategic and technological fundamentals that form the backbone of developing successful AI systems. These principles will support leaders in implementing AI including gen AI and agentic AI successfully across the various type of organizations.

CHAPTER 3

Building a Truly AI-Powered Organization: Strategy, Integration, and Transformation

Everyone wants AI. Few know what to do with it.
In the boardroom and strategy deck, the very mention of the term *artificial intelligence* is treated as a panacea, expected to drive exponential returns, cost optimization and spark innovation. But, in practice, many organizations struggle to move beyond pilot implementation and prototypes. Promising models rarely reach production stage, and even when they do, they don't deliver the expected value.

    The root of the problem isn't just technology, but everything around it, from misaligned teams, fragmented ownership, inadequate infrastructure to the absence of a coherent organization-wide data and AI strategy. As a result, many organizations remain in a cycle of endless experimentation without achieving any real impact.

CHAPTER 3   BUILDING A TRULY AI-POWERED ORGANIZATION: STRATEGY, INTEGRATION, AND TRANSFORMATION

This chapter starts by candidly examining the disconnect between AI's potential and the limited value most companies extract from it today. Understanding the root causes of failure is the first and most critical step. Once clarity is achieved, it becomes easier to develop an organizational roadmap that moves beyond isolated wins and impressive demos toward meaningful, scalable transformation.

Achieving this transformation requires three key enablers: leadership alignment, adaptable data and tech infrastructure, and fostering culture that embraces experimentation and continuous learning.

This holistic approach to building AI is essential because, in today's world, AI is no longer a technology tool. It has become the operating system on which the modern organizations think, make decisions, and evolve. To unlock its true potential, AI must be embedded into the DNA of the organization.

From Hype to Reality: Navigating AI's Adoption Challenges

AI has tremendous transformative potential, but it is elusive. To tap this tremendous potential, AI leaders must think beyond technical hurdles. They must elevate their role from just being product leaders responsible only for shipping features to strategic integrators and operational leaders who drive alignment across the organization.

By anchoring AI initiatives in business values, ethical principles, and cross-functional collaboration and prioritizing an organization's AI readiness, product leaders can transform AI from an aspirational vision into an organization's operating system of growth. For product leaders this is a generational opportunity, not just to ship smarter products but to shape business models and systems that define tomorrow's business.

CHAPTER 3   BUILDING A TRULY AI-POWERED ORGANIZATION: STRATEGY, INTEGRATION, AND TRANSFORMATION

Why AI Initiatives Underperform

As per various industry reports, AI implementations face a failure rate of 85%.[1] Why does this happen? And most importantly, what can we product leaders do to break this cycle? This section will answer those questions and explore the root causes beyond code and tech infra of AI failure and will then define the critical role product leadership must play in turning AI into an enterprise value. The following are some major root-cause categories that hinder AI success.

AI Doesn't Fail Because of Models; It Fails Because of Misalignment

Most failed AI projects show a clear pattern. It's not just technical complexity or bad data that leads to the failure; it is the breakdown of alignment between different teams, goals, and current technical capabilities of the organization and between vision and execution

The moment a product gets greenlit to use AI for solving a customer problem, all teams jump into execution, and the machine learning team builds a model trained on siloed or outdated data. Business, product, and machine learning teams all look at the different metrics that this solution will impact. Product teams struggle to integrate this model into their workflow. The engineering team struggles to prioritize the required infrastructure. With all this confusion, gradually executives lose interest, and the projects die a slow death.

This is not just a one-off story. This is an archetype of AI failure, and it starts at the very beginning of the discussion.

The Hype Trap: Overpromising and Underdefining

Another major reason behind the high failure rate of AI projects is that these projects are conceptualized on hype.

CHAPTER 3 BUILDING A TRULY AI-POWERED ORGANIZATION: STRATEGY, INTEGRATION, AND TRANSFORMATION

The mythical aura around AI often leads to inflated expectations. The stakeholders are promised human-like intelligence, rapid ROI, or full automation. AI is frequently positioned either as a sliver bullet or as a status symbol between teams. Despite this excitement, the implementation teams and product leaders often overlook the practical realities of AI. The real-world AI is "narrow or context based" and probabilistic in nature, and it has deep dependency on "quality and quantity of data" and "thoughtful" integration with technical workflows. Ignoring these factors during the planning and execution is a common reason why AI projects fail.

The first task for a product leader is to ensure that realistic expectations are set for stakeholders and that the implementation teams understand the practical limitations and nuances of AI.

Vague Goals = Unusable Models

Even when the right problem is chosen for AI deployment, it is vaguely defined. An example of a poorly defined goal is "This AI/ML implementation will help improve retention" or "This AI/ML implementation will make smarter recommendations."

The success of any AI implementation depends heavily on how clearly the problem space has been defined. AI needs specificity for choosing the data on which algorithms will be trained, for the choice of algorithms, and for the evaluation metric. A well-scoped AI problem should be connecting a business need to a set of data inputs, model outputs, and measurable results. Efforts should be made to translate the problem statement from a strategic viewpoint to a mathematical viewpoint. When it is missing, teams develop the solution in the dark. The following is the AI-specific version of the previously mentioned problem statement.

Retention problem statement: To solve this problem, multiple AI models will be required. A separate model will be required for identifying high-risk customers, and another model will be needed for recommending

CHAPTER 3 BUILDING A TRULY AI-POWERED ORGANIZATION: STRATEGY, INTEGRATION, AND TRANSFORMATION

personalized marketing offerings. These two models will have to connect to a CRM system to send communications to the customer. For the sake of simplicity, let's assume that the organization needs AI help only in identifying high-risk customers.

Here is the problem statement for developing identifying high-risk customers: Develop an AL/ML model that identifies high-risk users daily based on behavioral signals including shopping behavior, customer care interactions, product experience, and other relevant data

Later in this chapter, we will define best practices for translating a business problem into an AI/ML problem statement.

Data That Can't Deliver

AI doesn't just require a large volume of data, it needs to be relevant high-quality data. In many discussions we have observed that teams tend to focus on storing millions of data rows, without considering how much meaningful and diverse information those rows contain. While long data (row of data) represents multiple transactions, it is the variety and relevance of information represented by columns, also called *broad data*, that provides real insights and adds value to AI models. These columns contain critical information about each transaction. Additional forms of data such as text, images, and audio should also be considered for designing a robust AI model.

Other factors that are often overlooked but severely affect the probability of success of AI initiatives are fragmentation of data, poor data labeling, and poor data quality.

Product leaders should ensure that investments are made in developing a robust data ecosystem that supports high-quality data generation, storage, and ease to access. Today the availability of high-quality data is not just a nice to have. It is the table stakes for any AI product development and success.

Chapter 4 and Chapter 5 will help you understand more about developing a robust data platform and how data quality can be improved.

CHAPTER 3   BUILDING A TRULY AI-POWERED ORGANIZATION: STRATEGY, INTEGRATION, AND TRANSFORMATION

Shipping the Model Is Not the Finish Line

For most product managers, an AI product's active lifecycle ends once it is deployed. In reality, most models that go live never deliver the promised impact because of various factors ranging from not being able to integrate into real workflows to unmonitored model performance to not being able to AB test. Collectively, these issues lead to trust erosion over the time, and gradually these solutions fall off practical business use. Models that are not trusted or aren't used are just shelf wares.

Product leaders must recognize that deployment is not the end goal of an AI model; it's just the beginning of its real-world journey. To increase the likelihood of success, leaders should adopt a methodological end-to-end approach that spans planning, developing, and post-deployment activities of the AI initiative life cycle. Product leaders must also navigate the organization changes that might occur post AI model deployment.

Refer to Chapter 6 and Chapter 7 to read more about the best practices.

AI Is Not Software; Stop Managing It Like It Is

Traditional product management puts emphasis on planning and predictability, whereas AI product management demands something different: flexibility, exploration, and feedback. AI models are not deterministic just like traditional product features; their behavior changes with each customer they interact with. While it is difficult to completely scope and determine the quality of model output at the start, following best practices will help leaders a lot in navigating this flexible space.

Trying to force fit AI product management into the mold of conventional product delivery often results in unrealistic expectations, half-baked launches, and missed opportunities. AI product leadership is as much about learning as it is about delivery.

CHAPTER 3   BUILDING A TRULY AI-POWERED ORGANIZATION: STRATEGY, INTEGRATION, AND TRANSFORMATION

No One Owns the Outcome

Often AI products sit between multiple teams such as data engineering, machine learning, product management, engineering, and business. With too many teams involved and no clarity on common goals, the ask from each team becomes very ambiguous, and this ambiguity drags the down the AI initiative's performance.

With multiple teams working in silos and no single owner of the end-to-end solution, the AI initiative often becomes directionless. This reminds us of the parable of two blindfolded people and an elephant, where each person describes the elephant based on their own experience with complete disregard to each other's perception and overall bigger picture. In the same way, teams optimize for their own individual goals while the broader goal, i.e., delivering business impact, gets lost in translation.

Without a shared understanding, conflicting interpretations arise, and the true propose of the solution is lost. To avoid this product, leaders should lead the entire solutioning and take full ownership of the business outcome. Accountability in AI projects is not negotiable; it is essential for meaningful success.

Siloed Efforts, Isolated Teams

Another reason behind subdued AI performance is that often organizations try to develop AI initiatives in isolation confined only to the machine learning team with very little integration into product, tech, and business. This lack of cross-functional collaboration deprives teams from organizational readiness, which severely impacts the initiative's performance, and even the best of AI ideas and implementations go down the drain.

Organizations should ensure that every AI initiative is led by a product leader whose top priority must include gathering executive sponsorship, fostering cross-functional collaboration, and preparing organization prepared for successful implementation.

CHAPTER 3   BUILDING A TRULY AI-POWERED ORGANIZATION: STRATEGY, INTEGRATION, AND TRANSFORMATION

AI readiness is organizational, not just technical.

Ignoring Risk, Bias, and Ethics

Ethics in AI is often overlooked or treated as an afterthought. However, neglecting it poses significant regulatory risk and reputational risk. These risks impact customer trust and ultimately revenue.
Managing these risks should be the core responsibility of the product leaders. In the modern world, AI must be designed not just to work but to work responsibly.

Chapter 8 will help leaders deepen their understanding of ethics in AI and explore practical ways to embed ethical principles into real-world applications.

The Path Forward: What Product Leaders Must Do

Product leaders operate at the intersection of vision and execution. Their mandate makes them responsible for defining and executing the capability roadmap that aligns to the strategic goals of the organization. When it comes to leading AI, this role becomes even more critical and complex. Unlike traditional product management, successful AI implementation is not just delivering features. It is about orchestrating a symphony of interconnected systems, people, and process all working under a common strategy toward achieving a common goal. To get it right, product leaders must operate across three dimensions: strategic focus, technical fluency, and organizational influence.

Anchor AI initiatives in Business Value

AI product work does not start with algorithms and data. It starts with asking better questions. Product leaders must perform following tasks:

CHAPTER 3 BUILDING A TRULY AI-POWERED ORGANIZATION: STRATEGY, INTEGRATION, AND TRANSFORMATION

1. Clearly define business goals to be achieved by the AI initiative. Identify north star metrics (business metrics) that will be impacted by the initiative.

2. Translate a business problem statement into an AI problem statement. Identify what type of output the AI model will produce.

3. Identify the appropriate accuracy metric of the model and how changes in this metric are expected to impact the business north star metric. This helps in linking model performance to real-world business outcomes.

4. Once the north star metric (NSM) is determined, it should be tracked across various seasons for at least two business cycles. This helps in understanding how the external factors influence the NSM. If these factors are beyond control, then methods should be developed to negate their impact while assessing the AI model effectiveness. Otherwise, efforts should be made to capture the data of these factors for more accurate analysis.

5. Rather than starting with a narrow proof-of-concept mindset, begin by planning how the model will be integrated into the real-world technical ecosystems. Assess and identify the required changes within the existing infrastructure to support this model. Clearly define a deployment strategy and data management approach before initiating the POC. This comprehensive and proactive approach will prevent the project from becoming stuck in "proof of concept purgatory."

CHAPTER 3   BUILDING A TRULY AI-POWERED ORGANIZATION: STRATEGY, INTEGRATION, AND TRANSFORMATION

A strategic AI product leader does not chase use cases. They prioritize them like investments based on factors such as ROI, feasibility, and strategic alignment.

Develop Technical Fluency Without Becoming an AI Engineer

Software product managers can benefit greatly by drawing inspiration from product managers in industries like automotive and FMCG. Product leaders in those domains typically possess a well-rounded understanding of technology, design, and business needs. This holistic approach enables them to develop a business-viable, technologically viable, and desirable product. While we are not arguing that product leaders should attain the proficiency level of a software engineer or a machine learning engineer, building an application and concept-level understanding of both trades would help.

- Product leaders should understand the complete lifecycle of a data project. (Refer to Chapter 6.)

- Product leaders should understand what makes a use case data ready; they should understand concepts like data quality, volume of data, variety of data required for solving a use case, availability of data, difference between required frequency of data available, and current frequency of data generation.

- Product leaders should have a basic understanding of the types of learnings in AI including definitions, trade-offs, scenarios in which each should be applied, and how to measure accuracy.

- They should also understand the basics of MLOps including concepts like retraining strategies, deployment strategies, AB testing, etc.

- Finally, product leaders must develop a deep understanding of how to build trustworthy AI. This includes not only grasping the key concepts but also knowing how to operationalize these abstract concepts within their product development lifecycle. They should also develop the ability to define clear metrics and mechanisms to measure the progress. See Chapter 8.

Architect Cross-Functional, AI-Ready Teams

Often AI projects die a lonely death. To avoid this, leaders must act proactively to build cross-functional teams that operate in a collaborative style with each team's success metric tied to a common goal. At the onset of a project, this team should be formed with members from following expertise:

- Product managers: To define business goals and success metrics
- AI/ML team: To build model and evaluate performance
- Data engineering: To ensure data pipelines are well established
- Engineering and MLOps: To deploy, monitor, and maintain model in production
- Design: To shape up user experience in AI workflow
- Legal and compliance to manage any risk and regulatory constraints

Additionally, they should champion AI product readiness at the organizational level.

CHAPTER 3   BUILDING A TRULY AI-POWERED ORGANIZATION: STRATEGY, INTEGRATION, AND TRANSFORMATION

Why AI Products Are Different

AI is transforming the way digital capabilities including products and services are conceived, developed, and evolved. Unlike traditional software, AI-powered products are probabilistic, adaptive, and data-driven and often come with new types of dependencies and ethical requirements that were traditionally very rare.

In the previous section, we examined key factors behind the failure of AI products. At the core, these reasons stem from the organizational inability to recognize the different needs of AI software from traditional software. This disconnect often results in misalignment across teams, goals, and execution priorities. In this section, we will explore the differences between AI software products and traditional software to help product leaders make informed decisions while planning and prioritizing AI initiatives

Uncertainty Is Built into the Product

Traditional software operates through explicitly defined rules and deterministic logic. Developers define these rules and logic in a system using a programming language, and the software system is expected to behave exactly the same as programmed. On the other hand, AI systems are dynamic, and their behavior is dependent on the data on which it is trained. They learn patterns from the training data and use this knowledge to generate output based on probabilities. This behavior is emergent and often less predictable.

Behavior: Static vs. Evolving

In traditional software applications, the behavior remains predictable and static unless the software code is updated. The AI systems are diverse in behavior; in theory, almost every input data record may generate a different outcome because it would trigger a lesser applied pattern that the model learned during training. When an AI model is retrained, this diversity in outcome may further increase due to data drift or any new data pattern found in new dataset. All these factors can either improve or worsen the model performance.

Continuous Learning Replaces Static Releases

Software engineers have full control over the traditional application via programming languages codes and static configuration. In contrast, the AI system's behavior is driven by training data, hyperparameters, and algorithm architecture. Developers have less control over how the system will behave. To bring some level of predictability, product managers must shift from feature specs to data specifications, ensuring that the training data is a true reflection of the real-world in which product will operate.

The Business Case Is Less Predictable

Typically, a business case for any software development is built around driving efficiency, saving costs, or targeting new revenue avenues. Because traditional software development is more predictable, the ROI is more or else certain, provided the problem statement has been well researched. Whereas in AI products impact depends on data availability, model accuracy, and user behavior variability, which might not very evident early during the development, leaders must embrace iterative discovery, experimentation, and hypothesis-driven planning for AI investments.

CHAPTER 3 BUILDING A TRULY AI-POWERED ORGANIZATION: STRATEGY, INTEGRATION, AND TRANSFORMATION

Value Creation Shifts from "Only" Code to "Data and Code"

In traditional software applications, value is created by writing efficient code and implementing well-defined features. In AI while strong programming code is important, an equally important asset is data. Its quality, relevance, and adequate real-world representation directly influence the model outcomes. This requires a shift in investment from coding talent alone to resources for managing data lifecycle effectively.

User Trust: Predictability vs. Explainability

With traditional software, user trust is earned through consistent and predictable performance. AI systems, however, may produce varying outcomes in similar situations due to hidden data patterns and model complexity, making their decision-making feel opaque. Building trust in AI systems requires not only consistency and predictability but also explainability, fairness, transparency and thoughtful communication of levels of uncertainty to users.

Dealing with Deterministic vs. Opaque Failures

Traditional software malfunctioning can be traced to a specific bug or broken logic path. AI systems often fail in nondeterministic and nonreproducible ways, making the quality assurance procedure harder. Bugs may stem from data biases to underrepresented edge cases or subtle shifts in data patterns. Product leaders must work closely with data engineers and machine learning engineers to analyze and mitigate these failures.

UX, Brand, and Trust Are Intertwined

Traditional UX is shaped primarily by designing intuitive interfaces and reliable functionalities. In AI this concept is further extended to model performance and how well uncertainty is handled in interface design. Another critical aspect of designing AI products is including explainability and other ethical principles to ensure that users trust the information presented to them. Strategically, organizations must view trust and explainability as brand-critical features.

Ethics and Bias: Optional vs. Essential

Ethical considerations are minimal in traditional systems. In AI they are unavoidable. An AI product that exhibits biasness due to any reason could harm the impacted user category and cause severe damage to brand reputation. To build trustworthy AI, an active ML governance practice should be established whose primary responsibility is to oversee the compliance of trustworthy principles across all steps of the product lifecycle.

AI Demands a Product Culture Shift

Finally, AI demands a tectonic shift in an organization's approach toward developing digital capabilities. Organizations must move beyond shipping features to building intelligent systems. This includes treating data as an organization's asset and considering data governance and ethics as one of the strategic functions of the organization; shifting from deterministic thinking to more experimental and probabilistic decision-making; and lastly developing a culture of cross-functional collaboration.

CHAPTER 3   BUILDING A TRULY AI-POWERED ORGANIZATION: STRATEGY, INTEGRATION, AND TRANSFORMATION

Table 3-1. Key Differences Between AI Products And Traditional Software

Dimension	Traditional Software	AI-Powered Products
Logic & Execution	Rule-based, deterministic	Data-driven, probabilistic
Behavior	Static unless manually changed	Adaptive and potentially evolving
Control	Fully controllable via code	Behavior dependent on training data and tuning
Debugging	Root-cause traceable	Opaque failures; non-reproducible issues
User Trust	Easy to build trust through consistency	Requires explanation and transparency
Testing	Functional and integration tests	Statistical validation; edge cases are hard to catch
Dependencies	Logic, API contracts	Data pipelines, labeling quality, model weights
Product Lifecycle	Feature-centric	Model-centric (with lifecycle: training, tuning, decay)
UX Considerations	Interface-driven	Experience depends on AI accuracy and response variation
Ethics & Bias	Usually, low risk	Requires active governance to avoid harm/bias

Product Management Best Practices for AI Products

Table 3-2 outlines best practices that product management professionals should adhere to while developing AI products to achieve tangible, long-term benefits.

Table 3-2. *Product Management Best Practices for AI Implementation*

Focus Area	Best Practice
Discovery	Identify high-impact business problem. Start by translating business problem into well-written AI problem statement. Validate technical feasibility study including availability of data.
Data Strategy	Treat data as a product. Define KPIs for quantifying quality, and utilization of data. Develop feedback loops for collecting user feedback.
Cross-functional Planning	Avoid developing AI projects in silos. Align engineering, data, AI, and UX team early.
Metrics Design	Don't rely on accuracy alone for evaluating product success. Translate a model metric into a business KPI, e.g., conversions, revenue impact, trust score.
Monitoring	Set up real-time monitoring for input drift, model performance, and user overrides.
User Experience	Design with uncertainty, ambiguity, and explanation in mind. Communicate risk clearly.
Ethical Design	Embed Trustworthy AI practices into all streams of AI product lifecycle from ideation to design, data evaluation to post deployment.
Roadmap Planning	Plan for multicycle development including data acquisition, model iteration, UX improvements, policy updates.
Stakeholder Alignment	Work on creating AI awareness across leadership teams and nontechnical teams to set the expectations.

CHAPTER 3   BUILDING A TRULY AI-POWERED ORGANIZATION: STRATEGY, INTEGRATION, AND TRANSFORMATION

From Pilots to Platforms: Navigating AI Maturity

Up to this point, we have discussed the common reasons behind the failure of AI initiatives, how AI products differ from traditional software, and best practices that product leaders can adopt to mitigate the risks of AI failure. After analyzing the failure reasons, we identified that the key theme behind the high failure rate of AI products is not rooted in technology but in organizational challenges. As we saw in Chapter 1, AI-first companies succeed by deeply embedding AI into their strategy and operations from the ground up. Building on that pattern, this section will focus on identifying internal gaps and areas of improvement that may be holding organizations back from fully harnessing AI value. Let us first start by defining AI maturity.

AI maturity refers to the extent to which an organization is prepared to develop, deploy, and operationalize AI in a trustworthy and business-aligned manner. AI maturity encompasses how technical capabilities, strategic alignment, cultural readiness, and operational excellence work together to deliver an AI product with these characteristics.

From a technology perspective, AI maturity reflects the robustness of the technical infrastructure, including the data platform and the maturity of DevOps practices. DevOps is critical because machine-learning deployments are a specialized extension of traditional DevOps, commonly referred to as MLOps. It also considers the technical team's proficiency in the technologies required for the development, deployment, and maintenance of AI solutions.

From a strategic perspective, AI maturity indicates how well the AI initiatives are aligned with the broader business strategy.

From a product leadership perspective, AI maturity reflects the ability to leverage AI to meet user needs in an ethical and trustworthy manner.

CHAPTER 3   BUILDING A TRULY AI-POWERED ORGANIZATION: STRATEGY, INTEGRATION, AND TRANSFORMATION

Why AI Maturity Assessment Matters

The maturity assessment helps uncover tech infrastructure gaps and suboptimal workflows. It also highlights strategic gaps like a difference in department priorities, isolated AI implementations, and lack of organizational support in terms of executive support and resources. A comprehensive AI maturity assessment is a holistic diagnostic approach that examines various aspects of an organization such as strategy, governance, technology, and culture to identify gaps that could hinder success AI adoption

The outcomes of this study are diagnostic and foundational in nature, which helps in creating effective execution plans and making informed investment decisions directed toward ensuring a maximum return on AI investments. The following are the key benefits.

Clear Understanding of Current AI Readiness

- Offers a comprehensive snapshot of where the organization stands in its AI adoption and readiness journey across multiple dimensions like leadership, culture, strategic planning, data readiness, talent pool, technology infrastructure and governance policies

- Identifies readiness level of foundational elements such as data architecture, executive support, and AI literacy across technical and non-technical teams

- Identifies maturity-level disparities between different departments of the organizations, which enables leadership to channelize efforts where required

- Helps identify the strengths of an organization that can be leveraged to accelerate AI adoption and which weakness require immediate attention

CHAPTER 3   BUILDING A TRULY AI-POWERED ORGANIZATION: STRATEGY, INTEGRATION, AND TRANSFORMATION

Alignment Between AI Efforts and Business Strategy

- Ensures that AI initiatives are coupled with an organization's core strategic objectives
- Empowers executives to clearly differentiate between early-stage experiments and mature transformation focused AI projects enabling realistic ROI expectations and better-aligned resource allocation
- Helps in bridging the gap executive's vison and operational capabilities by identifying gaps and aligning resources around common goals
- Facilitates prioritization of use cases for AI implementation with maximum potential and strategic relevance, rather than chasing use cases that are technically exciting to solve

Visibility into Capability Gaps

- Exposes critical gap in technology infrastructure that can hamper AI initiatives success
- Helps in talent management plan by identifying talent gaps or reskilling needs especially in key areas like data engineering, AI, MLOps, governance, etc.
- Helps in identifying operational inefficiencies that might impact scaling AI solutions, including siloed workflows, bad data management process, or lack of lifecycle management practices

CHAPTER 3 BUILDING A TRULY AI-POWERED ORGANIZATION: STRATEGY, INTEGRATION, AND TRANSFORMATION

A Tailored Roadmap for Scalable AI Adoption

- One of the key outcomes of this exercise is creation of a prioritized, sanctionable roadmap that connects the current maturity level to the organization's strategic objectives. This roadmap spans multiple departments, reflecting an interdisciplinary approach. It outlines a multiyear plan with distinct short-term, medium-term, and long-term objectives with well-defined milestones and assigned ownership.

- Ensures that every step mentioned in the roadmap is timeboxed and tied to a specific objective with a KPI and progress indicator to track success.

- Helps leaders plan a budget effectively by helping fair resource allocation across foundational improvements and strategic use cases.

Confidence in Use Case Deployment

- Enables informed selection of use cases for AI implementation by assessing strategic alignment and organizational readiness status including technical feasibility, data availability, talent availability, and risk mitigation abilities

- Helps in establishing a framework for ongoing reassessment of use case readiness in line with evolving organizational capabilities

Stronger Risk and Governance Posture

- Facilitates development of Trustworthy AI frameworks aligned with international best practices, organizational risk tolerance capacity, regulatory needs with clear accountability, and policy enforcement

- Facilitates the development of Trustworthy AI frameworks aligned with international best practices, organizational risk tolerance, and regulatory requirements, ensuring clear accountability and effective policy enforcement. This approach reduces risk exposure by embedding trust, transparency, and compliance into AI systems throughout their design and deployment.

Cultural Readiness for AI Transformation

- Evaluates organizational readiness for AI adoption, by measuring awareness, trust in technology, and engagement level. It also identifies early adopters and areas of resistance. These insights help guide targeted communication and training for increasing AI acceptance within the organization.

- Measures current AI literacy level across different functions. It helps in identifying customized training and cross-functional collaboration needs.

- Promotes leadership alignment on change management processes and principles required for AI success and fosters future-ready culture that integrates AI as a core aspect of company's operations.

CHAPTER 3   BUILDING A TRULY AI-POWERED ORGANIZATION: STRATEGY, INTEGRATION, AND TRANSFORMATION

AI Maturity Scoring and Evaluation Framework

Now that we have outlined the need for and importance of conducting AI maturity study, let us now understand how to structure the study for maximum impact. While various approaches exist, our experience as seasoned professionals with a broad perspective on technology leads us to believe that any assessment aimed at evaluating organization's readiness for AI adoption should cover the following key elements.

Strategic Factors for Scoping an AI Maturity Assessment

1. Granularity at which the assessment will be conducted. i.e., at the company level or at the individual department level or at the team level should be guided by few strategically significant factors. These factors ensure that the assessment reflects the realities of the organization and delivers actionable outcomes.

    a. **Organizational Size and Structure:** Larger organizations with decentralized operations often have significant variations in AI maturity level across departments or business units. In such cases, department-level/business unit or team-level assessments are recommended to capture the difference. Once each business unit/ department is assessed, a comparative report should be created at the corporate level to give the executive group a clear understanding of where the group stands at each touchpoint.

CHAPTER 3   BUILDING A TRULY AI-POWERED ORGANIZATION: STRATEGY, INTEGRATION, AND TRANSFORMATION

It is wiser to conduct a company-level assessment for midsize to smaller companies who tend to have more centralized decision-making and shared technology function including data teams.

    b. **AI Adoption Footprint:** When multiple departments/business units are running AI programs independently, then a department/business unit level study should be conducted. We recommend collating these departmental results on an organization level to draw a clear picture of AI maturity.

On the contrary, if AI adoption is still in the nascent stage, then a broad company-level assessment is recommended.

    c. **Strategic Business Goals:** The granularity of a study should align with what the organization wants to achieve with AI. If the goal is to achieve an enterprise-wide transformation, then always start with a broader organization-level study followed by a deeper department-level study. If the focus is on a specific function, then always conduct an assessment for the specific department to identify specific interventions.

2. The assessment should cover readiness across the following five dimensions:

    a. **Strategy:** Assesses whether the organization has a clearly defined AI strategy aligned with business goals, with leadership commitment and sufficient investment plans in place

b. **Capability:** Evaluates the technical foundation including infrastructure, technology maturity, data management maturity, tool utilization, talent and working methodology, required to build and scale AI

c. **Governance:** Evaluates presence of control, policies, and organization's structures required for facilitating the development of Trustworthy AI

d. **Human values:** Focuses on measures taken for fostering talent management, skill development, and other softer aspect such as collaboration between teams

e. **Adaptive mindset:** Evaluates measures taken for making an organization culturally ready for adopting AI, including openness to change, innovation, experimentation and continuous learning

AI Maturity Framework: Strategic Pillars, Focus Areas, and Evaluation Criteria

To effectively assess an organization's AI readiness, the maturity model is structured into three interconnected layers.

1) Strategic pillar
2) Focus areas
3) Evaluation criteria

This structure ensures that the readiness study provides outcomes that are both strategic and operationally actionable. Another advantage of this structure is it provides a complete picture of where an organization stands and what needs to be done to attain the strategic goals set.

CHAPTER 3   BUILDING A TRULY AI-POWERED ORGANIZATION: STRATEGY, INTEGRATION, AND TRANSFORMATION

At the highest level, the strategic pillars represent broad organizational domains that are critical for a successful AI transformation. These pillars are

- Strategy
- Capabilities
- Governance
- Human values
- Adaptive mindset

The second layer of this structure is created by breaking down each strategic pillar into focus areas that capture specific functions within that domain. At this stage, the focus areas under each domain are identified, which are important for AI program success.

Each of these areas is further detailed into measurable and actionable subitems also referred to as *evaluation criteria*. These are the criteria on which each focus area under a domain is evaluated. The researcher responsible for conducting this evaluation will frame questions and conduct a study around these evaluation criteria.

The following are the definitions of each focus area, highlighting what it evaluates and how it contributes to the organization's AI transformation journey. These definitions serve as a foundation for deeper evaluation at the evaluation criteria level where specific measures are used to access maturity.

Strategy

- Strategy, vision, and purpose: This focus area evaluates how clearly the organization has defined the vision for AI implementation and how that vision is aligned to business strategy. This focus area also evaluates the level of executive commitment and allocation of organizational resources including finance toward the realization of that vision.

CHAPTER 3   BUILDING A TRULY AI-POWERED ORGANIZATION: STRATEGY, INTEGRATION, AND TRANSFORMATION

Governance

- Leadership, governance, and trust: This focus area examines the executive sponsorship in the form of degree to which they are involved and resources committed toward AI success. The presence of an AI governance structure including but not limited to the RACI matrix, AI board, and risk management framework that enables development and deployment of Trustworthy AI.

- Ethics, compliance, and responsible AI: Examines the presence and effectiveness of responsible AI frameworks and the degree to which AI development meets regulatory compliance.

Capability

- Data maturity: Measures the processes and governance structures involved in managing the data in an organization from data generation in transaction system to storage in centralized systems to the ability to utilize data. This focus area also evaluates the quality of data available in the organization

- Technology maturity: This focus area evaluates how mature the technology utilization is in an organization. It covers process and policies governing engineering practices. This focus area also evaluates different categories of tool utilizations required for developing and managing AI initiatives.

- AI operationalization and integration: Assess the degree to which is AI is embedded in business processes and how many AI initiatives are scaled across use cases and are delivering meaningful business impact,

- Talent and skills readiness: Reviews the availability and development of AI skilled resources not only in technology domain but also in business units. Also evaluates policies and strategies for talent retention and development.

Human Values

- Ethical data usage: This focus area examines how well ethical principles are defined and embedded in data management practices.

- Culture and human-AI collaboration: This focus area examines the degree of organization's willingness to accept AI, level of trust in AI, and development of workflows where AI and humans can interact.

- Stakeholders and user inclusion: Evaluates how stakeholders are engaged in the process of co-designing AI systems and how feedback is collected about the AI product performance and how that feedback is processed and implemented.

Adaptive Mindset

- Organizational agility and transformation: Measures an organization's ability to adapt to AI-driven changes and its ability to evolve business model accordingly

- Learning systems and continuous improvement: Evaluates the mechanisms employed to continuous feedback collection and model improvement. This focus area also includes evaluation of how well the existing process in the company collects learning made from AI systems and how these learnings are passed on to other departments.

CHAPTER 3 BUILDING A TRULY AI-POWERED ORGANIZATION: STRATEGY, INTEGRATION, AND TRANSFORMATION

Table 3-3 captures this framework and serves as a practical guide to assess and advance AI maturity across all levels of the enterprise.

Table 3-3. *AI Maturity Study Framework*

Strategic Pillar	Focus Areas	Evaluation Criteria
Strategy	Strategy, Vision & Purpose	Defined AI vision Linkage to business strategy Executive alignment AI investment strategy
Governance	Leadership, Governance & Trust	Executive sponsorship Governance structure Accountability models Trust & risk frameworks
	Responsible AI & Compliance	Regulatory compliance Fairness & bias governanceExplainabilityAI audit readiness
Capability	Data Maturity	Data availabilityData quality Data governanceMetadata & lineage
	Technology Maturity	Model development lifecycle MLOps infrastructure Monitoring & retraining Tooling strategy
	Operationalization & Integration	Use case scaling AI process embedding Impact measurement AI-as-a-service models
	Talent & Skills Readiness	AI skill depth Functional collaboration Hiring strategy L&D programs

(*continued*)

Table 3-3. (*continued*)

Strategic Pillar	Focus Areas	Evaluation Criteria
Human Values	Ethical Data Use	Bias management Data minimization Consent handling Data ethics reviews
	Culture & Human-AI Collaboration	Employee trust Role augmentationChange readinessHuman oversight in AI
	Stakeholder & User Inclusion	Customer co-design Internal stakeholder engagement Feedback loop integration
Adaptive Mindset	Organizational Agility & Transformation	AI experimentation culture Agility of business model Innovation alignment Governance flexibility
	Learning Systems & Continuous Improvement	Model improvement feedbackKnowledge capture Human learning loop Cross-domain insights

Scoring Methodology

The AI maturity evaluation framework consists of five distinct levels on the maturity scale. Each level represents a distinct stage in an organization's journey of attaining highest level of Ai maturity. These levels are as follows:

1. Initial: At this stage, companies have a very ad hoc approach toward AI. The company has no strategy alignment between AI implementation and business goals. The company does not have any infrastructure to support AI development.

2. Developing: The organization has started making early-stage efforts toward adopting AI. The consistency in efforts made and output received from AI

implementation is still lacking. AI-related activities are scattered delivering limited value. Data and AI governance structures required for oversight are missing.

3. Established: AI practices and processes have been defined clearly, but these practices are not scaled at the organization level. AI initiatives are now better aligned with business objectives. AI governance processes are still evolving.

4. Advanced: AI adoption has scaled organization-wide. AI governance frameworks are well established. AI initiatives are demonstrating tangible benefits on business performance. Teams collaborate with each other. The company has now created the culture, which is welcoming to AI implementation in key business processes.

5. Optimized: This is the highest level of AI maturity level. AI is deeply embedded as a core enabler for business. Continuous improvement and innovation drives AI efforts. All governance processes are mature enough to ensure all AI initiatives are truly trustworthy.

The maturity assessment uses a structured multilevel scoring technique to provide a nuanced and actionable report of organizational readiness.

Each evaluation criteria includes one or more specific questions designed to probe key aspects of maturity in that area. The surveyor should rate each question on a 1 to 10 where 1 represents minimal and 10 indicates highest level of optimization.

For each criterion, the score of all its questions is averaged to produce a single score, which will represent criterion score. The criterion scores are again averaged to generate a score of each focus area. This score reflects the overall maturity level of that focus area with respect to AI adoption.

CHAPTER 3 BUILDING A TRULY AI-POWERED ORGANIZATION: STRATEGY, INTEGRATION, AND TRANSFORMATION

To calculate a score for a given strategic pillar, scores for all focus areas under that strategic pillar are averaged, representing a maturity level across that broad organizational domain.

Finally, to calculate the overall AI maturity score at the organization level, the score of all strategic pillars is averaged. This hierarchical scoring model ensures that the study covers both detailed actionable insights at the micro level and provides a broad strategic view at the organizational level.

Table 3-4 recommend s best practices while rating each question on the criterion level.

Table 3-4. AI Maturity Scoring Best Practice (Per Evaluation Question)

Score	Maturity Level	Strategic Meaning & Use Guidance
1–2	Nonexistent/Ad Hoc	No formal practice. Activities are unstructured or individual-driven. No awareness or governance.
3–4	Emerging	Early-stage awareness or pilots exist. Efforts are informal, inconsistent, and lack coordination.
5	Defined	Capability is documented and used in pockets. Limited governance and measurement in place.
6	Structured & Managed	Standardized processes aligned to business needs. Governance emerging; early KPIs tracked.
7	Integrated Across Functions	Practice is applied organization-wide. Cross-functional collaboration and governance are active.
8	Enterprise-Wide & Measurable	Fully scaled and aligned with strategy. Impact is measured; Responsible AI is enforced.
9	Proactive & Benchmark-Aligned	Leading practices benchmarked externally. Predictive governance and external audits in place.
10	Optimized & Industry-Leading	Continuous improvement and ethical design embedded. AI shapes strategy and influences the industry.

Always calibrate scores 7+ against evidence, and not anecdotes. Avoid awarding 9–10 unless the capability shows enterprise-wide impact, continuous refinement, and responsible scaling.

Table 3-5 maps maturity scores to defined stages of AI adoption. This scoring scale helps organizations interpret their results and understand where they stand on the AI maturity curve.

Table 3-5. *AI Maturity Score Classification Table*

Score Range	Maturity Level
1.0–2.9	Initial
3.0–4.9	Developing
5.0–6.9	Established
7.0–8.9	Advanced
9.0–10.0	Optimized

Building an Enterprise AI Strategy Framework

After assessing your organization's AI maturity, the next step is to translate insights gained into an actionable AI strategy. In this section we will explore a structured framework for building an effective AI strategy. We also explore how to connect this strategy with broader product and technology agendas. A well-defined AI strategy isn't just a guiding principle for driving investments; it is the foundation for building scalable, ethical AI capabilities. Before we explore the framework for developing a good and effective AI strategy, let's start by understanding what the characteristics of a good AI strategy are.

CHAPTER 3   BUILDING A TRULY AI-POWERED ORGANIZATION: STRATEGY, INTEGRATION, AND TRANSFORMATION

What Is a Good AI Strategy?

A good AI strategy is a clear value-driven plan that helps leverage AI to solve high-value business problems. The following are the characteristics of a good AI strategy:

- A good AI strategy is aligned to the strategic objectives of the organization.

- A good AI strategy is highly focused and ruthlessly prioritizes use cases that have the potential of delivering high business value, is feasibility operationally and technically, and is financially viable.

- It helps develop AI as a capability and not just as a group of isolated AI projects mushrooming in the organization.

- It understands the importance of data and helps develop data infrastructure in the organization to enable trustworthy data generation, storage, and access.

- It helps in embedding governance and ethics in AI development to ensure all AI developed in the organization is trustworthy.

- A good AI strategy enables organizationally embedded AI development, which is supported by leadership, and helps in developing team capabilities and enables cross-functional collaboration.

Good AI Strategy Example

Strategic Objective: Establish AI-driven operational efficiency across customer-facing functions, beginning with customer support.

CHAPTER 3   BUILDING A TRULY AI-POWERED ORGANIZATION: STRATEGY, INTEGRATION, AND TRANSFORMATION

Near-Term Goal: Demonstrate value through an AI-powered ticket triage and response system, targeting a X% cost reduction over Y months.

Foundational Enablers:

- Leverage two years of support ticket data for model development.
- Build reusable pipelines and APIs for future support automation projects.
- Ensure all models meet organization-wide standards for explainability, fairness, and compliance.

Long-term Vision: Scale similar AI capabilities across other departments (e.g., sales support, HR inquiries) to drive consistent, responsible automation.

What Is a Bad AI Strategy?

A bad AI strategy is often tech driven, is vague, or is misaligned with organization's strategic needs. It has the following characteristics:

- Is not focused and often use cases are shortlisted because they were interesting
- Promotes AI as the panacea, a fit cure for all organization's problems
- Prioritizes projects with short-term wins over long-term benefits
- Underestimates the importance of embedding AI in an organization's fabric
- Pursues moon shorts without conducting a feasibility and viability study

CHAPTER 3   BUILDING A TRULY AI-POWERED ORGANIZATION: STRATEGY, INTEGRATION, AND TRANSFORMATION

After understanding what makes a good or bad AI strategy, let us now move forward and explore to how to develop a good AI strategy.

Strategic, Actionable AI Strategy Framework: A Six-Layer AI Strategy Framework

AI strategy is not just a vision; it is a structured, actionable plan that acts as a bridge between what a business wants to achieve and the AI capabilities of the organization while considering the maturity of the organization. It outlines how the organization will invest, scale, and govern artificial intelligence to drive a measurable business value over a period of three to five years. Table 3-6 shows the layers of a complete AI strategic framework.

Table 3-6. *Layer AI Strategy Framework*

Layer	Name	Goal	Output
1	Business Alignment	Know why you're investing in AI	Strategic goals tied to AI outcomes
2	Capability Check	See if you're ready to build and scale AI	Gap analysis of data, tools, skills
3	Use Case Selection	Choose valuable, feasible AI projects	Prioritized list of AI use cases
4	Execution Planning	Set up people, process, and tools	Delivery model, MLOps, governance plan
5	Investment Model	Budget and staff the right initiatives	AI investment roadmap
6	Learning & Feedback	Improve based on real-world results	Metrics dashboard and strategy review loop

CHAPTER 3 BUILDING A TRULY AI-POWERED ORGANIZATION: STRATEGY, INTEGRATION, AND TRANSFORMATION

Layer 1: Business Strategic Alignment: Understanding the "Why" of AI

The first and most important step in creating an actionable AI strategic framework is to ensure that it is aligned with the business goals of the organization. Without a clear link to business objectives, AI efforts can quickly become a bunch of visionless experiments with little or no association to what truly matters. It is this layer that helps explore the real reason behind the AI implementation. In simple terms, it answers the question, why are we doing this and what do we hope to achieve.

The objective of this layer is to align AI initiatives with the enterprise vision and business outcomes.

To achieve this objective, the following four major activities should be planned:

1. Define a company vision and link it to AI: Product leaders should work with business leaders to define how AI will contribute to the organization's growth in the next three to five years. This is done by identifying top areas of strategic importance for the organization and then defining on a high level if AI can contribute. This step plays a very important role in making AI a part of leadership's roadmap and not just a product and tech-only activity.

2. Build an AI investment thesis: In this activity, leaders define the rationale behind investing in AI and answer the following questions: Will AI help in driving growth? Or improve margin, market share? Or will AI help in creating a new product offering? This activity helps in setting the expectations from AI implementation and helps while prioritizing and funding individual AI initiatives.

3. Goal identification: Identify a list of three to five concrete strategic outcomes that will drive AI capability development. This activity provides clarity and direction for a team working while developing and identifying AI use cases.

4. AI value mapping: Identify how AI can unlock value across identified functions. At this stage, a high-level mapping is done between AI values and goals. For example, if one of the goals was to reduce revenue seepage by improving the trustworthiness of the platform, then the AI value will be: Develop fraud detection system to detect fraud in real time.

Deliverables from Layer 1

The following are the deliverables from layer 1 that act as the guiding map for all AI activities in the organization.

1. AI strategic brief: A one- to two-page document summarizing the business-aligned AI goals.

2. Executive sponsorship: By signing the Ai strategic brief, leaders commit their resources to support AI development.

3. AI value map: A list of high-level problem statements often at a time when AI can create measurable value. These problem statements are mapped to AI value themes to create a high-level view of how AI will help.

Review Mechanism: An AI strategic plan should be reviewed at the same frequency at which a company's OKR achievements are reviewed, as both are now linked together.

CHAPTER 3 BUILDING A TRULY AI-POWERED ORGANIZATION: STRATEGY, INTEGRATION, AND TRANSFORMATION

Layer 2: Capability Foundation (The "Can We?" Layer)

After the strategic goals for AI are established in layer 1, this layer aims at finding the answer to the question, "Do we have the capability to achieve the strategic goals?"

The objective of this layer is to evaluate internal readiness spanning technical, talent, data, and organization alignment required for executing effectively and for achieving the defined strategic AI goals. It ensures that teams don't just get to development stage without completing the necessary groundwork. This step helps in mitigating the risk of failure, delay, or potential regulatory or ethical issues during the implementation or post deployment.

The following are the set of core activities performed in this layer:

1. Evaluation of data infrastructure readiness: This step evaluates the quality, availability, and completeness of data. It also evaluates the state of the data ecosystem of the organization, which means whether the data architecture supports a unified view of data or not. Has the data ecosystem implemented concepts like lineage, observability, and data contracting to boost the quality and availability of data?

2. Evaluation of machine learning supporting platform and tools: This step evaluates how mature the machine learning development and deployment ecosystem including platforms and policies are. This includes automated MLOps procedures, availability of AB testing platform, policies, and tools to manage the reuse of features that are developed for various models. The core objective here is to understand how capable an organization is to execute complex AI projects.

3. Talent model maturity: This evaluates the availability of the right talent required to develop and operate complex ML applications. It also evaluates AI literacy levels across teams including executives and business. The core idea behind this step is to understand if AI is embedded in the culture of the organization or is localized to a specific team.

4. AI Governance level: This evaluates if the organization has any controls and structures to ensure that all AI developments are trustworthy. It evaluates what all frameworks and policies are in place to assess and manage risk associated with AI development including regulatory and ethical.

This layer provides the true state of internal capabilities of the organization required for achieving the strategic goals. It enables leaders to assess readiness across critical dimensions such as talent, data, infrastructure, and technology. By revealing capability gaps, this layer helps decision-makers determine where targeted investments should be made and in what form, either through in-house development, purchasing market-ready solutions, or forming strategic partnerships with technology providers.

Key Deliverables from Layer 2:

- AI capability audit report: Snapshot of current state and gaps across infrastructure, tools, and teams
- Build/buy/partner strategy: Recommendations for each capability domain

CHAPTER 3   BUILDING A TRULY AI-POWERED ORGANIZATION: STRATEGY, INTEGRATION, AND TRANSFORMATION

Layer 3: Strategic Use Case Portfolio: The "What" Layer

Once an organization has aligned on the strategic goals it aims to achieve through AI investment (layer 1) and has evaluated its capability to support those investments (layer 2), the next logical question it must answer is, what should we develop to effectively meet the strategic goals tied to the investment?

Layer 3 of this framework introduces a disciplined approach to choosing use cases. This layer ensures that use cases are worthy enough to solve and ensures that AI efforts are business led and not technology driven.

The objective of this layer is to identify and prioritize use cases that can balance value, feasibility, and financial viability. Under this layer the following activities are performed.

- Map strategic goals to business domains: Once strategic goals for AI have been identified, apply the value chain analysis framework to map business functions based on where they contribute to value creation with respect to the specific goal. This helps in identifying which functions play a critical role in achieving the strategic goals and enables focused use case identification.

- Identify and evaluate AI use cases: Based on the previous mapping, identify potential use cases from identified business functions. Each of these use cases will be evaluated to determine their desirability on the following factors:
    - Potential to deliver business value.
    - How feasible they are from technical and organizational standpoint. If a use case has a potential to deliver high business value but the

organization does not have the required technical and organizational structures to support it, then that use case should be deprioritized, and efforts should be made to first develop technical and organizational competency required for enabling that use case. Once that level of capability is achieved, then that use case should be prioritized.

- Financial viability of the use case: In this step each use case should be evaluated to check whether the development of this use case makes financial sense or not. If a use case that is feasible to develop and has the potential to be delivered but the cost incurred is way more than the potential benefit, then that use case will become a financial liability for the organization. The leader should not shortlist these use cases.

- Each of the use case should be evaluated on all three mentioned criteria and then should be shortlisted and prioritized.

- Once each of the use case is screened on the previous criteria, some organizations classify use cases into the following categories based on the potential to deliver business value, associated risk, and how feasible they are.

  - Flagship/Core: These are the use cases that are focused on delivering value to the core operations of the company. Examples are cost optimization projects focused on reducing recurring costs.

CHAPTER 3   BUILDING A TRULY AI-POWERED ORGANIZATION: STRATEGY, INTEGRATION, AND TRANSFORMATION

- Strategic Bets: These are use cases that are focused on transforming the way an organization functions or that have the potential to open additional revenue lines for the organization.

- Experimental: These projects are high-risk projects where the company does not have the technical and organizational resources but they have the potential to deliver high business value. An example is the application of Agentic AI for automating procurement.

• Another factor to consider here is how long the use case will take to fully develop and become available for the customers to use. Ideally, use following timelines to classify use cases into different categories:

- Short-term projects: If the duration is less than three months, then a project is classified as a short-term project.

- Medium-term projects: If the duration is between three to six months, then a project is classified as a medium-term project.

- Long-term projects: If the duration is greater than six months, then a project is classified as a long-term project.

• Use evaluation and time to market criteria to prioritize and create an order of execution of these use cases. Care should be taken to create a blend of all three categories of projects while deciding the priority and execution order.

111

CHAPTER 3 BUILDING A TRULY AI-POWERED ORGANIZATION: STRATEGY, INTEGRATION, AND TRANSFORMATION

Key Deliverables from Layer 3

- By applying tools and techniques of this layer, the organization can define a program-level AI roadmap with the following:

  - Three to five flagship initiatives: These are high-value strategic use cases that have high technical and organizational feasibility and are financially viable.

  - Five to ten high ROI supporting use cases: These use cases are typically directed toward solving current operational challenges that business is facing today.

Layer 4: AI Investment Model: The "With What?" Layer

To scale AI effectively, organizations must stop treating it as a tech playground. Instead, AI must be approached like any other strategic capital investment, where resources are allocated based upon expected return, risk profile, and alignment with the core strategy.

Layer 4 focuses on turning AI into a disciplined function complete with a business case, ownership model, and ROI tracking. This layer ensures that AI projects are not funded in isolation but receive portfolio-driven investment funding with defined goals to achieve.

AI is often underfunded or inconsistently funded because in most cases technology teams drive and lead AI initiatives. To prevent this, AI initiatives should always be led by a product leader who can elevate AI from tech playgrounds to a strategically managed portfolio-driven function.

Objective: Treat AI as a capital investment with expected ROI.

CHAPTER 3   BUILDING A TRULY AI-POWERED ORGANIZATION: STRATEGY, INTEGRATION, AND TRANSFORMATION

The following core activities take place in this layer:

- Viewing AI as a strategic class: Leaders must change their mindset and start viewing AI as strategic class alongside other capital initiatives. Once this mindset change is achieved, then AI investment decisions are like any other portfolio decisions, and project prioritization methods change from how interesting a use case is to more value versus risk comparison.

- AI portfolio management: Organize AI initiatives into a well-structured portfolio. A portfolio contains one or more AI implementations and associated technology development aimed to achieve one strategic goal. Classification can be done using the following three-tiered structure:

    - Core: This tier should contain projects aimed at developing core data and AI capabilities like data platform, technology infrastructure, and governance framework.

    - Enablers: These projects are aimed at developing reusable components, like shared data assets, MLOPS functionality, and other ML components that can be recycled across multiple projects. While the core project develops the infrastructure capability required for delivering AI, this layer aims at developing a standardized piece of technology that can be used anywhere.

    - Application: These are those projects that are directed toward solving business problems and developing AI applications by using capabilities developed by core projects and reusing enabling component developed in enabler layer

CHAPTER 3   BUILDING A TRULY AI-POWERED ORGANIZATION: STRATEGY, INTEGRATION, AND TRANSFORMATION

- Care should be taken to balance each portfolio by selecting all kinds of projects like short-term and long-term projects and high- and low-risk projects.

- Define clear ownership and accountability structures for each project and portfolio to ensure clarity on the role and responsibilities of stakeholders.

- Don't confuse the project categorization done in layer 3 with the portfolio structuring in layer 4. In layer 3, the focus is on identifying and prioritizing AI use cases based on execution readiness, potential business impact, and financial viability. The categorization is done to guide near-term planning and order of execution sequencing.

- In contrast, layer 4 focuses on managing these initiatives as a financial and operational portfolio. This categorization is done to achieve the following two objectives:

  - Ensure balance resource allocation between different project categories.

  - Introduce clear ownership and accountability structures including defining the progress of measuring KPI.

  - Define a clear funding model for AI: At layer 4, the objective is to institutionalize AI funding from ad hoc project spend to a formal capital investment governed by enterprise-level financial processes. The choice of funding level should be decided considering factors like organization's maturity, scale, governance structure, and how other strategic programs or portfolios are funded in company.

CHAPTER 3 BUILDING A TRULY AI-POWERED ORGANIZATION: STRATEGY, INTEGRATION, AND TRANSFORMATION

- Resource allocation and management plan: Investment in AI is not just about financial investment; it also about managing resources that includes talent, tools, and skills. In this step, an AI-specific resource management strategy should be created. This plan should include the following:

  - Talent management plan that should focus on hiring, upskilling, and retaining top talent
  - Procurement and lifecycle management of tooling licenses
  - Other hardware and software requirement management

Key Deliverables from Layer 4

- AI investment roadmap: A multiquarter plan for funding, owning, and scaling AI initiatives
- Build versus partner versus buy decision matrix: A decision matrix with rules to simplify build versus buy versus partnership decision considering long-term strategic goals and intention
- AI portfolio ownership models: Create a product management organization to manage the entire AI portfolio lifecycle. Define clear responsibilities for each role. Typically, the organization should be headed by a senior leader to manage the power dynamics of the organization. This plan should include the name and responsibilities of executive sponsors sponsoring the AI initiatives.

- AI investment scorecard: Develop dashboards or reports that link spending to the progress of each portfolio. It should also capture the impact of each portfolio and how it is contributing to the individual strategic goal of the portfolio.

Layer 5: Execution Architecture and Governance: The "How" Layer

In layer 4 the shortlisted use cases were classified into an organization-level portfolio with secured enterprise-level funding. The next logical step is to operationalize AI delivery of identified use cases at scale. This is the layer where AI moves from planning to tangible execution. This layer focuses on setting up an operating model including people, process, polices, and platforms required to deliver responsible and scalable AI.

This layer prepares strategies required to transform AI prototypes to production-ready capabilities that can deliver some tangible business value. Questions like "Why should we build AI?" and "What should we build using AI?" are answered in earlier layers. This layer answers the question, how do we deliver, scale, and govern trustworthy AI repeatably across the organization? This layer has the following core pillars for developing repeatable Trustworthy AI.

- People & Roles: Building the Right Cross-Functional Teams:

    The success probability of an AI initiative depends equally upon two important factors. One is of course the right technology including the hardware and software being used and the other equally important factor is human resource. It is very important to identify and create the right roles and define their responsibilities clearly and hire the right talent for each role. The cross-functional team should have roles to cater to all functions like business and digital strategy management, technology, and governance. The following are the major roles that we have observed in the industry:

- AI product managers serve as the bridge between business objectives and outcomes generated by AI initiatives. This makes them responsible for identifying the right use to technical execution in a timely manner.

- Machine learning engineers are tasked with developing AI models and experimenting. This group of engineers are responsible for developing AI which will solve business problems

- Machine learning operations engineers are responsible for deployment, monitoring, and maintaining machine learning models in production. They are also responsible for monitoring models in production for data drift and the model drift phenomenon.

- Data engineers design, build, and maintain the data pipelines and architectures that enable reliable the high-quality data flows required for developing AI models.

- Software engineers integrate AI models into applications, develop user interfaces, and design systems that interact with AI outputs.

- AI quality analysts ensure the AI model's performance is accurate and reliable. They design and execute test plans and validate model outputs. They also check AI models for any potential bias or any other ethical issue.

- Generally, the AI product manager doubles up as an ethics and risk officer. Under this role they identify potential regulatory issues and potential ethical issues that may impact the model. They are also responsible for ensuring that the AI model is free from any ethical or regulatory risk.

  - The AI product manager also acts as the change manager where they are responsible for driving adoption among users and various other stakeholders. They are also responsible for making the organization culturally ready for adopting AI in daily operations.

- Processes and delivery frameworks: Without standardized processes, AI delivery risks become chaotic and inconsistent with very high failure changes. To avoid this, this layer focuses on creating repeatable and reliable workflows that guide projects to discovery to post deployment. Chapters 6 and 7 of this book cover these processes and delivery frameworks in detail.

- Platforms and tooling: A robust technology foundation is essential for deploying AI across teams. This layer focuses on the plan for developing technology platforms that can support collaboration, scalability, monitoring, and compliance part of AI projects. Essential components include:

  - Development of organizational data infrastructure to support reliable data generation, storage, and easy access to quality data

- Development of MLOps platforms to automate workflows for model development, training, and monitoring

- Model registries and monitoring tools to help manage versioning of model and data to ensure performance replication.

- CI/CD for ML to automate testing, deployment, and retraining of model

- Feature stores to allow storage, sharing, and reuse of engineered data features across models

• Governance, risk, and ethics: As AI scales, governance has become non-negotiable. One of the key activities in this layer is development of a robust ethics and governance framework. Chapter 8 of this book will cover more about this.

• Scaling and enterprise integration: This step helps in developing a scalability plan for transforming AI into an enterprise-level capability.

Key Deliverables of layer 5

- AI Delivery Operating Model: Defines roles, responsibilities, and delivery approach

- MLOps Stack Design: Includes experimentation, deployment, and monitoring pipelines

- Model Risk & Governance Framework: Categorizes models and assigns oversight levels

- Change Management Toolkit: Upskilling plans, adoption playbooks, stakeholder guides

CHAPTER 3   BUILDING A TRULY AI-POWERED ORGANIZATION: STRATEGY, INTEGRATION, AND TRANSFORMATION

Layer 6: Strategic Feedback Loops

Layer 6 is the final and most critical stage in an enterprise AI strategic framework. While previous layers focused on defining why, what, and whether you have the right capabilities to develop AI, this section answers the question, "Are we improving over time and how we adapt faster?" AI systems are very dynamic in nature, and stakeholder expectations also constantly evolve. This final layer helps close the loop between AI performance and organizational learning ensuring that teams don't just deliver and instead they continuously learn from it.

The objective of this layer is to bring organization into a state of continuous evolution by enabling continuous learning, adaptation, and scaling. This layer has the following key components:

- Model performance monitoring plan: Creates plan for monitoring the performance of the model in production. This includes identifying key metrics that will define model performance. Metrics can identify data and model drift.

- Feedback to retraining plan: Develops a plan to identify signals based on identified performance and data metrics to retrain the model in production.

- Strategic alignment plan: Develops a plan to ensure continuous alignment with evolving business environments and current AI development. Plan to use insights from AI execution and model outcomes to realign goals, prioritizes, and resource allocation.

Key Deliverables of Layer 6

- AI Business Impact Dashboard: Linked to financial and operational KPIs

- Model Drift & Incident Logs: Automated alerts and retraining triggers

CHAPTER 3   BUILDING A TRULY AI-POWERED ORGANIZATION: STRATEGY, INTEGRATION, AND TRANSFORMATION

Structured Approach to AI Use Case Selection

In the previous section, we explored how to establish a clear AI strategy and embed it in the organization's digital strategy. The next critical step for us product leaders is to identify and prioritize use cases that align with the organization's goals where AI can make a difference. While the AI strategy provides an overarching vision, guiding principles, and capabilities required for achieving the vision, the success depends upon translating these overarching principles into concrete, high-impact initiatives. A structured approach toward use case identification ensures the optimum utilization of organizational resources and helps in building the momentum. This section will focus on how to discover opportunities where AI can deliver measurable business value. This section will also focus on what the best practices are for evaluating the potential impact of initiatives and how to conduct feasibility study. A good starting point is understanding how to identify a bad use case for AI.

What Is a Bad AI Use Case?

A problem or scenario where AI is unlikely to add any meaningful business outcome due to one or more limitations can be classified as a bad AI use case. These use cases typically suffer from issues like the unavailability of data, unclear objectives, low business impact, or simply AI being the wrong tool for the job. Just because we have data, not all use cases should be solved using AI. The following are the common characteristics of a bad AI use case:

CHAPTER 3   BUILDING A TRULY AI-POWERED ORGANIZATION: STRATEGY, INTEGRATION, AND TRANSFORMATION

Characteristics of a Bad AI Use Case

1. **Infrequent or One-Off Decisions**

    AI is most effective when it is used for automating task that have high-frequency repetition and occur at scale such as fraud detection, vehicle number extraction, or demand forecasting. Applying AI to a decision that happens rarely (just a few times in a business cycle) is not a very intelligent move. We are saying that because sufficient data will not be available for training the model. The cost involved in training, deployment and maintenance of AI model will far exceed the benefit.

2. **Problem Is Too Simple or Rule-Based**

    Often AI is seen as the panacea for solving all business problems. However, if a task can be solved easily by traditional programming approaches and has defined rules, then that use case doesn't necessarily need AI. Rule-based systems are simpler, faster, cheaper, and easier to maintain. AI in these cases will bring in additional overhead without significantly improving the output. Examples where AI is a misfit are automating password reset applications or an application monitoring and showing usage of individual users in a company.

3. **High Risk, Low Reward**

    Some use cases involve making very sensitive decisions like hiring, medical diagnosis, or legal judgements or any other regulatory decisions.

Implementing AI in such cases should be carefully considered as any oversight can invite serious reputational, legal ethical, or financial damage. If the potential benefit is small in comparison to the risk involved, we advise not to pursue it. In Chapter 8, we have quoted numerous examples where organizations have suffered significant damage by ignoring this point of view.

4. **Low Business Impact**

    Even if an AI application is technically successful and delivers desired results and does not fall under the mentioned reasons, it might still be a naïve decision to pursue it. If the implementation does not meaningfully affect the business metric, applying AI to low-value use cases just because it was an interesting use case without justifying investment cost is not a good idea. These use cases act as resource drainers for the organization, bringing down the overall efficiency of the AI program.

5. **Lack of Stakeholder Buy-In**

    Even the most efficient AI solution can fail if the people who need to use it don't trust and support it. If stakeholders are not engaged early in development lifecycle and their buy-in is not received, there is a high chance even a good use case for AI will transform into a bad one. Successful AI use cases require alignment across the business and technical teams.

CHAPTER 3   BUILDING A TRULY AI-POWERED ORGANIZATION: STRATEGY, INTEGRATION, AND TRANSFORMATION

6. **Lack of Quality or Accessible Data**

   While the previous points highlighted the strategic traits of a bad AI use case, this section will focus on one of the most critical but overlooked factor: data availability. From a senior executive's perspective, this issue is often underestimated mainly because of the high-level nature of executive decision-making. The practicality of accessing clean, relevant data for an AI use case often does not receive adequate importance. However, even a promising use case will fail if quality data is not available. Leaders should be prepared to re-evaluate projects that lack required data inputs, regardless of how attractive the use case seems to be.

Mental Models That Undermine AI Success

Selecting the right AI use case is critical for organization's progress, yet many organizations repeatedly fall into a similar pitfall. These mistakes often stem not from a lack of understanding of technology or business but because of underlying mental models that shape a leader's decisions. Understanding the role these mental models play in decision-making highlights why even most promising AI initiatives fail.

In this section, we will unpack the most common mental models behind poor AI use case choice and will also highlight common traps that leaders walk into due to these mental models. Recognizing them can help organizations make smarter choices.

1. **Choosing Use Cases Based on Hype, Not Need**

   Mental Model(s): Bandwagon Effect and Confirmation Bias

Leaders often get influenced by the bandwagon effect mental model, assuming that if a technology is popular, it will automatically benefit their organization, or if competition is doing it, we should also implement it. This creates pressure to adopt the technology without critically assessing their relevance. Another mental model that starts affecting leader's decision-making at this stage is "confirmation bias." Now leaders selectively seek out information that justifies jumping on the technology bandwagon and overlooks all other information. This often results in pursuing trendy solutions that lack real value or fit.

2. **Underestimating Data Requirements**

   Mental Model: Optimism Bias/Planning Fallacy

   While already under the influence of confirmation and bandwagon bias, leaders often overestimate availability of quality data without conducting a detailed assessment. This is often an outcome of another biasness, which influences the decision-making capability of a leader. This biasness is called *optimism bias*. Even if leaders acknowledge the availability of data as a challenge to make matters worse, leaders often underestimate the effort, time, and complexity required for AI initiatives. This mental model is called *planning fallacy*.

CHAPTER 3   BUILDING A TRULY AI-POWERED ORGANIZATION: STRATEGY, INTEGRATION, AND TRANSFORMATION

3. **Targeting Low-Value or Edge Use Cases**

   Mental Model: Availability Bias/Over-focusing on Feasibility

   One of the most common mental shortcuts in AI planning is availability bias. It is the tendency to focus on ideas that come to mind first or are already being circulating in industry circles or internally without considering their strategic value. This bias can lead companies to invest in technically achievable but low-impact initiatives or in shallow initiatives.

4. **Trap: Boiling the Ocean (Over-scoping Problems)**

   Mental Model: Analysis Paralysis/**Perfectionism**

   Infatuated by the charm of AI, often leaders try to solve large, complex problems at once, aiming for large comprehensive solutions instead of focused wins. This approach is often rooted in a mental model called *analysis paralysis*. Under the influence of this mental model, the leaders often overthink, overplan, and fear missing factors that might impact the success probability. All this leads to inaction and endless scoping discussions. This results in zero delivered value, shifts in priorities, and stalled momentum.

5. **Trap: Ignoring Change Management**

   Mental Model: Technology-Centric Thinking/ Neglect of Human Factors

   A common bias in AI implementation is the assumption that if a solution is technically sound, it will automatically be used. This biasness results from a mental model called *technology-centric thinking.* Under the influence of this mental model leaders often measure the success of an AI model only in technical terms rather than actual business impact and ROI or adoption.

This mindset makes leaders believe that AI is superior than humans and they ignore human factors that will impact AI adoption. Teams skip change management and stakeholder communication with a bold and flawed assumption that technology will speak for itself. But in the real world, even high-performance systems fail because users don't use them due to a lack of trust in systems.

Techniques for Spotting High-Impact, Feasible AI Opportunities

In the previous section, we explored how a flawed mental model often leads organizations to pursue bad AI use cases. To deliver high impact and strategically aligned AI investments, organizations must adopt a more structured and multiperspective approach for opportunity sighting. The following techniques form a comprehensive toolkit for discovering AI use cases that are not only feasible but also valuable:

1. **Business-Led Discovery: Start with Real Needs**

   All AI initiatives must be aligned with business objectives. To identify business objectives aligned with AI initiatives, product leaders must start engaging businesses teams early to surface their pain points. The following steps should be taken:

- **Interview business stakeholders:** One of the primary steps here is to identify key business stakeholders across the critical departments and talk to them. The conversation should focus on understanding operational inefficiencies, unmet goals, areas where decision-making is suboptimized, and workflows that are acting as bottlenecks. At this point, focus only on understanding the pain points without considering if AI will be of any use or not.

- **Align with Strategic Objectives:** Once pain points are identified, then map those to organization goals or OKRs. This step will help you weed out cases that are not aligned with business objectives.

- **Link to Product KPIs:** From a product point of view, all use cases must be mapped to a product KPI. This helps in understanding how an AI implementation will impact an existing product and how its impact will be measured.

While this kind of discovery will give you high-level use cases, multiple iterations of discussion should be done to nail down the exact use case. Once a final list of use cases is created and prioritized from business, due diligence should be evaluate the implementation feasibility.

2. **Data-Driven Discovery: Let the Data Guide You**

    Another approach that can be used for use discovery is a bottom-up approach in which teams analyze data from various sources to identify use cases. The following methods can be used:

- **Process Mining:** At times business users might overlook hidden anomalies in the workflow and business processes. To uncover, the data of critical processes should be mined and analyzed. Data sources like system logs, workflow data, and user journey data should be analyzed to identify inefficiencies. These inefficiencies can become your potential use cases.

- **Digital Exhaust & Behavior Patterns:** Use transaction data, clickstreams, or telemetry to detect anomalies or patterns that may be ideal candidates for AI use cases.

Once these use cases are identified, take them to business leaders and evaluate their fit in the overall organization's goals and OKRs. If any identified use case does not contribute to an organization's goals and OKR, ideally that use case should be dropped.

3. **Use Case Libraries & Industry Benchmarks: Don't Reinvent the Wheel**

    Another potential source of identifying potential AI use cases is reviewing proven use cases from your industry or adjacent sector. However, leaders must take a cautious approach here as what works for others might not work you at that point in time.

4. **Cross-Functional Workshops & Design Thinking**

    Facilitating joint application development sessions aka JAD sessions, as these are commonly known as by bringing in business, data, product and engineering teams together, can really help identify potential uses cases. Use tools like these:

    - Customer journey map, service blueprints, and "How might we" framework to identify friction points

- Ensure that end user perspectives are used in identifying friction points
- Verify the identified use cases with end users

By combining all these techniques, a power foundation for AI use case identification can be laid out that ensures that the use case chosen is not just technically feasible but also strategically aligned to deliver measurable business value for real-world adoption.

AI Use Case Feasibility Study Framework

Identification of use cases is just half of the battle. Before investing an organization's resources, product leaders must evaluate the feasibility of each use case. While doing so, one must consider factors that affect the development of the use case but also consider factors that influence the deployment and acceptance of the use case. This where the following mentioned feasibility study framework becomes handy. It will help product leaders access readiness across factors and allow the team to confidently move toward the implementation of the use case.

1. Before assessing the feasibility, ensure that a well-scoped and clear problem statement is defined (in the next section we will learn about best practices).

2. A feasibility study should be conducted by evaluating use AI uses on the following dimensions.

    a. **Data feasibility:** If the organization has enough quality data to train the model. To access data feasibility, the following evaluation criteria should be probed:

        i. Is the required data available?

        ii. Can this data be accessed easily and frequently?

CHAPTER 3 BUILDING A TRULY AI-POWERED ORGANIZATION: STRATEGY, INTEGRATION, AND TRANSFORMATION

      iii. What is the volume, granularity, dimensionality, and recency of data?

      iv. What is the quality of data?

      v. Is the data labeled?

Output: By evaluating data on the previous criteria, a data readiness score is generated, and a summary of data gaps and a mitigation plan to bridge gaps is generated.

b. **Technical feasibility:** Does the organization have enough resources, talent tools, and capabilities to solve this problem? The following evaluation criteria should be probed:

      i. Does the team have the skills to build, deploy, and maintain the solution?

      ii. Are required tools (e.g., cloud platforms, MLOps pipelines) already available?

      iii. Do we have required talent (data scientists, ML engineers, SMEs) to develop and support this?

Output: The technical ecosystem of the organization is evaluated on the previous criteria, and a technical feasibility score is generated. Another outcome of this step is the creation of a resourcing plan of talent and a platform.

c. **Operational feasibility:** How will this use case be integrated into existing technical and business workflows: Answers to the following questions must be gathered to gain clarity on how this model will operationalize in the future.

   i. In what format will the model produce output?

   ii. How will the output be consumed by the existing technology platform?

   iii. Who needs to approve or act upon AI's recommendations?

   iv. Where will this AI model sit in the overall business process?

Outputs: This step will have slightly different outputs from the previous steps. The first outcome is a set of modified process flow diagrams, one on the business side and other on the technical platform side. Other output of this step is the quantification of integration efforts.

d. **Regulatory/Ethical Feasibility:** This step evaluates if this use case is feasible from a regulatory perspective and the ethical considerations. The following criteria must be evaluated to check the regulatory and ethical feasibility of the use case.

   i. Is this use case legally allowed?

   ii. What are the data and privacy requirements of this use case?

   iii. Does this case involves processing of any sensitive data?

CHAPTER 3   BUILDING A TRULY AI-POWERED ORGANIZATION: STRATEGY, INTEGRATION, AND TRANSFORMATION

    iv.   What are the ethical requirements of this use case?

    v.   What is the risk of model misuse, potential bias, or potential noncompliance of the local laws?

The output of this step is a risk matrix to capture a list of potential risk, potential impact, and chances of translating risk in real threat. Another output is a list of regulations that will govern this use case. The final output is an ethical assessment sheet that covers the probable list of ethical concerns that can arise in this use case.

Feasibility Scorecard

We propose a two-step method for checking the feasibility of each use case. Each model is assessed first on Layer 1 criteria that evaluates an individual use case against kill switches or non-negotiable dimensions. These are those dimensions and criteria that are must-haves for any use case to pass through a feasibility study. Even if the use case fails on any one the dimension or criteria mentioned under this category, the use case will not move forward until that issue is resolved, no matter how well the use case scores on other dimensions.

Table 3-7 shows a list of kill switches or non-negotiable criteria on which each use case must be evaluated.

Table 3-7. Model Feasibility Study on Kill Switches

Killer Factor	Threshold Test	Pass/Fail
Regulatory/ Legal Risk	Will this violate GDPR, HIPAA, or local laws? Is this use case even legal?	
Ethical/ Fairness Risk	Could this use case reinforce bias toward vulnerable groups	
Data Availability	is the critical data available and accessible	
Model Failure Sensitivity	What is the potential risk of this model mis prediction or failure in terms of financial or reputational damage	

CHAPTER 3   BUILDING A TRULY AI-POWERED ORGANIZATION: STRATEGY, INTEGRATION, AND TRANSFORMATION

Only when a use case passes all kill switches is it evaluated on the second set of criteria. Table 3-8 is a multidimensional weighted scorecard that will help leaders make decisions for what to do with individual use cases.

Table 3-8. Multidimensional Weighted Scorecard for Model Feasibility Evaluation

Dimension	Sub-Factor	Weight	Score (1–5)	Weighted
Business Impact	Strategic fit, ROI, user impact	25%		
Data Readiness	Availability, quality, access	20%		
Technical Feasibility	Talent infra, tooling	17%		
Integration Complexity	Workflow fit, tech stack fit	15%		
Adoption Readiness	Buy-in, change needs, visibility	13%		
Org Capability	Talent, resources, ownership	10%		

Output Decision Matrix

Product leaders must use Table 3-9 to determine the outcome of feasibility evaluation of individual use cases.

Table 3-9. *Output Decision Matrix*

Layer 1 Result	Layer 2 Score	Recommendation
Pass	75–100	Proceed to prototype/MVP
Pass	50–74	Proceed with mitigation plan
Pass	<50	Re-scope or delay
Fail	N/A	Do not proceed (unless mitigated)

AI Use Case Prioritization Framework

With feasibility confirmed, the next step is to prioritize AI use cases to focus on those with highest impact and likelihood of success. We recommend a modified RICE framework to calculate the priority score of each use case. The use case with the highest priority score should be first picked for implementation.

Priority Score = *(business Impact × 2) + Tech Readiness + Adoption Readiness − Effort−Risk*

Defining Actionable AI Problem Statements

Once the identification of high-value AI use cases is done, the next step is to translate the problem statement from business language to machine learning friendly language. Any machine learning/AI implementation requires a clear, specific, and measurable problem statement to succeed. Many AI initiatives fail because the problem statement was vaguely defined that made execution difficult. In this section, we will explore best practices for translating business use cases into structured machine learning/AI problem statements.

CHAPTER 3 BUILDING A TRULY AI-POWERED ORGANIZATION: STRATEGY, INTEGRATION, AND TRANSFORMATION

Steps to Translate Business Needs into AI-Ready Problems

1. Start with a business objective and ask what decision or outcome we are trying to automate. An example is that we want to reduce customer churn.

2. Define that decision point in clear terms; i.e., the business outcome/ decision that we want to automate should be defined in simpler terms so that there is no ambiguity left. Example: Extending customer churn. One must ask questions like these:

    - What type of customer are we referring to? Examples are paid customers, freemium customers, all customers, active customers, etc.

    - What is the definition of churn? Examples are cancellation of subscription, inactivity, or no transaction in last X days (common for marketplace).

3. Clarify the decision point by breaking it down by asking what will happen if we get to know the outcome in advance. What sort of advance notice do we need?

4. Clearly define the expected output of the model. Is it a category name or probability score or a value or a cluster? For example, in our customer churn example, ask questions like these:

    - What is the expected output of the model? Is it a binary (churn/no churn) or propensity score (80% changes this customer will churn)?

This helps in fine-tuning the model output and helps in developing an integration plan with a larger technology ecosystem in which this application will reside.

5. Specify what data will be used to make predictions, at least on a dataset name level. For example, we will be using transaction data, login data, support ticket, and tenure of the customer data to predict customer churn.

6. On an AI program level, ensure that whatever output model is producing it is actionable and relevant systems and the team is considering it for actions. In our churn example, it could be if the propensity to churn is higher than 80%, then trigger retention offer.

AI Problem Statement Template

Business Goal: Reduce high-value customer churn

ML Problem Type: Binary classification

Prediction Target: Will a customer churn in the next 30 days?

Input Data Required: Purchase history, engagement, NPS, customer service calls

Actionability: Enable proactive offers via CRM within 24 hours of prediction

AI model Success Metric: AUC > 0.80, retention uplift ≥ 5%

AI project success metric: Reduction in customer churn as compared to last year's same month data

CHAPTER 3   BUILDING A TRULY AI-POWERED ORGANIZATION: STRATEGY, INTEGRATION, AND TRANSFORMATION

From Strategy to ROI: Financial Evaluation of AI Initiatives

According to industry reports, the average return on investment (ROI) for AI projects is approximately 6%, while top performing companies report an average ROI of 13%. The average cost of capital for most companies hovers around 10%. This indicates that many AI initiatives fail to generate sufficient returns justifying their investments.

Companies that achieve high ROI benefits from AI projects treat AI as a strategic enabler. They back these initiatives with full organizational support and follow a systematic, well-integrated approach in the discovery and development of AI solutions. As a result, such organizations realize ROI improvements of up to 30%.

One of the most critical factors contributing to the success of AI initiatives is selecting the right use case. This is a use case that not only contributes meaningfully to the business but is also technically and operationally feasible and financially viable. Choosing a use case that satisfies all the requirements not only increases success probability but also maximizes ROI.

What Is Financial Viability in the AI Context?

If an AI project can deliver sufficient economic value in terms of increase in revenue or a reduction in costs to justify the investment, then that project is considered financially viable. The investment includes initial development costs and ongoing expenses. In simpler terms, it answers the question, "Will this AI implementation deliver more value than it cost sustainably and at scale?" Evaluating financial viability involves considering the following factors:

- Initial investment: This is the development cost of the project. This includes cost of team members, tool cost, and infrastructure cost.

- Operational cost: This includes ongoing model management cost and infrastructure to support the model in production.

- Expected gain: This is the expected business value either in terms of increase in revenue or cost saving or productivity gain.

- Payback period: This is the time it takes for the project to recover an initial investment.

Why Financial Viability Matters Strategically

Financial viability isn't just a gating mechanism; it is a strategic signal. It not only determines which initiatives receive the funding but also which ideas scale, meaning which experiments mature into enterprise-wide transformation.

At its core, financial viability bridges the gap between the strategic intent of AI with the economic realities. The following are the benefits of conducting a financial viability evaluation of AI initiatives.

Driving Prioritization and Resource Allocation

Evaluating financial viability of each use case helps in optimum utilization of organization's resources. In an organization, resources are limited, and financial viability study provides a rational basis for prioritizing use cases across an AI portfolio. It helps categorize which projects justify scale, which capability development projects justify cost, and which projects should not be picked.

Without a financial lens, organizations risk spreading their resources thin. This evaluation keeps the transformation grounded.

Sustaining Long-Term Transformation

AI projects are not isolated micro-level initiatives; it is a strategic capability that can truly transform the way an organization functions and even open new business avenues. Financial viability is the method that ensures that use cases and foundational data and AI projects chosen are not only strategically valuable but also have the potential to scale at a sustainable cost. This paves the way for future AI innovations by boosting internal confidence and required infrastructure.

Building Confidence Across Stakeholders

AI transformation requires cross-functional team, and hence it requires buy-in from multiple stakeholders including executives, business leaders, and sometimes regulatory partners. All these stakeholders speak different business language, and financial viability is the shared language. A financial viability study helps leaders with diverse backgrounds communicate effectively and decide on a common list of initiatives that are financially viable and strategically contributing.

Balancing Innovation with Risk

An AI initiative with strong financial case doesn't always mean assured profit; it may also mean the following:

- Measured response to a risk greenfield project
- Clear milestones for staged investment for long-term strategic capability development projects

By conducting this study, organizations get answers to these questions, and it helps to strike a balance between innovation and prudence. In this sense, financial viability supports responsible innovation by encouraging bold decision without being too reckless. It also minimizes the risk of projects stalling midway due to lack of resources or unclear value.

CHAPTER 3 BUILDING A TRULY AI-POWERED ORGANIZATION: STRATEGY, INTEGRATION, AND TRANSFORMATION

Economic Fundamentals Behind AI Use Case Viability

Evaluation to access the viability of AI projects should be grounded in fundamental financial principles. These principles provide a structured way to assess risk, reward, and resource allocation. The following three principles play a very critical role.

The Economic Value of Money Declines Over Time

At its core, this economic principle states that the value of money decreases over time because of various factors such as inflation and the opportunity cost of not utilizing it effectively. In simple terms, this means that a sum of money today is worth more now than it will in the future. AI projects are often high-risk projects with delayed or uncertain results. A financial viability study must account for a lag in time between the investment and the payoff. This is achieved through discounting the future cash flow considering a delay in return and the risk quotient of the project to reflect the value at the current time. Doing this gives a fair view of the financial viability of the project.

Risk-Return Dynamics in Capital Allocation

At its core, this principle states that high-risk assignments must have high expected returns to compensate for the uncertainty and high failure risk. AI projects are high-risk projects but have the potential to deliver huge benefits. Because of this uncertainty, these projects must offer a higher return of investment than others.

Balancing Capital Investment and Opportunity Costs

In layman's terms, if a sum of money is invested in one initiative, then it means that the money is not invested in another initiative. This principle points out the lost opportunity cost or potential benefit from one initiative

CHAPTER 3   BUILDING A TRULY AI-POWERED ORGANIZATION: STRATEGY, INTEGRATION, AND TRANSFORMATION

simply because the other project was prioritized. It points out that initiatives should be considered carefully because if an organization's resources are allocated to an undeserving project, it will give the organization two kinds of losses, one from its suboptimized returns and another from the missed opportunity cost from a deserving project that was superseded.

While evaluating initiatives on the organization level, AI initiatives must not be evaluated in isolation but relative to other strategic options including doing nothing. Discounting rates and hurdle rates are used to compare the opportunity cost of all options in consideration. From a strategic perspective, a long-term data and AI capability development program may have a lower immediate impact but higher strategic value in comparison to an isolated AI implementation, but this project will have a higher strategic value and the leader should choose this initiative over short-term wins with limited scalability and reusability.

Assessing Financial Viability of AI Projects Through Key Financial Metrics

The following four fundamental financial metrics can be used to assess the viability of AI initiatives:

Return on investment (ROI): This is a simple metric that measures how much you will gain or lose in comparison to your investment. It is basic measurement to estimate the worthiness of an investment. If ROI is positive, then you are gaining, and if ROI is negative, you are losing on your investment.

$$ROI = \frac{\text{Net gain or loss}}{\text{Total investment}}$$

CHAPTER 3   BUILDING A TRULY AI-POWERED ORGANIZATION: STRATEGY, INTEGRATION, AND TRANSFORMATION

For example, if the total cost of developing and maintaining an AI project is $100,000 and it gives back $130,000, then the net profit = $130000 − $100000 = $3000.

$$\text{ROI} = \frac{3000}{100000} = 30\%$$

Net Present Value (NPV)

NPV calculates how much the future earnings of a project is worth in today's money after considering cost, because the value of money in the future will have less worth than it has today. To calculate the financial viability, first calculate the NPV using the following formula and then subtract the current investment:

$$\text{NPV} = \sum \frac{\text{Cash Flow in Year } t}{(1+r)^t} - \text{Initial Investment}$$

- **Cash Flow in Year t** = Money you expect to receive in each future year
- **r** = Discount rate (a percentage that reflects risk and the time value of money)
- **t** = The year (1, 2, 3, etc.)
- **Initial Investment** = How much you spend at the beginning

A simpler representation of the previous formula is NPV=(Value of all future money today)−(What you spend today).

This metric calculates the current worth of annual profit made in the future. If NPV is positive, the project adds value.

Internal Rate of Return (IRR): This calculates the average rate of return the project is expected to make over time. If the IRR of your project is greater than the discounting rate, then your project is financially viable.

Payback Period: This lets you know how long it will take to earn back the initial investment.

$$\text{Payback Period} = \frac{\text{Initial investment}}{\text{Annual profit}}$$

In plain terms, if your initial investment on an AI system was $200,000 and it saves $50,000 each year, then your payback period is 4 years ($200,000 divided by $50,000).

Discount Rate: This is the rate at which future money will convert into today's value. Because we know the future value of money is less than what it has today, we discount future cash flows to see its worth today. This helps compares costs and benefits over time. The discount rate can also have one or more references as mentioned here:

- Helps in representing the risk quotient in the financial calculation. Different discount rates are chosen to represent different risk profiles of the project.

  - A higher discount rate is used for risker projects; this makes future cash flow less valuable today.

  - For safer and stable projects, a lower discount rate should be used to make future cash flow more valuable today.

- It also represents the cost of arranging the capital (borrowing cost or opportunity) for a project.

Ideally a sensitivity analysis is conducted with two or three different discount rates varying from a low to moderate to high discount rate to understand how sensitive the NPV of your project is.

CHAPTER 3 BUILDING A TRULY AI-POWERED ORGANIZATION: STRATEGY, INTEGRATION, AND TRANSFORMATION

Practical Approach to Evaluating Financial Viability

Here is a practical way to apply financial metrics without overcomplicating the early stages of innovation:

- Use the ROI for a quick profitability check. Ideally, the ROI calculation should the first kill gate to identify AI use cases with basic financial promise.

- On the shortlisted initiatives, use NPV to evaluate if this initiative has any true financial value over time.

  - While calculating the total cost of ownership, include talent, licensing, and infrastructure cost; also include the cost of maintaining the solution post developed.

  - While calculating cash flow, include all the direct and indirect benefits translated into dollar values.

  - Choose a discounting rate worthy of representing the associated risk of the project. If you are not sure about risk profile, conduct a sensitivity analysis.

- Use the IRR for high-stake and high-risk projects. Projects that involve large capital investments and require C-suite approval should be evaluated using this technique. Product leaders must work with the finance department to model different scenarios and calculate the IRR. Once the IRR is calculated and normalized, use it to compare this project with other projects but similar capex requirements

CHAPTER 3   BUILDING A TRULY AI-POWERED ORGANIZATION: STRATEGY, INTEGRATION, AND TRANSFORMATION

Quantifying Expected Gains from AI Use Cases

We discussed various methods for calculating the financial viability of AI projects, but most of the methods we discussed rely on the potential benefits in dollar terms to evaluate the financial viability of an AI initiative. In most cases, if we are evaluating the financial viability of an AI initiative, then that initiative has not been implemented, which means the actual benefits are not available for reference. In such cases, the following practical methodologies that have tried and tested by us over the period can help in making an informed and educated estimation about the potential gains:

- **Use benchmarks and proxies:** Refer to similar case studies and projects in the industry to gauge the expected return. You can also refer to domain experts who have previously implemented similar use cases in the same industry.

- **Process simulation or modeling**: Create a rule-based simulation by mimicking the potential AI behavior and compare the performance of these rules with the existing system to gauge the expected return. You can perform an exploratory data analysis to identify patterns in data and use these patterns to predict the behavior. This prediction should be compared against the status quo system, and differences in performance should be considered as the expected return.

- **Execute a small pilot or proof of concept:** Another way of estimating the potential benefits of implementing a potential AI use case is to run a small POC by developing a lightweight MVP on a small set of data and test the model on real-time data to gauge the potential benefits.

After determining the potential benefit of an AI initiative using one or more of these methods, we recommend modeling the potential benefit value in three scenarios: best case, worst case, and expected case scenario. These scenarios should then be used to assess the financial viability, and doing so provides a more balanced view.

Operationalizing AI: Managing Change for Enterprise Adoption

In the previous section, best practices for aligning AI investments with the business objectives and evaluating the financial viability and expected ROI of initiatives were explored. You must understand that even the most valuable and cost-effective AI use case can fail in practice if it is not fully adopted by users and embraced by the leaders. Building the right AI solution is a technical and strategic challenge, but its adoption is not; it is a people challenge and should be handled softly.

This section will focus on how to manage the transformation required for successful AI adoption in the organization. Effective change management is the secret sauce that transforms AI from a mere technological implementation to a lasting competitive edge, by ensuring that it is properly adopted, trusted, and embedded into daily operations.

Before You Adopt: What AI Will Change in Your Organization

Before we design change management program for AI adoption, it is essential to understand the type of changes AI is introducing across the organization. AI disrupts the traditional ways of working, changing the decision rights and job responsibilities. Recognizing these areas of disruption is critical to preparing the organization for successful transformation. The following areas typically experience maximum shifts when AI is introduced:

1. **Decision-making:** AI is changing the way decisions are made from traditional human intuition, experience, or hierarchy to more data and insight driven. The traditional decision-making is increasingly getting influenced or even replaced by data-driven decisions. This creates cultural shifts that are harder to adopt, especially by those who were either making those decisions or by those following those decisions. A classic example is in sales forecasting, where AI may replace a senior sales leader's manual or semi-automated sales estimates for the next quarter with real-time more accurate forecasts, thus challenging the authority structures of the department.

    Roles and responsibilities: As AI systems take over repetitive tasks, human roles are evolving from execution to more supervision and oversight. Some roles may become redundant, while new roles may emerge. Even roles that may stay will see change in their day-to-day working and responsibilities. Without a proactive change management strategy, this shift may lead to confusion resistance and skills gaps.

    Processes: AI often alters business processes by automating the decision-making process at scale. This may eliminate intermediatory steps, speed up decisions, or require integration of new tools into existing workflows. If the business processes aren't redesigned around AI, then technology may sit unused for a larger period of time.

Success criteria: AI will fundamentally alter how success is measured. Traditional KPIs have typically focused on output quantity, process adherence, or lagging indicators. Once AI is introduced, an organization will move toward dynamic, real-time, and future looking metrics. The focus will shift from what happened to what's likely to happen and whether teams are effectively leveraging AI to change the outcomes.

Without managing these changes, organizations will not be able to adopt AI fully and will keep witnessing shadow IT and tool rejection, and all this will lead to AI underutilization.

Actionable Framework: AI Change Management Roadmap

1. **Start change management early—not after launch:**

   A change management exercise must begin early during the use case definition phase. Bring key stakeholders together by forming a use case squad in workshops, problem framing sessions, and MVP reviews. When stakeholders are a part of journey, they are far more likely to support and adopt the outcome.

2. **Start with a clear purpose and value alignment:**

   Ensure that all AI initiatives are aligned with the business objectives. Use this relationship to communicate the potential impact an AI initiative will have once developed and deployed to foster stakeholder buy-in.

3. **Create a stakeholder register:** Analyze all stakeholders to identify who will be impacted and in what sense. Classify stakeholders by the type of influence they have on this use case. Map stakeholders into different categories such as advocate, sceptic, neutral, and blockers based on the support you are receiving. Customize communication strategies for each stakeholder type.

4. **Launch AI showcases or "seeing is believing" demos:**

   Show before-and-after outcomes using actual data and workflows to reduce fear and ambiguity.

5. **Invest in people and skills development:**

   Develop and execute AI literacy programs in the organization to equip employees with knowledge to collaborate with AI. This program should not only cater to the theoretical knowledge of AI but also to the practical real-world knowledge that will help them execute their daily tasks.

6. **Promote transparency and explainability:**

   One of the biggest barriers to AI adoption is the perceived "black box" nature of machine learning systems. When users can't understand how an

AI makes decisions, they're far less likely to trust or act on its outputs, especially in high-stakes environments. To build trust, AI systems must be explainable, auditable, and accountable.

One of the biggest reasons people don't trust AI is because of its perceived black-box nature. When users don't understand how an AI makes decisions, they are far less likely to trust or act upon its recommendations. To build trust, AI systems must be explainable, auditable, and accountable. Another key point here is to design AI systems with a human in the loop; i.e., humans must remain actively involved in validation, decision-making, and the escalation matrix. Always allow users to question, override, or adjust AI outputs when appropriately reinforcing the idea that AI supports rather than replaces human judgment.

7. **Redefine roles and processes thoughtfully:**

    - Clearly articulate how existing workflows and job responsibilities will evolve.

    - Clarify how roles will evolve, for example, whether an analyst role can transform into the role of a model supervisor.

    - Link OKRs to AI usage and not just adoption metrics.

    - Recognize individuals who use AI systems effectively through performance metrics, bonuses, or visibility.

CHAPTER 3   BUILDING A TRULY AI-POWERED ORGANIZATION: STRATEGY, INTEGRATION, AND TRANSFORMATION

Summary

This chapter outlined the strategic tools and framework to enable product leaders to prepare their organization for AI transformation. The chapter began by exploring how to assess the current AI maturity level of an organization and how that assessment can inform the development of the AI strategy. How to align the AI strategy with a broader digital strategy to ensure coherence between relevant stakeholders was also discussed.

Methods for identifying use cases suitable for AI along with common cognitive biases and mental models that can lead product leaders to select ineffective ones were also highlighted in the chapter. A key aspect of prioritizing AI use cases is evaluating the financial viability of the solution through ROI analysis.

Finally, the chapter addressed how to design and implement a robust change management program to support adoption of AI.

In the next chapter, we will explore how to design a strategic data ecosystem that will become an essential foundation on which scalable and sustainable AI initiatives can thrive.

PART II

Building the Data Platform for AI Success

CHAPTER 4

From Strategy to Execution: Building a Data Operating Model for AI-Ready Growth

In the previous chapters, we explored best practices for designing AI programs strategically. We explored the importance of aligning an AI implementation with the organization-level AI strategy and further associating the AI strategy with the business strategy. One of the critical elements for designing an AI program with maximum chance of success is high-quality data that is easily accessible and trustworthy. In this chapter, we will explore how we can cater to this need for clean and reliable data.

Though companies invest significantly in an up-to-date technological infrastructure to create data platforms, they don't put into place the basic guidelines and governance infrastructure necessary to make them effective. In this disjointed manner, there is a proliferation of disparate data systems that cause enterprise-wide data isolation. These data isolations hinder cross-team collaboration that in turn restricts data utilization and ultimately the effectiveness of data applications, including the applications of AI.

CHAPTER 4   FROM STRATEGY TO EXECUTION: BUILDING A DATA OPERATING MODEL FOR AI-READY GROWTH

To achieve this, organizations require a well-planned and scalable framework to create a data operating model that links data efforts to the business. In short, it converts an organization's strategic intention into operational execution. Now, let us understand what a data operating model is and how it is different from a data platform.

What Is a Data Operating Model?

A *data operating model* prescribes guidelines for how different systems that are part of different data lifecycle phases interact and collaborate to provide good-quality and harmonized data. It lays down roles, procedures, and systems that assure data is not just accumulated but is also good for organizational use cases that often require good-quality data from different domains to be combined together. A good data operating model is extremely essential for AI readiness. Since AI readiness is not merely about embracing models but about building the foundation that can support them, it is extremely essential.

The Relationship Between a Data Operating Model and a Data Platform

While a data operating model is a relatively new term, there is another term that people often talk about, and that's a data platform. Let's now understand what a data platform is and how it is different from a data operating model.

A *data platform* as defined by leading technology companies is a technical solution that supports all stages of a data lifecycle starting from collection, storage, analysis, and governance. Here the emphasis is on "technology solutions," which primarily refers to the infrastructure and software solutions to support these activities. A data operating model is an abstract concept that defines how these various technology solutions work together in tandem to deliver high-quality unified and easy to access data. Just like an organization's operating model defines how various departments and teams across companies will work together in synchronization to meet organizational goals, a data operating model defines the rules of coordination between the various technological solutions to achieve the goals set. To put this into perspective, a data platform is one of the data solutions whose way of working is governed by guidelines set by the data operating model. Table 4-1 presents the key differences between data operating models and data platforms.

Table 4-1. Key Differences Between Data Platform and Data Operating Model

Aspect	Data Operating Model (DOM)	Data Platform
Definition	A set of principles, roles, processes, and governance models for managing data across the business.	The technical infrastructure and tools used to store, process, and serve data (e.g., Snowflake, Databricks, BigQuery).
Focus	Strategy, ownership, accountability, data product thinking, governance, quality, and alignment to business value.	Scalability, performance, compute/storage, pipelines, APIs, integration, data access.
Role	Defines how people and teams work with data—what standards to follow, who owns what, and how data is prioritized and activated.	Executes and supports the technical workload—data ingestion, transformation, querying, and serving.
Output	Aligned, governed, high-quality data products and insights ready for AI and business use.	Raw and processed data stored and accessed through technical systems.
Example	EDGE Framework, data mesh principles, domain ownership models.	Data lakehouse, CDP, data warehouse, ETL/ELT tools, observability tools.

CHAPTER 4   FROM STRATEGY TO EXECUTION: BUILDING A DATA OPERATING MODEL FOR
            AI-READY GROWTH

We have seen and consumed data from very different but connected perspectives and together have developed a data operating model that not only adds context to data efforts but also provides an architecture for how different data systems such as CDP, event management systems, transactional systems, centralization layer, and data products can work together. We call this operating model EDGE, for **E**nd-to-end **D**ata for **G**rowth and **E**nablement.

Introducing the EDGE Data Operating Model: End-to-End Data for Growth and Enablement

The EDGE framework is an enterprise data operating model that closes the gap between data and business goals by strategic and architectural coherence. It functions as a conceptual model that integrates business strategy, data architecture, governance, and consumption in a unified framework independent of specific technologies or platforms.

This model empowers cross-functional teams by defining consistent practices, standards, and semantic clarity by embedding governance mechanisms like data contracts, lineage, and quality checks at each layer. A key feature of this framework is its closed feedback system, where feedback from downstream systems informs upstream collection, modeling, and prioritization to ensure continuous improvements.

The EDGE framework provides a blueprint for designing and scaling enterprise data capabilities that are outcome-driven and operationally mature. It helps organizations move beyond fragmented data silos to a cohesive and business-aligned data ecosystem that fuels AI, analytics, and operational excellence.

EDGE comprises three sections each with two layers. Every layer is designed to achieve a specific milestone in the data value journey. The sections and the layers are outlined here:

CHAPTER 4   FROM STRATEGY TO EXECUTION: BUILDING A DATA OPERATING MODEL FOR AI-READY GROWTH

1. **Define & Align:** This is the first layer of the framework, and it focuses on aligning business and data teams together by identifying business goals and translating them into data needs. This layer has following two components:

   - **Business Context Layer:** This layer aims to align strategic business goals to data efforts.

   - **Analysis Planning Layer:** This helps identify data requirements aligned with identified business goals and define schema and business rules for identified data points.

2. **Build & Govern:** This is the layer where all core data collection and centralization activities take place. This layer defines foundational rules, structures, and pipelines for data collection, modeling, storage, and consumption.

   - **Data Collection Layer:** Defines rules for data generation and collection across systems while ensuring quality, compliance and traceability

   - **Data Centralization Layer:** Unifies and models transactional and user interaction systems for organization wide access

3. **Activate & Optimize:** This layer defines pathways through which applications can access data in a secure manner. It also helps in delivering measurable business impact and creating a feedback loop for continuous improvement.

   - **Data Activation Layer:** Defines ways in which data can be consumed by data products in real-time or batch applications

   - **Data-Driven Growth Layer:** Delivers measurable business impact and create a feedback loop for continuous improvement

CHAPTER 4 FROM STRATEGY TO EXECUTION: BUILDING A DATA OPERATING MODEL FOR AI-READY GROWTH

Figure 4-1 is a high-level diagram of the EDGE framework depicting how each layer and their subcomponents interact with each other. Please note that data lineage and observability solutions are a part of layers 3 to 6. For the sake of simplicity, they are not explicitly shown.

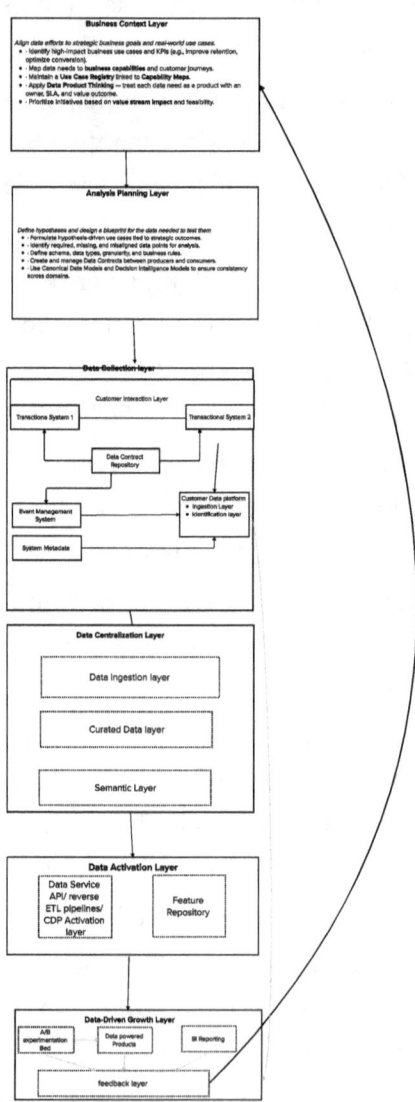

Figure 4-1. *EDGE framework diagram*

CHAPTER 4   FROM STRATEGY TO EXECUTION: BUILDING A DATA OPERATING MODEL FOR AI-READY GROWTH

EDGE Implementation Stack: Structure, Flow, Governance

In this section, we will explore each layer of the EDGE data operating model structure in detail.

Business Context Layer

The Business Context Layer is the first layer of the framework, aimed at aligning all data efforts with the strategic business outcomes and real-world use cases that generate measurable business impact. This layer focuses on identifying where new data needs to be captured with each use case tied to a defined success metrics. These metrics help in determining which data points are essential and in quantifying the success of the initiative. Use cases are often identified while defining AI strategy or to meet regulatory requirements, a process detailed in Chapter 3. Notably, this layer does not involve the handling of actual data but rather centers on discovering data needs in alignment with business goals.

The following are the actionable responsibilities of the product manager in this layer:

1. Partner with Business Stakeholders to Define Use Cases: Product manager should work with different business stakeholders in identifying and defining the use cases. This can be achieved by conducting one-on-one interviews or discovery sessions with various but connected stakeholders. Both approaches help in surfacing pain points and growth opportunities.

CHAPTER 4   FROM STRATEGY TO EXECUTION: BUILDING A DATA OPERATING MODEL FOR AI-READY GROWTH

2. Once use cases are defined, work with stakeholders to identify the primary KPI that would indicate business value. These metrics serve as the directional goals and not hypotheses, which will later guide hypothesis development and data planning in the next layer.

3. Build a value effort matrix to score and prioritize initiatives on business impact. To quantity the effort required for capturing the new data points, wait for the Analysis phase to complete.

4. Develop a use case registry to log and manage use cases. Each use case must contain:

    a. Business goal

    b. Hypothesis once analysis layer steps are completed

    c. Dependencies

    d. Priority

    e. Ownership (business and product)

    f. Required data points/domains (once analysis layer steps are completed)

Key Product Manager Deliverables

- Validated use case registry aligned to strategic goals
- Documented KPIs per use case with owners and targets
- Prioritized data use case backlog (impact/effort scored)

CHAPTER 4   FROM STRATEGY TO EXECUTION: BUILDING A DATA OPERATING MODEL FOR AI-READY GROWTH

Analysis Planning Layer

The Analysis Planning Layer translates business use cases into detailed data requirements by performing the following activities: identification of data gaps, planning data structures, quality rules and SLA, system dependencies, and ownership required for facilitating the data availability. To achieve these, the following actions should be taken:

1. Collaborate with data teams and business stakeholders to translate the business use case into a hypothesis tied to the business goal. Ideally, a thorough exploratory data analysis is conducted to identify factors affecting the use case. Once this analysis is complete, insights and business logic are merged to a draft hypothesis and then tested on new set of data. These steps ensure that the hypothesis that is created is fact based.

2. Identify the data points required for supporting this hypothesis. Of the identified list, the missing data points are also identified. This step also identifies gaps in the existing data points from an accessibility, structural point of view. The business logic used for calculating the data points is also evaluated under this step. The outcome of this step is a set of missing data points or existing data points that have structural, logical abnormalities. Data points that are inaccessible are also included in the list.

3. Map data needs to customer journey steps. This helps in understanding where the missing data point lies in the customer journey and how critical it is from a journey perspective. This can be achieved by asking diagnostic questions around the use case

and journey step. For example, if we are working on a business problem around improving new user conversion, then ask questions like these:

   a. Are we able to track the complete onboarding journey of a user or are some events/steps missing?

   b. Can we identify users who clicked on our promotion email and did not convert?

   Once missing data points are identified, document data dependencies for each data point.

4. For each data point, define the following:

   a. Schema details like data type (integer, number string, etc.), maximum and minimum permissible length. Name, nullability, uniqueness, sensitivity, etc.

   b. Granularity at which data will be collected, i.e., session level, daily, hourly, transactional level, etc.

   c. Business logic (e.g., "Active user = ≥ 3 sessions in 7 days")

   This helps reduce downstream confusion and enforce consistency across teams.

5. Define data quality metrics that are important for each data point. Define the SLA for each data quality metric.

6. Define data contracts, which will be enforced by transactional systems to standardize and enforce data definitions, quality, and behavior at the point of generation.

7. For each missing or new data point, estimate effort to collect that data point. Also rate how critical this data point is to the use case.

Key Outcomes of This Layer

- A list of data points that are either already existing, need to be newly captured, or require adjustments in their current implementation for each use case.

- Finalized data contracts for each new data points that should contain schema details, business rules, and granularity details.

- Finalized data governance details like data sensitivity, lineage, and quality requirements.

   Clearly defined ownership for each task. Table 4-2 maps various roles and their responsibilities.

Table 4-2. Data Roles and Their Responsibilities

Role	Responsibility
Data Steward	Defines semantics, data types, and quality expectations
Data Product Owner (DPO)	Owns scope, backlog, and contract negotiation
Solution Architect	Validates feasibility, scalability, and platform alignment
Engineering Manager	Provides effort estimates and platform constraints for new data fields

Data Collection Layer

The Data Collection layer is one of the core layers of the EDGE framework, responsible for capturing and ingesting high-quality data across enterprise systems. It ensures that every user interaction, whether from digital systems, internal systems, or external partners, is accurately stored, governed by contractual standards, and linked to a unified user identity. This layer establishes the technical foundation for downstream systems by providing a structured data ingestion pipeline by implementing governance and identity stitching from the point of data generation. By enforcing these practices early in data lifecycle, the data collection layer transforms raw signals into governed, business-relevant building blocks enabling the generation of enterprise-grade insights and powering scalable AI applications.

Let us now explore the core activities performed within this layer that enables it to capture, govern, and prepare data for downstream consumption.

1. The primary activity undertaken in this layer is capturing transactional data from various systems like CRM, ERP, billing, and order management, etc. This step also includes assigning clear ownership and SLAs against each data point being captured.

2. Another critical set of data that gets collected in this layer is behavioral and event-level data. In simpler terms, this is the data that gets generated when a user makes interactions like page visit, clicking a CTR, etc. It captures the sequence of steps a user takes before or after a transaction is made. This data adds context around the transactional behavior. This layer also helps in standardizing taxonomies, helping ensure consistency in how events are defined, named, and instrumented across systems.

3. Once transactional data and customer behavior data is collected, the next task is to identify guest users in the best possible way. This is accomplished by developing a subsystem under this layer called the customer data platform. Typically this platform has multiple layers from ingestion to identity resolution and the activation layer, but in the EDGE platform we have dissected this platform into its own independent layers and placed them according to their best fit. This layer will contain an ingestion layer that will be responsible for connecting to the event database, transaction database, cookie data, and other system metadata. This data will further be passed to a second layer, which is called the identify resolution layer. This layer uses the data collected from various sources and then applies deterministic and probabilistic methods to unify user identities across sessions and channels and help identify otherwise anonymous users.

4. Since this layer is focused on collecting various forms of data, it becomes very critical to establish a common definition and constraints for each data variable. In the absence of this common definition, various systems and channels may collect the same data in different formats and definitions, which may lead to inconsistencies across the system and leave data unusable at an organization level. To avoid this, data contracts must be implemented.

Data contracts can be defined as formal agreement between various data producers and consumers that explicitly define the structure, meaning the quality standards of data shared across systems, teams, and services. Data contracts also ensure that when two or more data producers capture the same data variables, they use the same definitions, standards, and business rules for shared data and ensure that data captured via different producers remains consistent. Adding data contracts to the data ecosystem can offer the following advantages:

- This helps in avoiding unexpected schema or logic changes that could disrupt downstream pipelines and applications.
- Clear contracts help in building confidence on data by ensuring that it meets quality and consistency standards.
- It acts as the common language and set of expectations between different producers and consumers.
- It helps shift data governance principles to the left, which means at the start of the data lifecycle. This makes data quality checks continuous and not just retrospective.

While there are many simpler ways to enforce data contracts, Google's approach toward data contracts is a very mature and holistic model. Rather than limiting contracts to simple schema definitions like field names and data types, Googles takes this concept a step further by embedding semantic meaning, business rules, usage expectations, and ownership metadata directly into machine-readable contract files. Google recommends storing individual data contract files centrally in a metadata catalog. This approach recommends that every time data is generated, ingested, and transformed, it should be actively checked for both schema validation and semantic validity.

> It further recommends integrating data contracts with real-time data pipelines to detect violations in real time. When violations occur, the system should automate actions such as halting the pipeline, logging the issue, or notifying stakeholders depending upon the severity of violation and data variable.
>
> This tightly coupled feedback loop ensures that producers are immediately accountable for any violations and consumers are protected against unexpected data changes or degradation.

5. EDGE prioritizes proactive data governance by embedding observability and lineage tracking from the start of the data lifecycle. This approach puts data trust and quality as a foundational element of the organization's data operating model and not just as an afterthought. Use data observability tools to monitor freshness, completeness, and schema drift of data. Data lineage tools should be used from the point data is captured in the system till it reaches the consumption stage. This will help Improve transparency and troubleshooting across pipelines.

6. The Data Collection layer's primary job is to collect data, and hence complying with regulatory data collection and storage requirements should start in this layer. While implementing this layer, personally identifiable information (PII) sensitive fields must be identified and tagged as early as ingestion.

CHAPTER 4 FROM STRATEGY TO EXECUTION: BUILDING A DATA OPERATING MODEL FOR AI-READY GROWTH

Measures should be taken to implement consent flags, opt-out forms, and other privacy signals in the forms and tracking used for data collection. The frontend layer of the application must be designed in a way that users feel informed about what data points are collected. The interaction layer must also provide easy-to-use controls to manage their consent and data preferences.

Building Blocks of Data Collection

Table 4-3 outlines the major and minor components that make up the Data Collection layer and their specific roles within the system.

Table 4-3. *System Components and Interactions of Layer 3: Data Collection*

Component	Role in Data Collection	Interaction & Flow
User Interaction Layer	Captures raw user behaviors and journey triggers (e.g., clicks, submits, scrolls)	Sends event data via event management system SDK to the event management system
Event Management System	Validates and structures raw behavior data using a standardized taxonomy	Publishes events to CDP ingestion and central data storage; enforces schema logic based on data contracts
Transaction Systems	Records structured, outcome-based data from systems like CRM, ERP, billing	Feeds completed transaction data into ingestion pipelines; aligned with behavioral events for journey tracking

(continued)

Table 4-3. (*continued*)

Component	Role in Data Collection	Interaction & Flow
CDP Ingestion Layer	Aggregates user signals and identifiers across all channels	Ingests events and transactions; passes data into the Identity Resolution Layer for stitching into unified profiles
CDP Identity Resolution Layer	Unifies anonymous and known users using deterministic/probabilistic matching	Outputs a unified customer ID tied to behaviors and transactions; writes identity mapping back to CDP and downstream systems
Data Contract & Schema Management	Maintains both technical schema definitions and contract-level data agreements	Used by all ingestion sources to validate data structure, enforce field types, ownership, freshness expectations, and business rule conformity
Data Observability Platform	Monitors data health metrics such as latency, completeness, and anomalies	Hooks into ingestion pipelines and warehouses; triggers alerts on schema drift, missing data, or quality violations
Data Lineage & Metadata Tool	Tracks provenance from source events through downstream sinks and transformations	Enables traceability for auditing or debugging; records transformation logic, ownership, and dependencies
Privacy & Consent Manager	Captures and enforces user consent and compliance preferences regarding data collection	Injects consent flags and PII indicators into data payloads; determines filtering or masking behavior downstream
Streaming / Messaging Engine	Routes event and transaction messages in real time or batch streams	Integrates with CDP ingestion and event management layers, feeding clean data into central storage systems

High-Level Data Flow in Layer 3:

1. When a user interacts with the digital platform, their actions are captured in the user interaction layer with the help of the SDK and sent to the event management system. Similarly, when the user completes the desired transaction line, the data is stored in the transaction systems.

2. Both behavioral and transactional data is fed into the CDP Ingestion Layer.

3. The Identity Resolution Layer stitches these signals into unified user profiles.

4. All transaction systems and event management systems are also connected to a central repository of data contracts to ensure data integrity at the source.

5. In the background, observability and lineage tools must be embedded in data pipelines starting the data origin point to monitor pipeline health and data quality.

6. The collected data flows into downstream systems via streaming and batch processing systems.

Ownership Roles:

- Engineering Manager: Owns data instrumentation and pipeline reliability
- Data Steward: Tags data sensitivity, validates contract adherence
- Platform Engineer: Implements observability, lineage, and ingestion tooling
- Compliance Officer: Validates privacy policies, ensures regulatory alignment

CHAPTER 4   FROM STRATEGY TO EXECUTION: BUILDING A DATA OPERATING MODEL FOR AI-READY GROWTH

Data Centralization layer

The Data Collection layer is the stage in the data operating model where raw data from various sources is ingested, processed, and centralized for enterprise-wide usage. The primary role of this layer is to ensure that diverse data is stored in a high-quality and integrated environment, allowing information from multiple sources to be combined effectively to address organizational challenges in a reliable, repeatable, and regulatory compliant manner. It enables downstream systems to have easy and standardized access to high-quality data, enabling a wide range of use cases from analytics to advanced AI-powered products to operational decision-making.

The Data Collection layer has three sublayers.

1. Ingestion layer
2. Curated Data layer
3. Semantic layer

Ingestion Layer

The Ingestion layer acts as the gateway to the data centralization layer, transforming data from decentralized sources into a unified, governed storage and processing foundation. This layer must be designed to balance scale, flexibility, and compliance to enable downstream layers to operate in a deterministic manner. This layer should support both streaming and batch data ingestion, depending on how and when each data variable is intended to be used.

The key responsibilities of this layer are as follows:

- Supports multimodal ingestion techniques like batch, streaming, and change data capture based on the nature and source of the data. This ensures that each use case is served with the right balance of latency and system efficiency.

- Enforces data contracts to maintain schema and business logic validity. Implement schema logs and versioning to enable traceability. These mechanisms protect downstream consumers from schema silent changes in the source systems.

- Respects personal identifiable information flagged data, consent signals, and opt-out states by handling data in accordance with PII policies and local laws. This embeds privacy-awareness early in the pipeline and avoids downstream risk.

- Stores operational metadata such as ingestion timestamp, latency, volume, error logs and retry mechanisms, and SLA compliance status. It allows data engineers to monitor ingestion health in near real-time and addresses bottlenecks or failures proactively.

The Ingestion layer must adhere to the following architectural principles to fulfil its core responsibilities:

- The Data Ingestion layer should be decoupled from the upstream and downstream layers via queues or buffers.

- The Ingestion layer should be designed to scale horizontally with an increase in data volume.

- It should have a resilient design to gracefully handle failures, reties, and schema drifts.

- The architecture of this layer should have latency awareness, and the decision to have real-time or batch loading must be user driven and not uniform.

- The Data Ingestion layer must be designed in a way that observability is engrained in the architecture and not as an afterthought.

Curated Data Layer

The Curated Data layer is positioned between the ingestion layer and semantic layer under data centralization layer. It is this layer where raw data is transformed into stabilized, enriched data and made analytically valuable, forming the single source of truth for all downstream data activities including analytics, AI applications, and decision-making. Although there are multiple architectural options for this layer, we recommend adopting the conventional lakehouse architectural approach. This approach is preferred because:

- It is easier to implement, is widely supported by technology platforms, and benefits from a large pool of available engineering talent.

- Can accommodate various data ingestion patterns seamlessly.

- Ensures data is easily discoverable and queryable across all stages of data lifecycle.

- Makes it easy to build clean and domain-specific datasets with local ownership in the form of data marts while preserving referential links across the entire organization. This enables both modular data management and unified enterprise-wide insights.

The Data Curation layer must have the following core subcomponents and design layers:

- Raw data storage zone: This layer serves as the long-term source of truth, capturing data in its original form or with minimal transformation applied. Transformation activities such as schema normalization, field-level tagging, flattening of records, etc., are applied. The data in this layer is in queryable format, which is in immutable format, partitioned by time, event type, or any other portioning approach that provides efficient retrieval. The primary purpose of this layer is to provide a stable foundation for all downstream transformations and curated datasets. Key characteristics of the raw storage zone include:

- This layer is designed as the raw storage zone of the lakehouse design pattern, which is optimized for traceability and not query performance.

- This layer preserves historical snapshots of incoming data for facilitating auditing, debugging, and reprocessing in case downstream system fails.

- Data stored in this layer is in a structured format, which makes it queryable with ease.

  One must not get confused with the Data Ingestion layer, as the Data Ingestion layer is simply an operational handoff zone. Its primary job is to reliably move data from the transaction systems to the data collection zone in the same format. Data here may not always be in a queryable format as it simply dumps data maintaining the same structure from the upstream system to the collection zone.

- Refined Data Zone: In this layer, data transitions from raw to a usable form. A wide range of quality, validation, and transformation procedures are applied to make data analysis ready. This may include:

    - **Cleansing**: Removing nulls, standardizing formats, resolving duplicates

- **Enrichment**: Joining with master/reference data, adding calculated attributes

- **Validation**: Applying business rules

- **Quality tagging**: Flagging data with quality scores or anomaly indicators

The outcome of this layer is high-quality and consistent data with uniform logic and comparability across teams. This data becomes the first shared data asset across teams.

- The modeled data zone plays a critical role in bridging the gap between semi-cooked data and refined data transformed into structured, business-aligned data models. This data is optimized and modeled for decision-making and machine learning applications. Data in this layer is not just of high-quality but also semantically enriched, reflecting real-world business entities and interactions in a reusable and consistent format. The data models encapsulate important business concepts and events such as customer, product, transaction, support, etc., in a structured representation. Each of these business concepts becomes a separate data model representing all the available information. For example, for the customer data model, it might represent attributes like demographics, behavior, lifecycle stage, etc. These data models serve as a single source of truth offering a simplified view of complex source systems.

To keep data models scalable and reliable, the following design principles must be followed:

- Data models created must be aligned with business domains like sales, human resources, operations, customer support, etc. This approach encourages increased data ownership in the product and business teams. The domain-led data model design also enhances clarity and collaboration between the platform team and product teams.

- Keys are distinct identifiers assigned to each record, essential for ensuring data integrity and reliability, especially when combining multiple datasets through joins. Special care must be paid by both product managers and data engineering team while assigning and creating these keys. Choosing a wrong key can have an adverse effect on the efficacy and effectiveness of data model.

- A robust data model zone must account for temporal dimension data. This helps downstream teams analyze how data has changed over time and draw historical trends. There can be various ways to represent temporal dimensions data, but the following are commonly used:

    - Slowly changing dimensions: How an attribute has changed over time

    - Event time recording: Timestamp of when an event as occurred versus when it was recorded in the system

To enable these design patterns, generally dimensional models designs like star schema or snowflake schema are used.

In simple words, a star schema has one big central table called a fact table that stores the fact or KPI you want to measure, for example, sales, revenue, etc. This table is connected to smaller tables referred to as dimension tables that describe the attributes against which the fact is measured. For example, if you are analyzing the sales of a company, then this model will have sales as the fact table and dimensions can be attributes such as geography, customer name, product category, date of purchase, etc. Now if you want to analyze the sales for each customer on monthly basis, you will join the sales fact table with the date of purchase and customer dimension using a common key.

A snowflake schema also has one big fact table representing the KPI that you want to analyze and is connected to multiple dimensions or parameters against which the KPI will be measured. The difference here is that dimensions are further split into subdimension tables representing additional information about the dimensions that otherwise one would store in a dimension table. These tables aim to reduce the duplication of information in a dimension table. Extending the example of a sales domain, a product dimension can have a subdimension of product category, and that information can be removed from primary dimension of product. Now if any downstream systems want to analyze sales data for each product, then the system will only join the main product dimension with the sales fact table. If the requirement is to analyze sales on each product category and product, then the system will first join the sales fact table with the primary product dimension and then the product dimension with its subdimension product category.

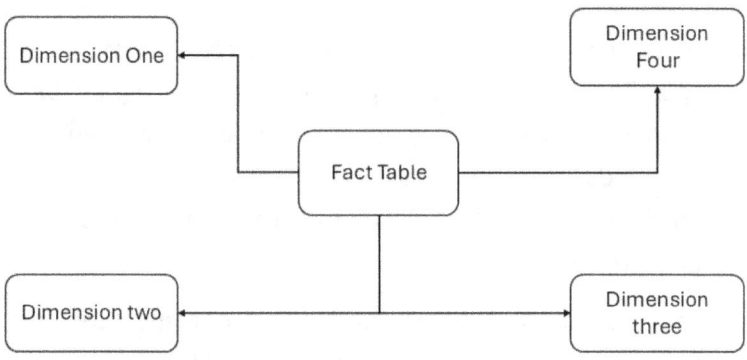

Figure 4-2. Star schema diagram

CHAPTER 4   FROM STRATEGY TO EXECUTION: BUILDING A DATA OPERATING MODEL FOR AI-READY GROWTH

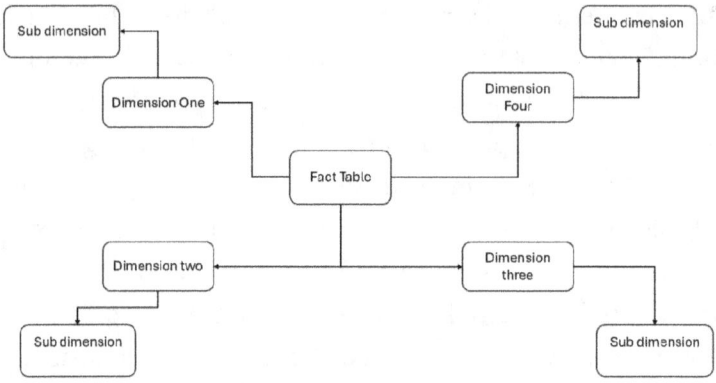

Figure 4-3. Snowflake schema diagram

- Data Marts (Domain Data Products): These are the ultimate output of this layer. These are modular and curated datasets designed for specific analytical or operational use cases. Unlike ad hoc datasets, data marts are directed toward a specific use case, governed by the SLA and owned by application owners aligned with the principles of data product thinking. Data marts can also be defined as a contract between the data platform team and business teams, ensuring expectations and responsibilities are clear on both sides. It has the following core features:

- All data marts are use case driven and serve a very specific purpose. For example, a data mart designed for marketing attribution would only contain information required for tagging a transaction to a specific marketing channel. It would not contain information around customer details.

- Each data mart has a data product owner responsible for maintaining its quality, updates, and SLA.

- Each data mart has an agreed-upon policy that governs its quality standards, schema change policies, consumption guidelines, refresh cyle guidelines, etc.

One must not confuse this layer with the data model zone layer. Table 4-4 presents the key differences between the two layers.

Table 4-4. Difference Between Modeled Data Zone and Data Marts

Aspect	Modeled Data Zone	Data Marts (Domain Data Products)
Purpose	Organizes and abstracts raw data into clean, reusable, and standardized data models representing business entities	Deliver ready-to-use data for specific business use cases, built on top of the modeled data zone
Audience	Data engineers, analytics engineers	Business analysts, domain teams
Focus	Business concepts (e.g., Customer, Transaction, Product)	Specific business questions (e.g., monthly revenue, churn Analysis")
Structure	Normalized dimensional models (facts and dimensions)	Denormalized, user-friendly tables or views
Reusability	High across multiple data marts	Often purpose-built for a single domain or report
Governance	Modeled with data platform in mind	Owned by business domain teams
Format	Dimensional (facts/dims)	Often flat/denormalized
Examples	Customer_Dim, Sales_Fact, Product Dim	Customer_360_Mart, Revenue_Snapshot, Marketing_Attribution_Mart

- Governance and Metadata layer: While all layers that were earlier discussed are horizontal each with a set defined outcome, this layer is unique as it spans all layers. This layer is responsible for managing data quality, regulatory compliance checks, and ownership throughout the data centralization layer. It is accomplished by implementing data lineage, schema versioning, sensitivity classification, maintaining ownership, and accountability details and validation of business rules.

Lifecycle of Data in Curated Data Layer

1. Data from the ingestion layer enters the raw data storage zone.

2. It is then cleaned, validated, and enriched into the refined zone.

3. Data is then transformed into business-aligned models in the modeled zone.

4. It is curated and published as a governed data mart ready for application-level consumption.

5. All steps are governed by metadata, lineage, and quality metrics.

    - **Semantic layer**: The semantic layer is the final sublayer of the data centralization layer. It is essential for unlocking the full potential of centralization. This layer embeds the business logic into usable code that can used for defining metrics and KPI into a governed abstract layer, ensuring that all downstream applications including dashboards and ML models use the same metric definition, leaving no room for confusion. By doing so, the semantic layer enables data democratization without sacrificing the consistency or governance. In EDGE, the semantic layer is not just a technical asset; it's a contract that protects trust and consistency in data as a product.

For a product manager, the semantic layer helps in the following ways:

- This is a data assurance system that solves the problem of inconsistent metric definition and interpretation across teams.

- This helps in reducing the ad hoc report requirements that the data engineering and analytics team get.

- This promotes self-service data access where metrics are always in a ready-to-use state across all type of access modes ranging from BI tools to Excel sheets to APIs.

- This enhances cross-functional alignment.

This layer has the following core responsibilities:

- The primary responsibility of the semantic layer is an abstraction of the business logic used for defining metrics. This logic is developed once in the semantic layer and can be reused across all layers. This ensures that all applications refer to the same source of truth.

- The semantic layer organizes metric models by business domain and not just technical subject. Each model is supported by rich metadata and documentation, making it easy for the users to discover required metrics through a semantic catalog or discovery interface. This enables self-serve analytics while preserving trust and governance. While representing data in business terms, this layer is also responsible for hiding technical complexities.

- This layer is also responsible for enforcing governance over metric definitions including version control, comprehensive description with associated filters, and dimensions for calculation, and the definition and enforcement of SLA.

Multichannel Metric Access via Semantic Layer

A downstream application can have diverse data consumption channels such as dashboards, Excel, API-based products, etc. This section will illustrate how the semantic layer defined in YAML enables consistent and governed access to business metrics.

API: Metrics can be accessed programmatically via an API, enabling seamless integration into internal tools, services, or external-facing applications. When an API request is made internally, the system references the YAML-based defined metric logic, builds a SQL query based on the requester's API request, and runs the query on the upstream database returning results. Internally, these results are transformed into structured JSON and return an API response.

Dashboard: A business user can connect dashboarding tools like Looker, Power BI, or Tableau to the semantic layer through the SQL interface, API endpoints provided by the data platform, or specific connectors. Once a connection is established, all the defined metrics and their associated filters and dimensions are visible to the user on the dashboarding tool. Once a user selects a metric with its dimensions and filters, it is then translated internally into SQL queries that are then executed on the targeted database. The results are returned to the dashboard tool. In this entire operation, the user is not required to write any SQL query.

Excel sheet: Excel sheet users can connect the semantic layer through data connectors or plugins such as ODBC/JDBC, power query, or custom Excel add-ins or DBT semantics. Once connection is established, all metrics and their dimensions and filters are listed on the Excel sheet (you can imagine the view as a pivot table) as the user makes selections internally and translates the selection into SQL queries, which then gets executed on the targeted database. The results are returned to the sheet.

Data Activation Layer

The Data Activation Layer is where data moves from storage to action. The objective of this layer is to make data available and usable across different operational systems. It enables batch versus real-time data movement from the centralization layer to the operational system. This layer also stores common AI/ML data features that can be reused across multiple applications. All the activities listed under this layer are accomplished either by enabling reverse ETL from centralization layer to front-line tools like CRMs, CDPs, ad platforms, and support systems or by designing pipelines that can serve multiple teams from the same trusted data layer.

Data-Driven Growth Layer

The Data-Driven Growth layer is the final and most important stage of EDGE. It ensures that all the data initiatives translate into meaningful and measurable business outcomes. It helps in completing the feedback loop between data, decision-making, and strategic growth. It is the layer where data is consumed by applications such as AI-powered products, analytics, etc., to drive real-world impact. This layer also closes the loop by feeding learning and outcomes back into the earlier layers of the data operating model for continuous improvement.

CHAPTER 4   FROM STRATEGY TO EXECUTION: BUILDING A DATA OPERATING MODEL FOR
             AI-READY GROWTH

Summary

This chapter explored how to develop a data ecosystem in the organization that can help the AI strategy to succeed. We explored how the concept of the organization operating model can be applied to data to develop a data operating model. This data operating model will define guidelines and SOPS for various data technology solutions to work and interact with each other. We also introduced a custom framework called EDGE that aligns data business goals, enables domain ownership, and democratizes the data access. EDGE ensures that at all stages all data activities are related to the business goals and the data maintains the highest standards set by the organization.

We also clarified the differences between data operating models and data platforms and how the data platform can be a subpart of the larger data operating model.

As organizations are increasingly using data for defining their product experiences, executive decisions trust in data has become the most critical trait. That is why in the next chapter we will focus on data quality. The next chapter will focus on data quality, not just as a technical concern but as a strategic enabler for building reliable AI and confident decision-making.

CHAPTER 5

Mastering Data Quality: Techniques and Best Practices for Product Managers

In the previous chapters, we have explored the strategic concepts around how to generate, store, and utilize data to help organizations achieve their strategic goals. Implementing these concepts will help transform data from being a passive sedimentary byproduct into a strategic asset.

However, for data to truly play that strategic role, it must be healthy, accurate, and trustworthy. This chapter focuses on this critical yet often overlooked aspect of the data lifecycle, i.e., how to maintain data quality throughout the lifecycle of data. Product leaders must consider maintaining data quality as one their core responsibilities because poor quality data undermines innovation, leads to poor decision-making, and hampers product development.

CHAPTER 5   MASTERING DATA QUALITY: TECHNIQUES AND BEST PRACTICES FOR PRODUCT MANAGERS

The goal of this chapter is to provide product leaders with the fundamental knowledge and actionable solutions to embed data quality management as a key step in the product lifecycle. It will cover the following:

- Fundamentals of data quality, including its attributes, maturity frameworks, and best practices to maintain data quality

- Fundamentals of data lineage in tracking data from generation to consumption, a role it plays in maintaining data quality throughout the data pipeline

- How to develop data pipelines that are transparent and auditable

- Best practices for maintaining data quality in AI applications

Before we start exploring these concepts, let's us first review why product leaders must bother about data quality.

Why Product Leaders Must Care About Data Quality

In an increasingly data-driven world, decisions, digital product offerings including personalized user experiences, and revenue models rely heavily on data. Poor data quality can adversely affect user trust, mislead product strategy, and eventually affect product usage. Poor data quality is often like a hidden malignant tumor whose symptoms are often dismissed as routine issues until they progress to a stage where recovery is unlikely. Data quality issues are often labeled as tactical issues receiving the lowest priority both from product and technology leadership. But in today's world, product leaders should adopt the mindset of a doctor who constantly advises for

CHAPTER 5   MASTERING DATA QUALITY: TECHNIQUES AND BEST PRACTICES FOR PRODUCT MANAGERS

routine health checkups, and in case of any anomaly, treating the root cause takes top priority, not just treating symptoms. The following are few astounding reasons that will make product leaders reprioritize their approach toward managing data quality:

- **Poor data = broken products:** From offering personalized user experience to recommending the right products to users to identify and preventing fraud, today most products are increasingly dependent on data. If the underlying data is wrong, it can affect the product experience, which will in turn affect the user trust. This loss in user trust will adversely affect product adoption. To summarize, in the era of increasingly data-driven products, the amount of trust you will have in your product is directly related to the amount of trust you have in your data.

- **Quality of data-driven decision-making is as good as data quality:** From experimentation to executive dashboards, decisions that predictive models make rely on data. If the input data is flawed, then the insights generated from these are too, which may lead to bad decisions. In today's modern world, the age-old proverb "Garbage in, garbage out" has become a reality.

- **Successful adoption of AI depends on quality of data:** As organizations race to adopt AI in their daily operations, often the focus lies on the volume and variety of data that is being captured and stored. But veracity or accurateness of data is often overlooked. All data applications including those powered by AI depend solely on the data. If the quality of training data is not high, then the data application will not deliver the correct output. Hence, protecting the quality of data in this era is mission critical.

- **Regulatory and compliance demand:** With AI taking central stage in industries, regulatory boards have started to form rules to ensure that data is collected and used in an ethical way. Poor-quality data including incomplete records can expose companies to regulatory risk.

- **Bad data contributes to hidden technical debt**: Often poor-quality data signals an underlying technical issue. Rather than addressing the root cause, teams typically resort to manual workarounds or isolated quick fixes. While these might temporarily improve efficiency, they accumulate over time, increasing technical debt and ultimately slowing down overall development cycle. As this debt grows, it becomes harder and costlier to manage.

As product leaders, we are accountable for the outcomes, and today these outcomes are increasingly dependent on data. By treating data as an integral component of the product, we will ensure it gets the due importance it deserves because today data quality is no longer just a technical concern; it has transformed into a product leadership imperative.

CHAPTER 5   MASTERING DATA QUALITY: TECHNIQUES AND BEST PRACTICES FOR PRODUCT MANAGERS

Introduction to Data Lineage and Quality in Modern Data Architectures

After establishing why product leaders must prioritize data quality issues, let's now move forward and understand what data quality and lineage mean in modern data architectures.

As data becomes foundational to product innovation, decision-making, and AI adoption, the enterprise architecture is moving from static monolithic siloed systems to modern distributed data architectures. These architectures are powered by real-time data pipelines, data lakes, and cloud-native platforms. While new enterprise architectures offer flexibility and scalability, they also introduce new complexities. The most critical challenges in this environment is ensuring data quality and traceability across increasingly distributed and fast-moving pipelines. This is where data lineage comes into the picture.

What Is Data Lineage?

Data lineage refers to the ability to track the data lifecycle from the point of origination or generation to the point of final consumption. If data originated from a single source is consumed at multiple applications each with separate transformation steps, then ideally lineage should track each journey separately. Ideally, data lineage should cover data from its point of origin, all intermediary transformation, storage, and eventual consumption. To summarize, data lineage implementation should be designed to answer the following questions:

- From where the data came from (point of origination also called *source systems*)
- What processing or transformations the data underwent
- Where was it consumed (end point systems or application where this data was utilized)

CHAPTER 5   MASTERING DATA QUALITY: TECHNIQUES AND BEST PRACTICES FOR PRODUCT MANAGERS

In the modern data architecture, data lineage enables transparency, traceability, and auditability of the data, which not only helps in maintaining the quality of data by enabling easy debugging and bug fixing but also ensures regulatory compliance.

What Is Data Lineage (and Why Does It Matter to PMs)?

Data lineage provides end-to-end visibility into from where data comes, how it changes, and where it flows across systems. It traces a data point's full journey, enabling product teams to understand how a metric works, how a feature in AI model is developed, and what contributes to a prediction of an AI model. From a product lens, lineage helps in the following ways:

- Builds trust in the metrics and data trends observed, enabling more confident decision-making.

- Helps diagnose the root cause when AI model produces unexpected results.

- Enables product managers understand how changes in upstream data systems can affect downstream systems, features, and user experience reducing risk of unintended consequences.

- Enhances explainability of data-driven decisions by providing clear visibility into how data is collected, transformed and used. This not only helps in regulatory compliance but also enhances user trust.

CHAPTER 5   MASTERING DATA QUALITY: TECHNIQUES AND BEST PRACTICES FOR PRODUCT MANAGERS

How Data Lineage Enhances Data Quality, Transparency, and Governance

Modern data systems consist of dozens of interconnected pipelines. A single faulty transformation, schema mismatch, or upstream change can propagate incorrect data downstream, silently breaking all data applications. Without gaining insights about how data is flowing and changes in these intermingled pipelines, teams are often flying blind or spending long hours diagnosing data issues. Data lineage is the cornerstone for developing a reliable, governed, and explainable data system. Let's now explore how data lineage enhances data quality, transparency, and governance of the modern data system:

- **Data lineage helps improve data quality with root-cause traceability:** In real-world scenarios, data quality issues are rarely localized. One flawed data source or transformation can impact multiple systems. Data lineage provides visibility into how and where data is generated and consumed, making it easier to do the following:

    - Identify source of quality issue

    - Evaluate the impact of schema or logic changes across systems

    - Trace the propagation of errors across downstream systems

- Building confidence in data helps executives and product leaders build trust in data metrics, which they use by providing explainability and traceability. This also helps business, analytics, and engineering teams interpret data in the same language, which again helps in reducing data quality issues as data definitions are agreed upon and interpreted in a common language.

- **Helps in strengthening data governance and compliance:** Data lineage helps in governing compliance to regulations by doing the following:

  - Analyzing impact before deploying data changes
  - Identifying where sensitive data resides or flows
  - Ensuring compliance to data retention and deletion polices

Key Challenges in Managing Data Lineage and Quality at Scale

While we have discussed the benefits of implementing lineage and quality solutions, scaling them in a large-scale environment introduces complexities that are often underestimated. In this section, we will explore most common challenges encountered, because understanding these challenges in advance will help product leaders prepare an effective plan to scale data lineage and quality solutions.

- **Tool sprawl and siloed systems:** Large organizations often use dozens of tools for data storage, ETL operations, and streaming data, and each tool may have their own metadata and lineage logic. While in isolation this may look good as these systems may provide an answer to questions that are of local importance, on an organization level, it leads to confusion and creates blind spots and poor traceability. This can also give rise to conflicting definitions of the same metrics or fields, which would not only corrupt data but also make it impossible to use for an organization-level use case. To avoid this, leaders must take the following steps:

- Consolidate metadata of all data fields into a centralized data catalog. Atlan, Collibra, and Datahub are some major metadata tools that can be used.

- Always use internationally accepted standards and platforms for managing lineage metadata storage and generation.

- Data ownership and system/process dependencies should be clearly identified and documented at an individual team level.

- **Dynamic, constantly changing pipelines:** Modern data architectures are always evolving, which includes not only additional data points being created but also transformation logic refactoring and schema changes. All these changes have the potential to impact downstream data applications adversely if done uninformed. Lineage solutions can direct these changes, but with partial or no integration with CI/CD workflows and schema validation tools, lineage systems are just reactive alerts rather than a proactive safeguard. To avoid this situation, a schema versioning practice should be implemented, and ideally data contracts should be used to handle schema declaration. Product managers must also ensure that automated alerts are set up for raising uninformed changes in production data pipelines.

- At scale, the implementation of real-time lineage tracking and quality check solutions across numerous data pipelines is very resource intensive and costly. To avoid incurring this cost, product managers must take following steps. Identify critical data assets and classify them into multiple tiers. Once asset tiering is done, then set up a tiered lineage tracking and quality monitoring framework. Tier 1 assets that are most critical must have real-time monitoring in place. Tier2 assets that have medium criticality should have a periodic batch check. You can split this tier into multiple subtiers and set the frequency from daily to weekly to monthly. Tier 3 is where not so critical assets are placed and should have on-demand monitoring.

Key Dimensions of Data Quality

In this section, we will explore data quality in detail including its definition, key dimensions, assessment methods, and key metrics.

Data Quality: Definition and Key Dimensions

In the convergence of AI and business strategy, data quality is the linchpin that determines the degree to which organizations can leverage data as the strategic asset. All AI models are only as trustworthy as the quality of data their train on, and poor data will lead to poor decisions regardless of how sophisticated the algorithm is. With the context now set, let's now proceed to understand what data quality is.

CHAPTER 5 MASTERING DATA QUALITY: TECHNIQUES AND BEST PRACTICES FOR PRODUCT MANAGERS

What Is Data Quality?

Data quality is defined as the degree to which data fits for its intended purpose whether it is about powering an AI/ML applications or analytics to support data-driven decision-making. This includes evaluating data on various features like accuracy, completion, consistency, and timeliness.

In a modern context where AI has taken precedence, data quality is not just about correctness; the concept is extended to include trustworthiness, timeliness, and usability throughout the entire data life cycle. Today data powers real-time decision-making; managing data quality is no longer be a reactive task addressed after the fact. Instead, it must be embedded into the design of the systems from the onset just like performance and security considerations. Ideally, modern data architecture must have following components:

- Data observability and monitoring tools to identify data issues early and raise alerts

- Data contracts between teams to define schema expectations and ownership

- Data lineage tools for tracking data throughout the lifecycle

- Incorporate data quality metrics into OKR and SLA

Key Dimensions of Data Quality

To understand and manage data quality, it is essential to understand the key dimensions on which the quality of data is evaluated. Each dimension defines a different aspect of data.

CHAPTER 5   MASTERING DATA QUALITY: TECHNIQUES AND BEST PRACTICES FOR
            PRODUCT MANAGERS

Accuracy

This dimension of data quality captures how well and truthfully data represents real-world events. Accurate data reflects the true value as it exists at the source without any distortion, errors, or bias. In simpler terms, it means that data should be factually correct, not made up, and not outdated. Data is representing what happened and not what we assume or hoped. The term distortion in this case means that while capturing data neither system or human responsible for capturing data made any mistake which would affect the accuracy of data.

For example, if a user performs a transaction of INR100 in an e-commerce marketplace but in one of the systems it gets recorded as $100 because the default currency was set in that system was dollar, the simple change in currency made this data point inaccurate. This dimension matters because inaccurate data will make AI models learn wrong trends, which will in turn make wrong recommendations ultimately affecting the efficiency of the model.

Common Causes of Inaccuracy

- Human data entry errors.

- Incorrect data mapping. Fields from the source system are incorrectly mapped to the wrong columns or formats in the target system.

- Lack of validation at the source.

- Stale or outdated information in the systems of record.

- Data loss in transition. When data is transmitted from one system to another, sometimes some part of data is either truncated or dropped off, or an entire data format gets changed due to differences in schema definition, differences in encoding methods (UTF-8 vs ISO 8859-1), timeouts, or any other technological issue.

CHAPTER 5   MASTERING DATA QUALITY: TECHNIQUES AND BEST PRACTICES FOR PRODUCT MANAGERS

How to Measure Accuracy

- The match rate is calculated as the percent of records that are accurate of the total records in the database.

- Error rate is the percentage of data fields of value that are incorrect of the total data fields.

How to Improve Accuracy

- All data entry points, i.e., the screen from where the user will enter data, should have validations in place.

Use authoritative references data from trusted verified source to validate user data like government identification, address, country codes, etc.

- Automate accuracy checks against external or internal master sources.

- Implement exception reporting to flag values outside of expected ranges.

- Allow users or systems to report and correct inaccuracies in data

- Implement data contracts for schema validation and checks in data pipelines

- Use tools to identify drift in source vs designation data

- Use automated data diffing tools to compare source and destination data for discrepancies.

Completeness

Completeness is defined as the degree to which a record is complete or contains every piece of expected information. In simpler terms, this means that a record has no missing column, and a dataset has no missing records, which is essential for a particular use case. Data completeness is critical for any data application as missing data can create biasness in models if missing data is from a specific user or product or geographic category. It can lead AI models to learn incomplete trends, limiting its ability to generalize, which may affect its performance in real-world situations both from efficacy and ethical point of view.

An example of database completeness could be imagining a dataset stores all logs of all activities of a server including the number of requests, status of requests, type of request, concurrent requests, etc., but accidentally the developer has coded the logic conditionally such that only error logs are stored. If you want to use this dataset for developing a prediction to anticipate the occurrence of a specific type of error, then this dataset is incomplete because it does not store normal observations that would have taught the model to differentiate between error trends and normal trends.

Common Causes of Incompleteness

- Poor form design: Forms used for data entry may lack proper design such as not enforcing mandatory fields or absence of inefficient validation rules. This can result in incomplete or inconsistent data capture.

- Data dropped during ETL or ingestion: Data may be unintentionally dropped during the ETL process due transformation error or schema mismatches. This can lead to incomplete records.

- Delays in system syncing or API responses: If two systems interacting with each other for data transfer have significant difference in latency or API response time, there is a high likelihood that some records may fail to synchronize properly. This may lead to incomplete or out-of-sync databases.

- Manual data entry errors or omissions: Manual data is prone to human errors such as typos, incorrect values, or missing fields. This may affect the completeness of the record.

- Misconfigured data pipelines: In some cases, validation used either at the input forms or at the start of a data pipeline is outdated or incorrect. This can result in valid data being rejected at ingestion, causing certain fields to miss or be incomplete in downstream systems.

How to Measure Completeness

- **Null Value Percentage:** This metric should be calculated on a column level. It is calculated as a percent of the missing values in key fields.

- **Record Completeness Score:** Proportion of fields filled out per record.

- **Coverage Ratio:** Number of expected records versus the number of actual records.

How to Improve Completeness

- On the user input forms, enforce validation to make critical fields mandatory.

- On the user input form, add data validation rules to flag incomplete or inappropriate before feeding them in the system.

- If a particular data field consistently has a common or expected value, consider setting that value as the default to reduce input error. Example, fields like citizenship or ISD code before the phone number can default to a country-specific value.

- Another way to improve completeness of records is by imputing them. Use statistical methods or business logic for identifying the imputation value.

- Create data contracts to define and enforce schema expectations with data producers.

Consistency

Consistency in data quality means that a data entity should have uniform values across different systems, format, and time. For example, a customer whose record is present in the CRM, discount management system, and order management system should have the same loyalty tier status in each system. Inconsistency in data is a critical issue as it only brings information asymmetry between different teams and systems but also makes decision-makings learn inconsistent trends that affect the quality of decisions.

Common Causes of Inconsistency

- Data pulled and merged from multiple sources each with a different schema or common transformation logic can lead to the same information presented in different formats.

- An individual source system evolves independently as it is governed and controlled by different teams. Sometimes these systems witness a change in a data schema such as the addition of new columns, renaming a field, or changing the data structure or length of the column independently without informing others this could lead to information asymmetry across systems. An example of such asymmetry can be the customer status field in one system is stored as "Active/Inactive" and in other system it is stored as "Enabled/Disabled."

- Lack of common rules across systems for defining and transforming data fields. If there is no common definition of critical data entities, then individual teams will define those entities as per their convenience, which would create fundamental inconsistencies in data entities. For example, if there is no common definition to determine how to classify a user active, then each team will use different benchmarks to classify user as active or inactive. Even if definition is decided but the data format including length and value type is not decided at the organization level, then information asymmetry can occur.

- Sometimes manual overrides and local edits done in a downstream system will create an information gap between upstream and downstream systems.

- Different systems using different reference data to look up value for the same data entity can lead to information asymmetry across systems.

- In ML training workflows if the labeling logic changes, then the same model with the same features might give different outcomes, which may affect the decisions .

How to Measure Consistency

- **Cross-System Comparison Rate:** Percent of values that match across defined systems

- **Conflict Rate:** Percent of records with mismatched or contradictory values

- **Format Uniformity Score:** Percent of records that adhere to a standard format or structure

How to Improve Consistency

- Establish common definitions and standardized formats for all critical data entities. This should also include standardizing date formats, currencies, conversion rate, encoding norms, etc.

- Designate clear domain ownership for key data entities such as customers, products, and employees.

- Automate data synchronization between different systems. This should include queue-based implementations that would ensure that data is not lost in transit due to different in latency between two systems.

- Conduct regular cross-system validation to identify any mismatches in critical data fields.

- To ensure consistency across systems, implement data versioning, like how software engineers use version control for code. Assigning version identification to datasets allows teams to track changes, verify which version is in use, and quickly identify systems that are referencing outdated or inconsistent data.

- All systems have same information use tools for data versioning just like software engineers use code versions to name code. This will help in identifying easily which system is *not* using latest versions of data.

Timeliness

Timeliness refers to the freshness of data. The data must reflect the most recent and accurate state of the real world when it is needed for operations. However, this definition is contextual to the use case for which data is required. Some use cases such as product payment fraud detection may require real-time data, whereas other use cases may tolerate some delay such as lead conversion model are generally executed in batch form and may tolerate some delay.

Common Reasons for Poor Timeliness of Data

- Most organizational data is processed in batches on fixed schedules. If the frequency and timing of these batch processes do not align with when the insight is needed for decision-making, it can result in stale or outdated information being used.

- If data ingestion and transformation pipelines are slow, fragile, or prone to error, then it might impact the timeliness of data. Additionally, if the pipeline's throughput does not match the velocity of incoming data, latency and backlogs will increase.

    This issue is especially common in a digital/ecommerce business where during peak periods such as annual sales and seasonal events, the data generation rate and volume exceeds the normal level exponentially. Pipelines not architected for handling a high rate of concurrency or horizontal scaling may impact all aspects of data quality including timeliness. Without the support of queuing mechanisms, load balancing, and auto scaling, these pipelines will struggle to manage load.

- Human dependency for data generation or data upload can impact timeliness of data.

- A common cause for stale data is misaligned SLAs and expectations. Different teams often maintain different SLAs and batch processing frequency or schedules. When data involves multiple teams, these inconsistencies can impact the timeliness of data.

How to Measure Timeliness

- Data latency (ingestion lag): Time difference between when data is generated and when it becomes available downstream.

- Data freshness: How recently the data was updated.

- Data delivery SLAs: Whether systems meet agreed delivery schedules

How to Improve Timeliness

- Define data freshness requirements for each use case. While defining freshness requirements, product managers must refrain from asking for real-time or near-real-time freshness for all use cases. Remember, real-time or near-real-time freshness comes with a very cost on infrastructure.

- Modernize data pipelines by using streaming or micro batch architecture for low-latency use cases.

- For use cases where data from multiple domains is used, define common SLAs on a use case level. Individual team/domain-level SLAs must be aligned with this overarching SLA.

Validity

Validity refers to the extent to which data conforms to the defined formats, values, business rules, and constraints. A valid data meets the eligibility criteria set by the system, product, and model that consumes it. This dimension of data quality focuses on the compliance of data with the expected structure and business rules defined, which includes the data type, range, and format of the data.

Common Validity Constraints

- **Format constraints** (e.g., regex for emails, date formats)
- **Type constraints** (e.g., strings, integers, Booleans)
- **Domain constraints** (e.g., only accepted values like country codes or status types)
- **Range constraints** (e.g., percentages must be between 0 and 100)
- **Mandatory fields** (e.g., user ID must not be null)
- **Referential constraints** (e.g., product ID must exist in the product table)
- **Business constraints**: These are domain-specific rules that define how data should behave logically in expectation with business logic and policies.

Common cause behind invalid data

- When validation logic on data entry points is inadequate and does not cover all types of previously mentioned constraints, there is a risk that the entered data may not remain valid or reliable across all systems or use cases.

CHAPTER 5   MASTERING DATA QUALITY: TECHNIQUES AND BEST PRACTICES FOR PRODUCT MANAGERS

- When there is no standardized schema, or agreed upon set of business rules to govern data across organizations, the individual team will have its own rules or key rules entirely. This may lead to data validity issues.

- When data needs to be transformed, the rules used in ETL or preprocessing may be incorrect due to a gap in understanding the underlying business processes. Alternatively, the logic may be developed overly aggressively (for example missing on boundary values). In both scenarios, the transomed data violates business constraints leading to data validity issues.

How to measure Data Validity

- Validity Rate (%): The percentage of records that fully comply with all defined format rules, constraints, and business logic.

- Rule-Level Violation Count: Tracks how many records fail each specific validation rule.

- Invalid Field Density: The average number of invalid fields per record.

How to Ensure and Enforce Validity

- Use data contracts and schemas.

- Set up pre-checks before loading them to production tables or models.

- Define exception handling policies to decide whether invalid data should be rejected, quarantined, corrected, or bypassed with defaults. These rules should also cover the tolerate rate of individual rule violation for each use case.

Uniqueness

Uniqueness refers to the requirement that each record in the dataset represents a distinct, nonredundant real-world transaction. Duplicity in records can adversely affect metrics, may introduce biasness in AI systems, and can degrade user experience. Ensuring uniqueness means enforcing one representation per event within the relevant context.

How to Measure Uniqueness

- Duplicate rate: This is calculated as a percentage of the number of duplicate records of the total number of records.
- Unique key violation count: This is the total count of violations of a business -defined uniqueness constraint.

Common Causes of Duplicity

- Absence of unique key constraints in database where data entry is stored.
- When data is integrated from different sources, improper use of join operations can result in record duplication.
- Fuzzy matching or poor merge rules.
- Accidental reprocessing of the same data multiple times.

Deduplication Strategies

- Define uniqueness criteria for relevant fields in the database.
- Use advance matching algorithms to identify records that contain similar information in slightly different variation.
- Design processes for reviewing and merging duplicate candidates manually or via rules.

CHAPTER 5   MASTERING DATA QUALITY: TECHNIQUES AND BEST PRACTICES FOR PRODUCT MANAGERS

Data Quality Management Framework: The PROMT Framework

In the modern world where decisions are increasingly data-driven, data quality is not just an abstract concept, it is a measurable discipline. Just the way programming codes are measured for performance, product performance is tracked against set business objectives, and data quality must be tracked, managed, and reported.

We present a multilayer framework with the name "PROMT Framework" to design and implement data quality and performance indicators across the data lifecycle from generation to consumption. Implementing a strategic multilayer approach will enable organizations to move from reactive firefighting to proactive data stewardship. In today's AI-driven organization, managing data quality is not just a best practice but survival. The PROMT framework outlines a multilayered approach for building effective data quality management that is as follows:

- **P**urpose-driven
- **R**esponsible and accountable
- **O**perationalized in lifecycle
- **M**easurable across layers
- **T**ool-agnostic and transparent

Let's now explore PROMT Framework in detail.

This framework is based on six foundational tenets.

1. **Product and strategy-centric quality:** Data quality must be assessed based on its impact on product experience and strategic business decisions and how it affects AI adoption in the organization, not just on technical correctness. This ensures that data is treated as a strategic asset across the organization.

2. **Layered metrics**: Data quality should be measured and reported across multiple layers. This will ensure that each stakeholder gets the insights needed for effective decision-making and action.

3. **End-to-end coverage**: Data quality should be monitored and maintained throughout the lifecycle from data generation to ingestion to consumption, with common SLA and rules governing data quality at the organization level.

4. **Tool agnostic but integrated**: Establish generic, holistic data quality management policies that serve as the strategic framework with tools actings as flexible enablers to implement and enforce these policies effectively.

5. **Actionability and accountability**: Data quality policies must establish clear accountabilities for all stakeholders, with clear metrics and SLA to measure performance and drive timely actions.

6. **Composite Indexing**: Synthesizes metrics into indexes (like DQI or Trust Scores) for executive visibility and prioritization.

The PROMT Framework Components Explained

The PROMT framework helps define data quality principles and translate them into actionable steps that ensure data is trustworthy and fit for purpose. This enables organizations to confidently drive data-informed decision-making and accelerate AI adoption. The following are activities that are required to implement the PROMT framework.

CHAPTER 5   MASTERING DATA QUALITY: TECHNIQUES AND BEST PRACTICES FOR PRODUCT MANAGERS

Align Data Quality Metrics with Strategic Objectives

Every strategic implementation that is data dependent may have a different definition or requirements around the quality of data required for its success. To achieve this, every dataset of the pipeline should have metrics directly tied to the use case that it supports. This ensures that data quality is defined, measured, monitored, and tailored to the needs of each use case, enhancing user confidence in the data and driving greater product adoption. For example, if a specific dataset or pipeline will be supporting a pricing model in an e-commerce marketplace, then timeliness, accuracy, and data drift might be the right set of major data quality metrics. Similarly, for executive dashboards timeliness, accuracy and completeness could be the top metrics. While this does not mean that for a specific use case only these metrics will be monitored, these metrics or data quality dimensions will take precedence over others.

Mapping datasets and pipelines to a specific use case also helps in identifying critical data assets and directing investment of organizations resources accordingly.

The product leader must undertake the following listed activities to achieve the desired outcome of this step:

- Map each dataset to the product feature, AI/ML implementation, or business goals it supports.

- Prioritize each data asset including the dataset, pipeline, and storage platform based on criticality of the use case it supports. Ideally on the completion of these two steps, you have a list of data assets in ascending relevance and criticality.

- Collaborate with various stakeholders including business, product, data engineering, AI team, and software development to define quality goals for each asset, for example, freshness, accuracy, data drift, consistency, etc.

- Once the quality metrics are identified, define the success criteria for each quality metric. For example, for a product recommendation feature for an e-commerce marketplace, completeness of the records with the agreed on fields should be above 98%.

- Prioritize development, monitoring, and measurement of quality checks based on user impacts and strategic visibility.

- Prioritize quality checks based on user impact and strategic visibility (e.g., production versus internal datasets).

Define Multitiered Performance Indicator Layers

Once all the datasets have been identified and prioritized in order of their relevance and their core data quality metrics have been identified, structure these metrics into three layers. Each layer caters to a defined set of stakeholders, so the metrics are tailored to meet their unique needs and expectations. Each layer builds trust by showing the right data quality view to the right set of stakeholders driving clarity, accountability, and traceability. Table 5-1 presents the purpose and target audience for each layer.

CHAPTER 5 MASTERING DATA QUALITY: TECHNIQUES AND BEST PRACTICES FOR
           PRODUCT MANAGERS

Table 5-1. *Define Multitiered Performance Indicator Layers*

Layer	Audience	Purpose
Operational Metrics	Data engineers, software engineers	Detect technical issues (drift, freshness, lag)
Product & Model Readiness	Product managers, AL/ML engineers	Validate data fitness for features or models
Executive/Strategic KPIs	CDOs, exec sponsors	Track SLA adherence, domain health, trust scores

The following steps must be completed to achieve the desired outcomes of this step:

- Start by defining and documenting what data quality means at each layer, by understanding how data quality affects their area of responsibility or the strategic objective they are trying to achieve using this data asset. The following are sample metrics for each layer:

    - **Operational metrics :** These metrics are used for detecting data quality issues at the most granular level in real time across the pipelines and system.

        - Ingestion Success Rate : % of successful data loads over total runs

        - Null Value Rate: % of records with missing required fields

        - Schema Conformance: % of records adhering to schema definitions

        - Data Freshness Lag: Time since last successful update versus expected SLA

214

- Row Count Drift: Variance in expected versus actual row volume

- **Product Readiness**: Ensures that dataset used for a product feature, personalization systems, or decision intelligence meets the required data quality standards.

  - Feature Drift: Change in distribution of features over time

  - Label Coverage: % of training records with valid target labels

  - Class Imbalance Ratio: Distribution skew across classification targets

  - Outlier Rate: % of anomalous or extreme values in key features

  - Business Logic Violations: Records failing domain-specific validation rule

- **Executive/strategic KPIs:**: Track data health as a strategic asset and measure its impact on business outcomes and innovation readiness.

  - Data Quality Index (DQI): Composite score across dimensions (accuracy, completeness, etc.)

  - SLAs Met on Critical Tables: % of critical datasets meeting quality and timeliness SLAs

  - Model Launch Delay Due to Data Issues: # of launches delayed due to poor training data

  - Data Incident Frequency: # of major data quality failures impacting business ops

- Trust Score for Key Dashboards/Models: User-rated or automated score measuring perceived data reliability

• Align metric granularity with decision-making timelines (e.g., daily for ops, quarterly for strategy).

• To drive accountability and action, metrics must be visible and accessible across teams. Design monitoring and visualization mechanisms for each layer of metrics. To ensure each metric gets due attention from the stakeholders, the reporting setup should be segregated by domain, asset, or product. While reporting for operational metrics should be in real time supported by alerting system, dashboards to the product and exec stakeholders should be scheduled weekly/monthly.

• Establish a cadence for reviewing executive and product KPIs (e.g., monthly data health scorecards).

Embed Data Quality Across the Data Lifecycle

Ideally data quality must be measured and monitored across the lifecycle of data and not just at the ingestion or consumption. This helps in capturing the fault near its point of origin, making it easier and quicker to debug and fix. The following are some of the most common checks that should be performed at each stage:

- Ingestion: Schema, null checks, duplicate check

- Transformation: Business rule validations

- Model training: Feature drift, leakage checks

- Dashboards: Completeness, freshness scoring

Another critical task apart from embedding data quality metrics across the entire data lifecycle is creating common SLA policies where business logic is applied across the lifecycle to enable consistency and governance at scale.

The product manager should perform the following activities to achieve the desired outcomes:

- Define the schema validation logic including nulls, duplicate checks on transaction systems, or systems where data is being generated.

- Work with the engineering team including data to create a data catalog to centralize the definition, declaration, and business logic required.

- For each use case define the specific transformation logic required to make data usable for a certain context.

- Work with the quality analysis team to develop a test case suite to test how accurately the transformation logic has been applied in reality.

- Define the data quality entry criteria for each use case including AI/ML projects.

- Define checkpoints in your product lifecycle to verify data quality compliance.

Define Policies, Then Choose Tools

In many organizations, the adoption of tools is often seen as the silver bullet for solving all data quality issues without defining what quality means. This tool-first approach often results in fragmented and superficial implementation, vendor lock-in, and inconsistent application of technical and functional standards across domains. As a result, it not only limits the full potential of data but also erodes executive confidence in both data and the team responsible for managing it.

To develop a scalable and future-ready data quality capability, organizations must develop tool-neutral policies as the strategic foundation of data quality practice. The organizations must treat tools as the plugin or delivery mechanism used for achieving the strategic outcomes. Becoming policy first in data quality has the following advantages:

- When policies are centrally defined, they remain consistent across teams regardless of tech stack or individual team preferences.

- A policy-driven model enables portability across different tools without affecting the quality of the rules implementation.

- It enables larger stakeholder participation by allowing nontechnical users to contribute to policy creation without needing to understand tools internally.

- As your org grows and new domains and use cases are added, consistent policies reduce duplication and governance overhead.

Table 5-2 presents critical information about the components that a Data Quality Policy document must contain.

Table 5-2. *Essential Components of Data Quality Policy*

Component	Example
Quality Dimensions	Accuracy, Completeness, Timeliness, Validity, Consistency, Uniqueness
Thresholds or Rules	"Product ID must be present in 99% of records in production datasets"
Criticality Tiers	Define critical, important, and optional fields with different thresholds
Use Case Mapping	E.g., "Timeliness is essential for fraud detection but not for HR reports"
SLA Guidelines	E.g., "Customer data should not be more than 10 minutes delayed"
Roles and Responsibilities	Articulates responsibilities of each stakeholder involved
Violation Response Plan	Auto-alerts, data quarantine, incident tickets, or triggering a rollback

The following activities must be untaken by the product manager to realize the full potential of this step:

- The product leader must work with various stakeholders including business, governance, and engineering to define tool-agnostic data quality policies.

- The product leader must work with various stakeholders in developing data quality standards including definition and calculation logic and threshold logic across domains regardless of tech stack.

- Drive adoption of common metadata standards to unify quality monitoring across tools.

- The product leader must work with individual teams adopting shared tooling without being locked into specific vendors. Let the organization's policies guide the evaluation and utilization of tools and not the other way around. Avoid premature vendor lock-in by piloting solutions with open standards.

- Once the data quality policies are developed and approved on an organization level and individual domain level, track the policy coverage.

- Review the policy effectiveness periodically.

Assign Ownership, SLAs, and Escalation Paths

Metrics without ownership are just noise. To attain sustained data quality, clear accountability must be assigned. This includes defining and assigning ownership, designing a service or domain-level SLA, and establishing agreed-upon escalation workflows to ensure timely resolution when issues arise.

To make data policies actionable and accountable, the following elements are critical:

- Every critical dataset or domain must have an accountable owner such as data product manager or domain lead.

- Service-level agreements must define acceptable thresholds for key data quality dimensions, tailored to each use case.

CHAPTER 5 MASTERING DATA QUALITY: TECHNIQUES AND BEST PRACTICES FOR PRODUCT MANAGERS

- To enforce accountability data quality, metrics must be embedded into the team's performance KPI or OKR.

- Escalation paths must be defined to ensure timely triage, incident management, and resolution when thresholds are breached.

The following actions must be taken by the product leader to implement this step:

- Assign data owners at the domain or data asset level. Use the RACI model to clearly identify the roles and responsibilities of each stakeholder. The RACI metric must be made visible in the metadata catalog centrally.

- Define the SLA threshold on each data set for each data quality dimension. The threshold value must be based on the data's downstream impact.

- Set up escalation protocols such triggers on which escalation can be made. Route for each escalation matrix, etc.

- Integrate quality incident reporting into existing incident management systems and share quality dashboards with leadership on a regular cadence.

- Include data quality KPIs into your product team's OKRs or sprint goals.

Roll Up Metrics into Indices for Strategic Visibility

To make sense out of low-level operational metrics or even product/domain-level data, quality metric requires fundamental understanding of data domain. This lack of knowledge will act as a barrier to stakeholder engagement and data adoption. To solve this problem, low-level technical metrics must be synthesized into a generic metric that can interpreted and consumed by nontechnical stakeholders for key decision-making. This can be achieved by calculating an index that defines the overall health of the data. The data quality index can be defined as a composite score that summarizes the overall quality of a data asset based on various key dimensions on which data quality is evaluated. Think of this index as a credit score that does not replace the need for details but gives a high-level indication of overall financial health. The data quality index helps in transforming raw facts into actionable business intelligence.

Product Manager's Actionable

- The product leader must maintain a data quality index for each product area.

- Efforts should be made to baseline the data quality index (DQI) by first calculating the overall data quality index across the entire data ecosystem. The scores of individual data assets or products should be then compared against the baseline to identify gaps and prioritize improvements.

- Develop a visualization for DQI using relevant representations. We have seen heatmaps and scoring dashboards work best.

- Incorporate DQI trends in product reviews and OKR tracking and in postmortems of data-related incidents.

- Use index scores to prioritize quality debt remediation and platform investments.

Steps for Defining Data Quality Index (DQI)

- Identify relevant dimensions for quantifying the data quality of each data asset. Commonly used dimensions were listed and described earlier in the chapter.

For AI/ML use cases, additional dimensions such as representativeness or data drift may also be relevant.

- For the identified dimensions, quantify the method to calculate a measurable metric.

- Not all dimensions are equally important for a use case. Depending upon the importance of each dimension, assign the weight to each dimension.

- Multiply the score of each dimension's metric by the assigned weight.

- Calculate the overall DQI by aggregating the weighted scores for each dimension. This is a simple formula:

  $DQI = w_1 * Accuracy + w_2 * Completeness + w_3 * Timeliness + w_4 * Validity + ...$

- Once the DQI is calculated, set thresholds and targets for each data asset. Use these thresholds to enforce quality gates in pipelines or to trigger alerts. The following is an example:

  \> 90: Excellent: Data is reliable for all use cases

  70–90: Moderate: May require caution or selective use

  < 70: Poor: Requires immediate remediation

Table 5-3 maps each step of the PROMT framework to its corresponding core tenet.

Table 5-3. *Mapping PROMT Steps to Foundational Principles*

Step	PROMT Tenet	Outcome
Tie Metrics to Use Cases	Product & Strategy-Centric Quality	Data aligned to impact
Multi-Layered Metrics	Layered Metrics	Clarity by stakeholder level
Lifecycle Integration	End-to-End Coverage	Quality embedded across ingestion to consumption
Tool-Neutral Policies	Tool-Agnostic but Integrated	Strategic control, flexible implementation
Ownership + SLAs	Actionability & Accountability	Defined accountability and measurable responsibility
Composite Quality Scoring	Composite Indexing	Strategic visibility and prioritization

Designing SLAs for Data Quality

While discussing the PROMT framework, we explored the importance of defining SLA for data quality key dimensions on the data asset or domain level. Equally important is to define the SLA in the right manner as defining SLA for data quality are no longer optional; they are essential for ensuring trust, accountability, and operational continuity. An SLA helps in defining quantifiable, enforceable expectations as key data quality indicators and helps align stakeholders across different teams. They are not just technical guardrails; they are the strategic commitments that protect decision quality and user trust. Placing data quality SLA on par with infrastructure SLA organizations reinforces the idea that data is a product and its quality is not negotiable. Table 5-4 presents key components of data quality SLA.

Table 5-4. *Core Components of a Data Quality SLA*

Component	Description	Example
Metric/ Dimension	What aspect of data is being measured	Timeliness, accuracy, schema conformance
Threshold	Minimum acceptable value	≥ 98% completeness
Frequency	How often it is monitored	Hourly, daily, per batch
Scope	What data asset it applies to	transactions_fact, customer_profile_dim
Owner	Responsible team or function	Data Engineering, Source System Owner
Escalation Path	What happens on failure	PagerDuty alert → Slack → Jira ticket
Reporting	How and where results are surfaced	Data quality dashboard, executive reports

How to Design SLAs in Practice

1. Identify critical data assets that are tied to key business goals or offerings.

2. Define the core data quality metrics for each dataset. Ideally three to five metrics should be identified. These are metrics that are very critical for the usability of that data asset.

3. Set realistic but high standards, but don't go for perfection, i.e., 100% achievement, unless mandated by regulatory compliance. Use the baseline data to determine an achievable target.

4. Instrument your pipelines by using data observability tools to automate monitoring and alerting.

5. Define who gets alerted when an SLA violation is reported, how it's tracked, and how incidents are resolved and reviewed.

6. Report SLA performance to stakeholders by including SLA dashboards in product reviews and model readiness gates. Show trends, not just breaches.

Measuring Data Quality: Practical Approaches

So far in the chapter we have explored the fundamentals of data lineage, data quality including key dimensions, the PROMT framework for implementing organization-level policy, and data quality governance. Now it's time to understand how to measure data quality in the real world. Measuring data quality is fundamental to managing data quality. Without continuous monitoring, organizations risk flying blind and reacting to data problems only after they cause downstream impact. This section will describe methods for measuring data quality.

Data Profiling: This is the method in which datasets are scanned systematically to gather statistics and summary about the data stored, structure, and the relationships in between. It helps in identifying dataset's incompletes, uniqueness, schema validity, and outliers

Data Quality Metrics: This is a simple method of measuring data quality in which metrics are defined and tracked over the period. This method helps in getting answers to already defined data quality metrics.

Data Contracts: Data contracts are formal agreements between data producers and consumers that define expected structure, content, and quality of data. By explicitly setting rules and expectations, data contracts enable automated validation and monitoring of data quality.

Data contracts help in monitoring and managing data quality in the following ways:

- Data contracts help in establishing measurable quality standards.
- Data contracts can be easily integrated into pipelines to validate data at the source or during transformation.
- Violations of contracts can trigger alerts, logs, or quality scores.
- Contracts help track SLA adherence and other KPIs.

User Feedback Loops: By incorporating user feedback from end users, analysts, product managers, and AI/ML engineers can help in identifying quality issues that automated methods may miss.

Effective data quality measurement is not a one-time activity, but a continuous automated process integrated into every stage of the data lifecycle. By combing profiling, metrics, data contract implementation, data lineage, and user feedback data, product teams can ensure that their data stays reliable and actionable fueling confident decision-making.

Profiling and Quality Checks in Large Datasets

Today organizations generate and store data at an unprecedented scale. This data often resides in distributed storage systems and streams through real-time processing engines. Traditional profiling methods struggle to perform under these conditions. Therefore, the following methods should be applied to effectively profile large datasets:

- Incremental and partition aware profiling: Rather than profiling entire datasets repeatedly, focus on incremental data that is new or changed. It can be further optimized by leveraging partition keys to profile new partitions only. Another way to implement this is by using change data capture to isolate updates. Compare data profiling results with historical data profiles to identify changes in trends.

- Sampling strategies: Employ random sampling techniques to create a statistically representative snapshot of data and employ data profiling techniques on that sample. Always employ a sampling technique that ensures that all segments in data are represented in the same proportion in which they are in present in the master data.

- Distributed computing for parallel profiling: A distributed computing framework such as Apace Spark can be used to parallelize profiling computations. This is a very computationally expensive procedure.

- Schema and metadata-driven profiling: Automate data profiling using metadata definitions. Schema registries can be used to detect changes in schemas and trigger re-profiling.

- Use machine learning models to detect anomalies in complex datasets.

Data Quality Challenges in Unstructured and Semi-Structured Data

In modern data-driven systems, unstructured and semi-structured data is being used frequently especially in AI applications. Free text, audio, video, and images can be classified as unstructured data. Data in XML, logs, and JSON format is called semi-structured data. While these data formats capture rich data and enhance the understanding of the real world in a far better way, maintaining data quality standards in these formats poses unique and complex challenges.

Product leaders must collaborate closely with data teams to enhance their approaches in monitoring, measuring data quality, and ensuring this new type of data can deliver value to the organization.

In this section, we will explore the key challenges in monitoring, measuring, and quantifying data quality metrics for unstructured and semi-structured data.

- These types of data formats often lack a standardized schema, which means the structure in which data is placed in a record is not clearly defined. This issue is especially more pronounced in unstructured data, where even the start and end points of an individual record is not easily identifiable. In contrast, semi structured data does offer a basic structural framework, but it can still present challenges. Subrecord-level fields may be dynamic, nested, or vary in number from one record to another.

  While structured and unstructured data is often rich in information and has the potential to unlock valuable insights beyond what structured data can offer, their complexities make it difficult to enforce validation rules. This complicates the detection missing or malformed data and hinders the application of uniform quality control measures.

  To manage schema-level complexities of unstructured and structured data, a schema inference tool such as AWS Glue, Apache Avro, and JSON schema validator can be leveraged. These tools help in dynamically inferring and enforcing data structures. These tools analyze the data to automatically detect the schema, enabling validation, transformation, and quality control measures without requiring manual schema definitions.

- Parsing and normalizing data from sources such as logs, images, or any other unstructured data often involves complex steps. These preprocessing steps may vary depending on the type of data. For logs and unstructured text, natural language processing (NLP) tasks like tokenization, lemmatization, topic modeling, or name entity recognition are essential, but these tasks are prone to error, especially when dealing with noisy, ambiguous, or multilingual data. These challenges make it difficult to standardize and normalize data, which in turn complicates downstream tasks.

   One way to address this problem is to establish standardized parsing and normalization procedures supported by version-controlled preprocessing logic. This approach helps in standardizing common tasks, enabling the systematic identification and handling of edge cases that may arise after applying the initial logic. By versioning preprocessing procedures, the newly discovered edge cases and patterns can be incorporated into the shared logic repository, allowing continuous improvement and refinement of the original standards over time.

- Often the meaning or intent behind unstructured inputs is contextual and subject to interpretation, which makes data labeling for ML applications subject

CHAPTER 5   MASTERING DATA QUALITY: TECHNIQUES AND BEST PRACTICES FOR PRODUCT MANAGERS

to human judgment variability. To address this, inter-annotator agreement metrics such Cohen's Kappa must be used to assess and monitor label consistency. Implementation of active learning techniques can help in identification ambiguous or uncertain samples, which can be subject to further analysis and review, improving label quality and model performance over time.

- Semi-structured formats such as API structures often change over the period of time. Without robust data observability, these changes in structure and semantics may go unnoticed, adversely affecting data pipeline and model performance. To avoid this schema drift, detecting tools and log schema versioning tools must be implemented.

- Since it is difficult to implement data quality control checks on unstructured data, often unstructured data sources contain redundant and noisy content. The redundancy and noisy content increase the overall volume of data, giving the impression of high data availability, but it severely compromises the quality of data that adversely affects the performance of value extraction procedures while increasing the processing cost.

To handle redundancy and noisy content in unstructured sources, implement techniques like noise filtering, NLP-based relevance detection techniques like regular expression matching, stop word removal, topic modeling for meaningful information extraction, and data quality improvement for downstream processing.

- Most off-the-shelf observability tools available cater to structured data due to nonavailability of industry-wide standards. To overcome this limitation, extend observability by defining custom metrics like text length distributions, missing captions, resolution checks, and maintain metadata catalogs with lineage links.

Table 5-5 contains a product manager's action items for handling challenges in managing data quality in unstructured and semi-structured data.

Table 5-5. PM's Action Items for Handling Challenges in Managing Data Quality in Unstructured and Semi-Structured Data

Challenge	PM Actionable
Changing formats or schema drift	Build contracts with upstream API teams and define expected JSON/XML patterns
Noisy or redundant unstructured input	Introduce content quality thresholds and automatic filtering before model input
Poor labeling consistency	Standardize labelling guidelines, measure annotation quality, and revalidate periodically
Tooling gaps	Champion investment in data profiling tools that support nested and non-tabular formats

We explored earlier why conventional data quality measurement and error validation techniques are not appliable in unstructured and semi-structured data. Table 5-6 presents lists of techniques applicable for unstructured and semi-structured data quality validation.

Table 5-6. *Techniques for Data Validation and Error Detection in Unstructured and Semi-Structured Data*

Technique	Semi-Structured Data	Unstructured Data (Text/Image/Audio)
Schema Validation	JSON Schema, XSD	Not applicable
Pattern Matching	Regex for fields	Spell check, language detection
NLP Checks		NER, sentiment, topic modelling
Image Validation		Image resolution, format, color depth, corrupt file detection, dimensions and aspect ratios
Audio Validation		Audio sample rate, silence durations, clipping, or noise levels
Duplicate Detection	Hashing, clustering	Hashing, clustering, Fingerprinting, clustering techniques like Minhash
Statistical Checks	Field distribution analysis	Feature distribution, outliers
Metadata Verification	Lineage, checksum	Provenance, processing status
Human-in-the-Loop	Annotation audits	Manual review, active learning

Data Quality for AI/ML Workflows: Ensuring Data Quality for Model Training and Testing

Developing a high-performance machine learning model requires data that is not only correct but also has good quality. Failure in any of the key data quality dimensions can lead to model underperformance including low efficacy and biasness. This section outlines the set of activities that must be performed to ensure trustworthy and production-grade ML data pipelines.

- Collaborate with various stakeholders including the AI team, data team, and business teams to define what quality means for each feature and label. The definition should include feature formats, cardinality, allowed ranges, label freshness, accuracy, leakage prevention, and coverage requirements (e.g., minimum % of non-null rows).

- Perform data profiling of the training data to analyze distributions, outliers, null patterns, and relationships between different variables.

- Validate feature consistency across time, environment, and data sources.

- Ensure the temporal integrity of data is preserved by ensuring the correct time-based structure of data throughout the data pipeline and modeling process.
    - Preserve the chronological order of events and data points.
    - Avoid lookahead bias, where future data is incorrectly used to predict past or present data.
    - Properly handle timestamps and ensure lag features are correctly aligned with the target variable.

- Check label quality and balance. This can be achieved by auditing for the following:
    - Incorrect labels due to manual errors or delayed pipelines

- Label leakage (e.g., target data used as input feature)
- Class imbalance (especially in binary or multi-class classification)

- To address class imbalance in training data, apply techniques such as SMOTE, class reweighting, and over- or under-sampling.

- Apply bias and fairness audits. Chapter 8 of this book will cover this step in detail.

- Use data versioning tools to track and version training data and the associate model version with training data version to ensure predictability and reproducibility in the ML model outcome.

- Define the acceptable latency for features and labels. To report an SLA breach, develop trigger mechanisms.

- While designing a deployment strategy and roadmap, work with dev ops or MLOps engineers to identify and design quality gates to embed quality checks in CI/CD pipeline.

Data Quality for AI/ML Workflows Implementing Quality Gates in AI/ML Model Development Pipelines

To enforce data quality standards across the data pipeline for developing trustworthy AI, automated quality checkpoints must be embedded into the AI/ML workflow. These automated checkpoints are called *quality gates*. They help prevent poor-quality data from contaminating model performance. Quality gates act as the defensive line for trustworthy AI by converting theoretical quality checks into auto-enforceable rules preventing silent failures, accelerating feedback loops, and improving the probability of aligning models with business, ethical, and regulatory expectations.

Quality gates are an abstract concept that define best practices for ensuring data integrity and reliability. They are technology agnostic and can be implemented using any suitable tool or platforms, depending on the existing system architecture and requirements. The following are some commonly used quality gates that we have found personally useful.

Schema Validation Gate

A schema validation gate is like a security guard at the entrance of a data pipeline. Its job is to check that very piece of incoming data follows the defined rules. It checks the schema conformance of the incoming data with the quality check rule. The following are the major checks it performs:

- Incoming data contains all the required fields as described in the quality check rule.

- The data fields of the incoming data have the right data types and match with the data type including formats and length mentioned in the quality check rule.

- Whether incoming data contains any field other than those mentioned in the quality check rule.

- If the incoming data contains optional fields, they must still meet defined quality and format rules.

Data Distribution Gate

The data distribution gate ensures that the training data aligns with the expected data distribution by comparing the data distribution of the incoming data with the expected data distribution. Simply put, the data distribution gate checks whether the overall shape or pattern of the data looks as expected; it is like checking that the ingredients in a recipe are in the right proportions or not. The data distribution gate is important because if the distribution of data changes too much, then it will adversely affect the AI model's performance. This gate helps in catching these sudden changes in the data before they may cause bigger problems. The following are some common checks:

- Mean/variance, range checks
- Distribution distance (e.g., PSI, KS test)
- Detection of outliers or anomalies

Label Quality Gate

The label quality gate prevents corrupt, missing, wrong, or future leaking labels from contaminating model training. This quality gate checks whether labels in the incoming data are correct, consistent, and complete. This gate performs following checks:

- Label presence/completeness
- Temporal leakage (e.g., future labels used in past features)
- Label skew or inconsistency

Bias and Fairness Gate

A bias and fairness gate is a checkpoint in the data or model pipeline that evaluates whether the system behaves equitably across different demographic groups (e.g., gender, race, age, location). It helps identify and mitigate any systemic bias that may lead to unfair or discriminatory outcomes. The following checks should be performed:

- Statistical checks to detect imbalances in model predictions across subgroups

- Auditing training data for over- or underrepresentation of certain populations

- Assessing fairness metrics, such as disparate impact, equal opportunity, or demographic parity

Model Readiness Gate

This is a final validation checkpoint before the model is pushed to the production environment. This quality gate determines whether an AI model is ready for deployment based on the defined set of technical and business and ethical criteria. The following checks should be performed:

- Completeness and accuracy of training data

- No major feature drift

- SLA-compliant freshness of source data

- Meets robustness, explainability, and fairness standards

- Test performance meets defined KPIs

- Has monitoring and rollback plans in place

Mapping Data Quality Gates to ML Lifecycle Stages

Table 5-7 maps data quality gates to relevant AI project lifecycle stages.

Table 5-7. Mapping Data Quality Gates to ML Lifecycle Stages

Stage	Quality Gate Type	Integration Point
Data Ingestion	Schema & Completeness Gate	ETL pipelines
Feature Engineering	Consistency & Validity Gate	Feature store population scripts
Model Training	Drift & Label Validation Gate	Training pipelines (CI/CD)
Model Testing	Fairness & Accuracy Gate	Automated test suites
Model Deployment	Readiness & Recertification	Model registry or release pipeline

Product Manager's Additional Responsibilities

The product manager will be playing an integral role in ensuring that defined quality gates are well placed throughout the data pipeline. Apart from overseeing the operationalizing these gates, they have the following additional responsibilities

- Work with different stakeholders to define pass/ fail criteria for each quality gate.

- While writing PRD, ensure that requirements for developing these quality gates are clearly mentioned and the relevant time is allocated to the teams for implementation.

- Review gate outcomes during model readiness reviews and track failures or manual overrides.

- Work with technical teams for including gate compliance metrics in ML product dashboards for visibility.

- Develop an escalation and resolution workflow for acting on quality gate failure reports. Ideally the following are the best practices for defining escalation and resolution workflow:

  - If a quality gate failure is reported, then auto fail the downstream pipeline.

  - Log issue immediately into incident tracker tagged with severity and relevant owner. Efforts should be made to automate this step.

  - Send auto alert to responsible team.

  - Develop mechanisms for allowing limited and controlled retries or allow override with competent approval.

Connecting Data Lineage to Model Explainability and Fairness

Modern AI systems are deeply embedded in the product experience and strategic decisions, and hence explainability and fairness have emerged as non-negotiable requirements and not just technical nice-to-haves. Data lineage plays a very pivotal role in enabling both. It offers transparency into questions like where data originated, how it was transformed, and how it was used in feature engineering, model training, and decision logic. Data lineage provides the crucial traceability required, explaining the model behavior, validating assumptions, and ensuring ethical and fair outcomes. It is the connective tissue between raw data and responsible AI. In this section, we will explore how data lineage can be connected to AI models for enabling better explainability. Before that, let's understand how data lineage helps. Table 5-8 explains how data lineage can help in operationalizing trustworthy AI principles.

Table 5-8. Enabling Responsible AI Through Data Lineage

Concern	How Lineage Helps
Model Bias	Trace biased predictions back to upstream data sources or features
Drift and Anomalies	Identify when new or changed data begins to affect model output
User-Level Explanations	Show how specific inputs flowed into a decision
Audits and Compliance	Provide documented trails for how and why decisions were made
Accountability	Pinpoint responsible owners for flawed or unfair data pipelines

Best Practices for Leveraging Data Lineage for Model Explainability

- Enable record-level tracing from feature calculation to transformations to upstream data source by integrating lineage metadata into Table 5-8 model training pipelines. It will enable backtracking of model predictions to raw inputs.

- Use data lineage metadata to map features with sensitive attributes to data sources including Table 5-8 transformation they underwent. Once mapping is achieved, audit how these features contribute to predictions or fairness issues.

- Use data lineage to track fairness across all steps of the data pipeline, which should include data insertion, major or minor transformations, and consumption and not just at entry and exit of the data pipeline.

- To enhance model transparency, combine explainability tools with feature provenance from data lineage logs. This integration will enhance local model explanations by providing contextual information about each feature origin. This added context supports clearer, more trustworthy insights into how and why a model makes specific predictions.

Emerging Trends in Data Quality: Synthetic Data and Data Augmentation

As businesses expand their use of AI, it has become increasingly challenging to guarantee an adequate amount of high-quality training data. This is where synthetic data and data augmentation techniques help. These techniques help in addressing data gaps, biasness issues, and other generalization problems.

Synthetic Data: What Is It?
Artificially created data that replicates the statistical characteristics of real-world datasets is known as *synthetic data*. The following techniques can be used to create it:

- Simulation models based on real-world business rules can be used to create syntenic data.
- Advanced generative algorithms, like GANs, VAEs, or diffusion models, can be used to produce data that replicates distributions in the real world.
- Large language models can be used to generate synthetic data.

Practical Use Cases Where Synthetic Data Can Be Used

- Augmenting sparse or imbalanced datasets to improve model learning

CHAPTER 5   MASTERING DATA QUALITY: TECHNIQUES AND BEST PRACTICES FOR
            PRODUCT MANAGERS

- Simulating rare events, such as fraud or anomalies, for more resilient training

- Producing privacy-preserving substitutes for sensitive data in regulated domains like healthcare and finance

- Evaluating model stability by introducing edge cases and uncommon data patterns

What Is Data Augmentation?

This technique involves transformation or enhancement of existing data to improve its quality, robustness, and diversity without collecting new data. Common techniques include the following:

- For image data, generally image rotation, injecting noise deliberately, scaling, color filtering, and contrast adjustment are few techniques.

- For text data techniques like synonym replacement, paraphrasing, back translation, injecting noise in the form of spelling, and punctuation errors are used.

- For structured data techniques like scaling, transformation, SMOTE, and noise injection by modifying numerical values by adding small random noise, random placement of null or out of distribution value, and bootstrapping are used.

Practical Use Cases Where Data Augmentation Can Be Used

- Helps in training data enrichment by expanding limited datasets with realistic variations to help models learn broader patterns

CHAPTER 5 MASTERING DATA QUALITY: TECHNIQUES AND BEST PRACTICES FOR PRODUCT MANAGERS

- Helps in correcting class imbalance by generating additional samples for minority class

- Helps in missing data imputation by simulating missing values to maintain data integrity during model training

- Helps create rare or extreme case scenarios to evaluate to model stability under edge conditions

Table 5-9 compares capabilities of synthetic data and augmented data.

Table 5-9. Synthetic Data vs. Augmented Data: Capability Overview

Benefit	Synthetic Data	Augmented Data
Fills data gaps	Yes	Yes
Preserves privacy	Yes (fully synthetic)	No (based on real data)
Improves model generalization	Yes	Yes
Reduces labelling cost	Yes	Yes
Simulates rare/edge cases	Yes	Yes
Introduces variation for training	No	Yes

Summary

This chapter began by highlighting why data quality should be treated on par with other strategic initiatives by discussing real-world challenges especially for product managers working in AI-driven companies. We also discussed foundational concepts like key dimensions of data quality and PROMT framework to guide structured data quality management. The relevance of data lineage in data quality management was also discussed. The chapter also touched upon practical techniques to support implementation across large-scale datasets and ML pipelines. Modern techniques like synthetic data generation and its relevance to data quality and augmentation were also discussed.

By mastering these techniques and best practices, product managers can actively contribute to building trustworthy, scalable, and AI-ready data systems.

PART III

Navigating the ML Landscape: Essential Knowledge for AI Product Leaders

CHAPTER 6

Introduction to Machine Learning for Teaching Systems

While it is worth discussing successful implementations of artificial intelligence, discussing failed implementation and the reasons behind them gives us wisdom of what not to do. A comparative analysis of applications that failed and succeeded will give us a perfect blueprint of how to develop an implementation plan for guaranteed success. Let's explore IBM Watson, for example.

In the past decade, IBM Watson[1,2,3,4] has been hailed as a radical force in artificial intelligence, redefining AI capabilities and the way AI was perceived to be used. From beating a world chess champion to winning *Jeopardy*, it was setting new milestones every day. Then IBM Watson forayed into healthcare with the noble intention of helping cancer patients with personalized care. But things didn't go per plan for Watson. During the early implementation phase, reports of unsafe and incorrect cancer treatment recommendations started surfacing; another major issue that was reported was its inability to explain its predictions. All this led to product withdrawal. Now let's analyze what went wrong with Watson.

CHAPTER 6   INTRODUCTION TO MACHINE LEARNING FOR TEACHING SYSTEMS

- IBM trained Watson on a limited dataset from very specific medical institutions that limited its exposure in terms of type of patients, variety of disease, and symptoms. It also failed to align itself with local healthcare norms of real-world cases.

- Watson did not involve domain experts like oncologists and other doctors while designing and developing the product, which led to usability issues and also prevented Watson from understanding the complexity of cancer.

- IBM used a lot of synthetic data that restricted its ability on real-world data.

Fast-forward to 2025. IBM has made significant progress in addressing the root causes behind Watson's initial healthcare implementation challenges. By emphasizing AI ethics and adhering to other best practices, IBM is emerging as a leading force in the AI space. This incident not only highlights common reasons that can hamper the success of any AI initiative but also depicts how thoughtful strategic changes and targeted resource (which includes financial and organization both) allocations can revive the most complex projects.

What makes this example even more relevant from a product leader's perspective is that Watson never lacked technical prowess; in fact, it was a marvel at that time (this was way before ChatGPT and other major LLMs became common). However, it faltered temporarily due to softer challenges like misaligned expectations between the product and other departments, insufficient data exposure, limited understanding of local healthcare, and data regulations. These are precisely the types of issues that we as product leaders should anticipate and address.

As product leaders, it is our responsibility to bridge the gap between organizational strategy, technology, the business, and users. We should work toward defining the product vision, which is not only ambitious but

CHAPTER 6   INTRODUCTION TO MACHINE LEARNING FOR TEACHING SYSTEMS

also grounded. It is our responsibility to ensure that product teams get cross-functional alignment and navigate ethical and legal complexities. Our roadmap should not only list features that should be developed but also include solutions to real-world challenges and constraints, implement best practices around AI development, promote scalability and should be able to adapt to feedbacks received. This is a powerful reminder that AI product success relies not only on building the right solution but also on other softer aspects that often get ignored.

The IBM Watson case clearly shows the importance of diverse and representative training data, strong ethical oversight, and stakeholder involvement for developing a successful AI product.

Challenges That Detail AI Implementation

Here are some of the common pitfalls and challenges faced in AI/ML projects, which would prevent these projects from being deployed in production.

Data Quality and Availability: Often the AI/ML projects are initiated without confirming if the data required for the project is available in the right quality or amount. The project delivered based on poor data cannot give accurate insights or predictions. Here are some common issues with data:

1. Inaccurate or missing data

    Erroneous data: This includes incorrect entries and ambiguous values, such as misspelled names or incorrect dates.

    Missing values: This includes None or NaN values or data points that are unavailable.

Inconsistent data formats: Nonstandard data can be difficult to process or be consumed properly with other standard data. For example, the dates may be available in different time zones or in different formats like DD/MM/YYYY in some tables and MM/DD/YYYY in others.

2. Stale and obsolete information: Data can become irrelevant due to changes in trends or macro factors. For example, the historical sales data for the products during the COVID lockdown is not irrelevant in post-COVID times. Another example is data collected for user engagement; for example, time spent by a user on a product detail page cannot be used for any predictive modeling if there has been significant change in the user's circumstances.

   1. Redundant records: Multiple entries for the same entity or event, leading to inconsistencies and bias. For example, near-real time updates for the rainfall data being sent by a remote meteorological station can lead to inconsistencies and data redundancies because there are multiple rain gauges installed. If one gauge is not working, then data from another gauge can be used, but the two gauges may record different data values for the same timestamps. Such inconsistencies need to be resolved before using this data for any kind of predictive modeling.

3. Inconsistent data
   - Conflicting values: Contradictory information within the same dataset, such as different addresses for the same person. This can happen if the person

has changed cities or addresses over several years. This is a common data situation for the delivery or transport services.

- Nonstandardized units: Data that is measured using different units, making it difficult to compare and analyze.

4. Biased data

- Sampling bias: Data that is not representative of the entire population due to a skewed sampling process. For example, users randomly chosen for A/B testing for different search ranking algorithms could still be biased in terms of their user behaviors causing the unreliable results of the A/B testing.

- Measurement bias: Data collected in a way that systematically favors or disfavors certain outcomes. For example, if the search results of keyword-based matching are always shown before the search results by an AI-based model in ranking, then this data further used for any modeling will always favor the results of the keyword-based matching due to the position bias.

5. Noise and outliers: Inaccurate data points that do not follow the expected pattern. An example is the data collected by a malfunctioning sensor measuring the temperature of a large server room. Another example is that a sudden boost in sales of products on e-commerce platforms can be caused because of a social media campaign or discounts. A data scientist not aware of this event would not

be able to explain the sudden increase in sales, and this could cause inaccurate results from a sales forecasting model.

While data issues are an important reason for the failure of data science and AI/ML projects, the choice of model to be trained and the input features could also be a cause of failure even if there are no data issues.

Model Complexity and Overfitting: Overly complex models can perform well on training data but could fail to generalize in real-world scenarios due to overfitting. The overly complex models are very good at memorizing the training data , but they fail to perform for new, unseen data because they have learned the noisy patterns of the training data rather than the actual underlying patterns that can generally explain the data. This phenomenon is known as *model overfitting*, and it results in the inconsistent performance of a model in real-world applications.

Common example can be an image classification problem where an overly complex model with millions of parameters, trained on a large dataset of images, shows very high accuracy on training data. Despite showing high performance on the train-test set, the model may still struggle to classify images in the real world. This could be due to over training of the model to an extent that it fails to adapt to the smallest change in real-world data. Another classical example is the failure of IBM Watson.

Lack of Clear Objectives: Many projects are lost or abandoned eventually due to the lack of well-defined objectives or demonstrated value. The following are examples of such projects:

- Customer churn prediction: A company wants to implement a customer churn prediction model, but they did not start with a clear definition of a "churned" customer. The data science team defined "churn" in a way that was aligned with model training, while the business and product teams had their own definitions. This would lead to confusion among different teams and miscommunication regarding model performance and functioning. Ultimately, it could result in ineffective retention strategies that would also affect churn rate negatively giving the impression that model is not working well.

- Personalized recommendation systems: Stakeholders may influence the data science teams in order to achieve short-term goals without realizing that data science projects have long-term effects. Business team could ask the data science team to build a recommendation system to optimize revenue without analyzing the effects of such a system on user experience in the long run. Such a system would make very aggressive or intrusive recommendations that may negatively impact customer experience.

Designing Successful AI/ML Systems

In the previous section, we have identified the major reasons that can lead to AI product failure. Now, let's apply those learnings to the development of our AI implementation project plan. In this section, we will explore how to avoid the previously identified pitfalls while designing AI implementation project plan. We will also outline a comprehensive project plan based on our extensive experience.

CHAPTER 6   INTRODUCTION TO MACHINE LEARNING FOR TEACHING SYSTEMS

From Pitfalls to Best Practices: Smarter AI Implementation Planning

1. **Define Clear Objectives:** The objectives of the AI/ML systems must be defined such that they are specific, measurable, and achievable. For example, the objectives of the AI/ML system could be to improve customer engagement, increase conversions, reduce fraudulent transactions, etc. But the terms like customer engagement, conversions, and fraud must be well-defined and validated by all the stakeholders before starting the project.

2. **Data Collection and Preparation:** Data is the foundation of every AI/ML system on which the models are designed, trained, and evaluated. Data collection and preparation are critical stages to ensure data quality and availability, which can significantly influence the performance, accuracy, and reliability of AI/ML systems. The following steps should be followed properly during data collection:

    a) <u>Identifying data sources:</u> Data sources must be aligned with the problem definition and the objectives of the AI/ML project. Before making any plans for project execution, it is essential to know what data is available and how to access it. Temporary data storage is required where data can be pulled and temporarily stored while working on a project. So, data access should be available to download to temporary storage. This could be an issue while working with PII data.

CHAPTER 6   INTRODUCTION TO MACHINE LEARNING FOR TEACHING SYSTEMS

The following are different types of data sources usually available:

- Structured data that is generally stored in relational databases residing inside data warehouses on the cloud or on-premise databases. Examples are customer records, transaction logs, etc.

- Unstructured data consists of data in the form of text, images, videos, audio, etc. Examples include posts or comments on social media, emails, pictures, and audio-video recordings from video calls or demos.

- Semi-structured is another type of data, which is less common. Examples include user session streams from their activities on the social media apps or e-commerce websites. This type of data contains both structured and unstructured elements like user demographics, content (images or videos or audio) viewed or consumed, items clicked, time spent on page views, etc.

- External data sources could be web scraping and public datasets that could be of any type: structured or unstructured.

- With the advancement in technologies, synthetic data has emerged as a new source of data. Synthetic data can be generated using generative models like generative adversarial networks (GANs) for computer vision or large language models (LLMs) for natural language. Such techniques are used to increase variability in training data where it is not possible or is too costly to gather more data points. Synthetic data mimics the statistical properties and patterns of real-world datasets. A healthy mix of real and synthetic data must be used as too much synthetic data may result in overfitted models.

b) <u>Ensuring data quality and relevance:</u> The success or failure of any AI/ML project is highly dependent upon the quality and relevance of the data used in the project. Data quality depends upon its accuracy, completeness, consistency, and timely updating. Data accuracy is reflected in its ability to explain the real-world actions or events without any significant errors. Data completion means there must be minimum missing values, and all the necessary information must be recorded in the data. Data consistency ensures uniformity in formats, units, naming conventions, etc., across different datasets so that data collected from different sources can be used without any conflicts. Data must be timely, refreshed, or updated so that the data available remains relevant to the context of the problem.

c) Data preparation consists of a series of processes to convert raw data available from different sources into a useful form that can be consumed for model training and evaluation. These processes are as follows:

- <u>Data cleaning:</u> In this step, the inaccuracies, inconsistencies, and incomplete records within the dataset are identified and rectified. Imputation is used to handle the missing values in the data. Missing data is filled with mean, median, mode, or more sophisticated interpolation methods. The incomplete records (those with missing values for several factors) are deleted. Data deduplication is performed to remove redundant records so that the model is not biased or influenced toward duplicate data points. This step also involves elimination or correction of anomalies like typographical errors, inconsistent naming conventions, and outliers.

- Data transformation: This step involves transforming data into a suitable format that can be consumed effectively by the AI/ML models. As a rule of thumb, numeric fields in the data are normalized by adjusting the range of values across fields to ensure uniformity; i.e., no field should have very high or low values that may influence model's results. Normally, values for all fields are normalized between 0 and 1. This is particularly important for algorithms sensitive to the scale of feature values (e.g., k-nearest neighbors, support vector machines). Other techniques like embedding techniques like one-hot encoding, label encoding, etc., are used to convert categorical variables into numerical form so that these can be consumed by the AI/ML models.

- Feature engineering is an important type of data transformation technique for improving accuracy of the ML systems. With this technique, new features (or fields) are created from the existing data, which helps the AI/ML models to capture underlying patterns and relationships in a better way. Feature engineering may involve steps like data aggregation, extracting components like day of week, month, day of month, etc., from date and time fields, or generating new fields by interaction of two or more existing fields.

- But a very large number of features or flat data (large number of columns and a smaller number of rows) is also not good for model training due to the curse of dimensionality. Thus, dimensionality reduction algorithms are applied to reduce the number of features in the data. This may involve feature selection or feature transformation with algorithms like Principal Component Analysis (PCA).

- Data integration: In this step, data from multiple sources is combined to get a more comprehensive view. This involves merging datasets by aligning and combining them based on common keys or identifiers. This also helps in identifying conflicts in data from different sources. The conflicts can be in the form of discrepancies in definitions, formats, or units across different sources. These conflicts are resolved before using this data for model training purposes, and the result is a consistent dataset free from any anomalies.

CHAPTER 6   INTRODUCTION TO MACHINE LEARNING FOR TEACHING SYSTEMS

- Data splitting: Before training a model, the training data is split into train, test, and validation datasets. The train set usually consists of 70% of the total training data, and it is used together with the validation dataset, which consists of 20% of total training data. The train and validation sets are mixed and split randomly to further generate multiple sets of train and validation splits. This process, also known as the k-fold technique, where k = total number of train-validation splits generated, is used to detect any bias in the train set that may have been introduced while splitting the data. For example, while splitting the data, it may happen that the number of male data points are more in the train set compared to female data points while it is opposite in the validation set. This would cause the model trained on the train set to be biased toward male and biased against female. If there is no bias, then the model performance on the validation dataset would not vary significantly. If bias is detected, then further investigation is required. The test dataset constitutes the remaining 10% of the total training data, and it is always kept separate and not used for model training. The test accuracy is used as a proxy for the model's performance in the real-world scenario.

- Data augmentation: This technique is popular among visual datasets like images or videos to expand the existing dataset artificially by applying transformations like rotation, scaling, cropping, pixel manipulations, etc., to the images in the existing dataset.

3. **Model Simplicity and Interpretability:** Start with simple models to establish a baseline and gradually introduce complexity. Setting a baseline means establishing a relation between the offline and online metrics. The offline metrics are those that are used to measure the model's performance during training and validation such as accuracy, RMSE, precision, recall, F1-score, IoU, etc. The online metrics are the business metrics that are used to measure the actual business impact like click-rate (CTR), conversions (CVR), add-to-Cart (ATC), etc. To establish a relation between the offline and online metrics, several A/B experiments are required with ML models of different offline metrics. Once the online metrics are available for the different ML models, a relation can be established between the offline and online metrics.

   Once this baseline relation is established, the following can be estimated:

   a) Level to which the online business metric could be achieved

   b) Offline metric required to achieve a certain level of impact on the online business metrics

   c) Choice of model for achieving the online business metrics

CHAPTER 6   INTRODUCTION TO MACHINE LEARNING FOR TEACHING SYSTEMS

4. **Continuous Monitoring and Evaluation:**
   Implement a feedback loop to monitor model performance and adapt to changing data patterns. Regularly evaluate the system's impact on learning outcomes.

5. **Scalability and Deployment:** Design with scalability in mind, considering cloud services and edge deployment for real-time processing and feedback.

Designing AI Implementation Plan That Succeed: Leason Learned

In the previous section, we explored best practices for AI implementation. Merely building an understanding of these practices is not enough; they must be embedded into a structured, repeatable process to be truly effective. Many AI implementations fail to deliver not because a poor AI model was developed but because of a lack of proper planning, improper change management, and misalignment with organizational objectives. In this section, we will introduce a robust AI/ML project management methodology that helps bridge the gap between execution and planning and helps deliver high-impact AI offering. This approach focuses on clear goal setting, stakeholder collaboration, continuous evaluation, and clear communication. This methodology ensures that the AI solution technically sound as well as strategically aligned. The life cycle encompasses the key phases shown in Figure 6-1.

CHAPTER 6   INTRODUCTION TO MACHINE LEARNING FOR TEACHING SYSTEMS

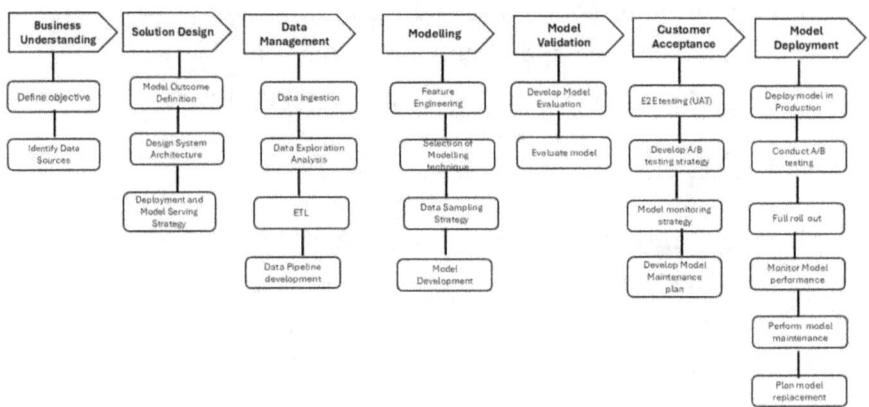

Figure 6-1. *AI project lifecycle*

Business Understanding

A deep understanding of business problems is the foundation of any successful AI/ML project. This phase involves defining the core objective that this AI implementation is expected to achieve, how this objective will be measured (metric), and the translation of business problem into a machine learning problem.

There are two major tasks in this stage.

1. **Define Objectives:**

    - Clearly articulate the problem statement this AI model will address. For example, a business problem such as "improve the search performance of my e-commerce platform" should be translated to "Develop an AI/ML model to improve search relevance by ranking results based on user intention and historical interaction data." This is a learning to rank problem, where models will learn to predict the relevance score of each result item for a given search query.

CHAPTER 6   INTRODUCTION TO MACHINE LEARNING FOR TEACHING SYSTEMS

- This translation ensures that business objectives are framed in a way that it makes problems very objective and precise, which makes selecting data and development of model and evaluation simple.

- Translating a business problem into an AI/ML problem requires asking direct and relevant questions. It's essential to understand the core drivers of the problem. How are these factors affecting the problem? How can these factors be measured, influenced, or predicted using data? By framing the problem in this way, we can determine whether it is suitable for a machine learning solution and, if so, what type of model is most appropriate.

- For our e-commerce example, the following questions can be good for starters:

  - What does "search performance" mean in measurable terms? Are we focusing on relevance, speed, user satisfaction, or conversions?

  - What user behaviors indicate a successful or failed search?

  - How can we define "relevance" for a search result?

  - Is the relevance of a result the same for all users? How do user preferences or profiles affect what's relevant?

  - What would a good outcome look like for the business?

CHAPTER 6   INTRODUCTION TO MACHINE LEARNING FOR TEACHING SYSTEMS

- **Key Decision Point:** Here we have a critical decision to make. If factors affecting the identified problem statement cannot be influenced by ML in cur6rent situation, we can make a decision to terminate the AI project and look for alternate solutions.

- Identify the key ethical guidelines that this AI model should follow (refer to Chapter 8 on trust AI).

- For identifying factors, the problem statement use a process flow diagram (in swim lane format). The process flow diagram shows how a particular task works. What are the different stages in the task, and how are they accomplished? (See Figure 6-2.)

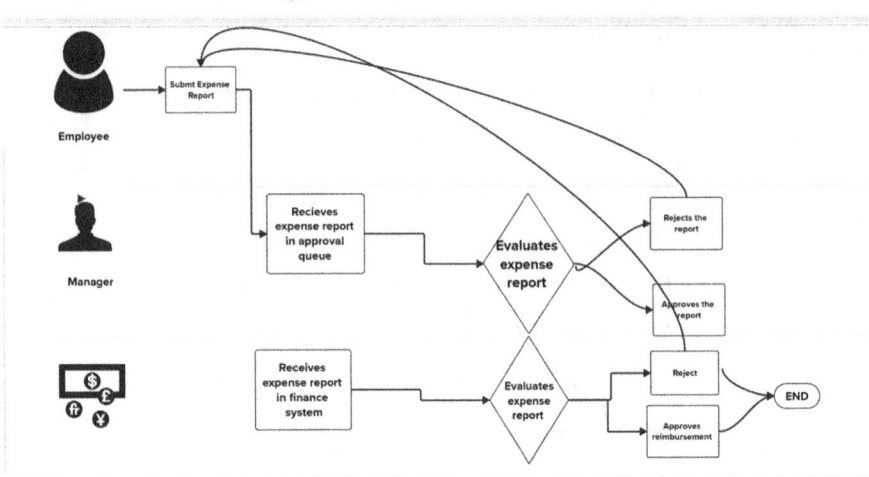

Figure 6-2. Example of a swimlane diagram

- Once factors affecting the problem statement are identified, another study should be conducted to identify the variables that will define the problem statement conditions in data. Ideally all factors affecting the problem statement should have some relationship with this variable.

- Define the business metric (also called the online metric), which can objectively define the performance of the problem statement.

- Once clarity on these pointers is achieved, define what type of ML problem this translates into. The following questions can help:

  - How much or how many? (Regression)

  - Which category? (Classification)

  - Which group? (Clustering)

  - Which option should be taken? (Recommendation)

  - In what order should the options be presented? (Ranking)

  - What is being said or meant? (NLP-Specific Tasks)

- Modifying the previous set of questions for computer vision domain, we have the following:

  - What is in this image/video/audio? (Computer Vision/Perception)

  - How much or how many? (Regression; for example, predicting the number of people in a room using video or photo)

- Which category? (Classification; an example is classifying photos into two or more categories such as identifying tumors by analyzing CT-Scan images)
- Which group? (Clustering; an example is a hotel booking app classifying and tagging images of room into different categories like bathroom, bedroom, etc.)
- Which option should be taken? (Example: Recommending similar product images in an e-commerce app)
- In what order should the options be presented? (Ranking; example: ranking search results based on visual similarity)
- What should be generated? (Generative AI; example: Creating realistic images from text prompts)
- Object detection (Identify and locate multiple objects such as using photos and videos to identify people who are carrying weapon)
- Optical character recognition (OCR; extract text from images)

- Identify data sources: Find the relevant data that helps you answer the questions that define the objectives of the project. Create process flow diagrams to understand what parameters are required to answer the question. Create a data flow diagram to understand how and what data sources will be required to answer the question. Create a data dictionary to document which data sources will be used and where they are located.

A data flow diagram (DFD) is a diagram that helps in visualizing the information flow within a system. With visual aids, it shows how the information enters and leaves the system and what changes the information undergoes at each step and where information is stored.

The outcome of this exercise is to identify the relevant data sources that the business has access to or needs to obtain.

Key Decision Point: Only if we have data for all factors affecting the problem statement should we go ahead with the AI project; otherwise, a project to capture data for missing factors should be started.

Solution Design

The objective of this step is to come up with the solution design required for developing a solution for the problem statement, which includes discovering answers to the following questions:

- What will be the outcome of the AI model? Will it be a category name, probability score, a continuous value or something else?

- What is the frequency at which model's outcome will be generated?

- How will the external system consume a model's outcome? Will it be through an API or via some intermediatory database or via some system event?

CHAPTER 6   INTRODUCTION TO MACHINE LEARNING FOR TEACHING SYSTEMS

- How will the AI model integrate with larger external systems?

- What are the changes/developments the external system has to make to accommodate the AI model?

- What are the constraints that can affect the AI model's performance?

- How will the AI model be deployed in production (more of this in Chapter 7).

- How will this new AI model serve the larger ecosystem (more of this in Chapter 7)?

- Identify key teams whose platforms will be impacted by this new model and establish a communication channel with them.

Key Decision Point: If the new AI model cannot be integrated with existing technical ecosystems, a thorough study should be conducted to identify the underlying reason. If the issue lies in the way AI model delivers its outcome, then it should be discussed and agreed upon before proceeding further. Conversely, if any changes are required in external systems, then a parallel project should be launched, after due approvals. If neither of these systems can be adapted to work together, then the AI project should be put on hold until a viable solution is established.

Data Management

- This step of machine learning/AI project lifecycle involves establishing connections with various data sources and extracting data from those sources.

CHAPTER 6   INTRODUCTION TO MACHINE LEARNING FOR TEACHING SYSTEMS

- **Key Decision Point:** If connection to any data source is not established due to technical challenges or any other challenges, a decision should be made by choosing either of the following choices:

    - Identify any alternative data source that can give the same data and use it as a replacement.

    - Drop that data source completely and proceed with the project lifecycle without that data source.

    - Stop the project until the deadlock is resolved.

- Evaluate data from all sources to identify and address issues that could impact quality, which includes missing values, data inconsistencies, redundancies, duplicates, outdated, or any other anomaly that affects the reliability of data.

    Key Decision Point: If data quality is too low, then a decision should be made to either replace the data from another reliable source and proceed or initiate a project to fix the underlying reasons behind bad-quality data. In this case, the AI project should be stopped.

- Develop a data pipeline to ensure a steady and reliable flow of data from the source to the AI application. The data pipeline should also include any transformation logic used to making data ready for AI modeling step consumption.

CHAPTER 6 INTRODUCTION TO MACHINE LEARNING FOR TEACHING SYSTEMS

Modeling

These are the major goals of this stage of ML/AI project lifecycle:

- Determine the optimal data features for the machine-learning model.

- Create an informative machine-learning model that predicts the target most accurately.

- Create a machine-learning model that's suitable for production.

- Evaluate the model on offline metrics that can capture model performance objectively.

- Develop and implement a comprehensive sampling approach that can segregate the data into training and testing datasets in a way that both datasets are equally representable.

Customer Acceptance

- Plan and conduct end-to-end testing of the entire workflow in which the AI model is integrated.

- Conduct test using near-real-time data to understand how well the AI model is performing on more recent data. If the model performs satisfactorily, then the solution is accepted; otherwise, it is not.

    Key Decision Point: If the model performs well and meets the acceptance criteria, then proceed with integration with the overall ecosystem. If the model performs poorly, then revisit the model assumptions, redefine features, and look at the model architecture before considering integration with external systems.

- Develop a comprehensive AB testing plan that will be used to test the model in the production environment (refer to Chapter 7).

- Develop a comprehensive plan on how the model will be monitored in production, including the key metrics to monitor.

- Develop a comprehensive plan outlining when and under conditions the model will be retrained; i.e., if it is a performance degradation, how will it be measured, or will it be scheduled periodically. Additionally, how will a feedback loop be established to capture feedback on incorrect or suboptimal decisions, and how will this feedback be incorporated in future models?

Model Deployment and Post-Deployment Strategy

1. Deploy the model in production as per the agreed upon deployment and model serving strategy.

2. Perform A/B testing for comparing the performance of the new model against the existing solution (refer to Chapter 7 for more details on AB testing).

3. If A/B test results are positive, then proceed with a full rollout of a new AI model; if not, then discuss and roll back a new solution.

4. Monitor the model continuously in production on the agreed upon metrics to detect performance degradation or any anomalies.

5. Regularly evaluate and fine-tune the model based upon new data, feedback loops, or any noted data drift.

6. Establish the criteria and a roadmap for replacing the model with a new version including retraining schedules, model benchmarks, and rollback protocols in case of underperformance.

Types of Machine Learning Algorithms

Popular AI/ML algorithms can be divided into the following categories:

1) Classical machine learning
2) Deep learning
3) Generative AI

Classical Machine Learning Algorithms

Classical ML algorithms are relatively simpler algorithms (compared to deep learning) that consume features directly to find patterns in the data using statistical techniques. The models trained using classical ML are smaller in size with fewer parameters (compared to deep learning). Also, the classical ML models can be trained and deployed using cheaper compute resources without needing any GPUs.

Classical ML algorithms can be further divided into two types, supervised learning and unsupervised learning algorithms. Both these learning types will be covered in detail including definition, evaluation criteria, challenges, and application scenarios

CHAPTER 6   INTRODUCTION TO MACHINE LEARNING FOR TEACHING SYSTEMS

Supervised Learning

To explain supervised learning, let's look at an analogy. You are a young parent, and you want to teach your little one the difference between oranges and guava. So, you will first show them both fruit and tell them which one is which and then tell them the color of orange, and its shape; it needs to peeled before eating, and then finally you will make them taste it. You will follow a similar approach for describing guava. After practicing this for some time, the child will learn to classify fruits between orange and guava. This type of learning is called *supervised learning*. In this learning, a machine learns how to predict the outcome (generally into known categories or values) using input data points. The output can be a category name (just like we had in our baby example), or it can be a continuous number like number of oranges sold in day.

The supervised learning algorithms learn the mapping of data points to the desired output. This mapping of input to output is the model that is then used to predict the output for the data points not seen by the model during training. Some common supervised algorithms are listed here:

- **Linear Regression:** Imagine you the owner of a juice house. You have observed that days that are relatively warm are days on which you sell maximum cooler drinks. Now you want to use this insight to predict how many cooling juices you will sell next month. You have data like average day temperature (during 9 a.m. to 6 p.m.) on a day level for the last year, and you also have data on the number of coolers sold each day. Linear regression is that type of machine learning algorithm that will help you find the relationship between these two variables, in this case the relationship between temperature and sales of cooler drinks. It does so by drawing a straight line that best fits your past data.

CHAPTER 6　INTRODUCTION TO MACHINE LEARNING FOR TEACHING SYSTEMS

This line shows how your sales will vary based on the average daily temperature. Once the line is created and a relationship is drawn, you can use this relationship to predict sales by simply putting in the temperature. In simple terms, linear regression helps, and you define the relationship between a set of input variables (can be more than one) and output variable. This relationship between the input and output variables is generally linear, and hence it is called *linear regression*. This relationship can be used to predict the value of the output variable based on the values of input variables supplied.

- **Logistic Regression:** Imagine you are doctor, and you want to predict if the patient has a specific disease or not based on their symptoms like fever, body chills, cough, fatigue, loss of appetite the and number days since they had first bout of fever. Historically the you have collected this data for all of your past patients. You want to use this data to build a system that simply gives you a yes or no (if a person has that disease, then yes and otherwise no). Logistic regression will help you build a relationship between the input variables like patient symptoms and output variables (disease output). The outcome of this algorithm is not a continuous variable like in the case of linear regression. Logistic regression generates the propensity (probability score) of having the disease, and then depending upon the probability cutoff decided by you, it will classify "Yes" or "No." It is a classification algorithm that predicts the class or category in which an observation will fall based on the value of its input parameters.

CHAPTER 6   INTRODUCTION TO MACHINE LEARNING FOR TEACHING SYSTEMS

- **Decision Trees:** A decision tree is a machine learning algorithm used to make predictions by splitting data into smaller subquestions. These subquestions are based upon certain deciding factors (input variables). It starts by asking a main question also called *root*. This is the input variable that is most important for getting the answer to your question and further keeps dividing the data based upon less important input values. Each input variable with certain value condition become a node. At each step, the algorithm chooses the best input variable to split in a way that it improves the prediction. Each input variable or node keeps on getting split till it reaches a final decision point. This is called the *leaf*. The value of the leave is the actual prediction. Decision trees can be used both for classification and regression.

- **Support Vector Machines (SVM):** Imagine a fruit seller wants to segregate apples and guavas on a table in a way that one side of the table has apples and the other guava. The fruiter seller is using a ruler to make the segregation. He wants to place the ruler in a way that it best separates apples from guava. He can place the ruler in many ways, but the agenda is to place it in such a way that:

    – All apples and guava are on opposite side of the table.

    – The gap between the closest fruit on both sides of the ruler is the maximum.

The ruler here is like an SVM decision boundary or hyperplane that is used for segregation. The fruits closest to the ruler on both sides are called *support vectors*. These support vectors define the line where it should go. In simple language, SVM is a machine learning method that finds the best line (or boundary) to separate two groups of things, while keeping the boundary as far away from both groups as possible. SVMs work well for smaller linear and nonlinear datasets and are also robust to outliers. But their computational requirements increase exponentially with an increase in data size, and they lack interpretability due to projection to a higher dimensional space.

Evaluation Metrics for Supervised Learning

Evaluation metrics help in assessing the efficacy of the machine learning/artificial intelligence model. However, which metric should be used for evaluating the efficacy of the model varies on a case-to-case basis. In this section, we will understand evaluation metrics that can be used for assessing the efficacy of supervised models. This section will also highlight scenarios in which each of these metrics should be used.

Classification Metrics

- **Accuracy:** Tells us how often the model predictions are correct. For example, a model with 90% accuracy predicts correctly 90 out of 100 times. This metric is relevant for a classification model to know how much the model can be trusted. The same metric could also be used for a regression model by defining an error tolerance of, say, 1% or 5% of the actual value. In other words, all the predictions within this much error will be considered correct. So, if a regression model is 90% accurate, then that means that its predictions are within this error margin.

CHAPTER 6   INTRODUCTION TO MACHINE LEARNING FOR TEACHING SYSTEMS

The accuracy can be computed using the test data or the production data with the known ground truth. It should be computed using unbiased data that is representative of the whole population or else the accuracy computed would be biased.

- **Precision** as a metric measures how many times the predicted positive was actually positive. To put it simply, out of all the times the model said "yes," how often was it correct? For example, your fruit classification model predicts apple 100 times of which 80 times the fruit was actually apple. So, the precision of your model is 80/100 =80%. Precision is more important in cases where correct prediction is more important than making incorrect predictions for the class. For example, a spam email detector needs to be precise in classifying an email as spam because otherwise an important email can go unread. So, it is still better to misclassify spam as email than to misclassify an email as spam. This is also referred to as false positive (FP) or false negative (FN) rate. Classifying an email as spam is FN, while classifying a spam as email is FP. Similarly, the correct classification of spam as spam is called a true negative (TN), and the classification of email as email is called a true positive. Precision can also be defined as the ratio of total true positive to sum of total of all TP+FP.

- **Recall**: Measures how many of the actual positive cases your model was able to identify correctly. Put simply, of the actual yeses, how many did the model predict correctly? Recall is better in cases where sensitivity to true positive is extremely high, for example, in fraud detection and disease prediction. The difference between precision and recall is this: use recall in cases where objective is to reduce false negatives. For example, in the healthcare world, misclassifying a sick patient as not sick is more dangerous than classifying a nonsick person as sick. Precision is important in cases where minimizing false positives is more important (spam detection).

- **Confusion matrix:** This is a table that summarizes the performance of a classification model for all the classes. It gives the class-wise precision values along the diagonals, and the nondiagonal terms tell which classes are being misclassified as which other classes. See Figure 6-3.

Confusion Matrix

	Actually Positive (1)	Actually Negative (0)
Predicted Positive (1)	True Positives (TPs)	False Positives (FPs)
Predicted Negative (0)	False Negatives (FNs)	True Negatives (TNs)

Figure 6-3. Confusion matrix

- **F1 score:** There are situations in which balancing the precision and recall are necessary. In those cases, an F1 score can be used. It is a harmonic mean of precision and recall. It is used in cases like credit risk scoring models where you don't want to approve loans for too many risky customers but you also don't want to miss on too many eligible customers.

CHAPTER 6   INTRODUCTION TO MACHINE LEARNING FOR TEACHING SYSTEMS

Regression Metrics

- **Mean squared error:** This is a way to measure how well a machine learning model predicts a continuous value. It tells how off the predicted value is from the actual. The bigger the number, the worse the model performance. It is measured as the average of the square of difference between the predicted value and the actual value.

$$\text{MSE} = \frac{1}{n}\sum_{i=1}^{n}(y_i - \hat{y}_i)^2$$

- yiy_iyi = actual value
- y^i\hat{y}_iy^i = predicted value
- nnn = number of data points

- **RMSE**: The root mean squared error tells how far off your model performance is on average; it is in the same unit as your actual value. RMSE is calculated as the square root of the MSE; it brings errors on the same scale as the actual value.

- **MAPE:** The mean absolute percentage error is the average percentage error between the predicted and actual values. It is easier to interpret errors in the form of percentage values as errors for different datasets can be compared on the same scale, i.e., 0 to 100%. But MAPE cannot be calculated when the actual value is zero, and it penalizes underestimation more than the over-estimation. Also, it is hard to interpret when errors vary too much across data points.

- **R²(R squared):** In simplest terms, it explains how much of the variation in the actual data the model explains. It is defined as the proportion of variance in the dependent variable explained by the independent variables. R-squared is also known as the coefficient of determination, and it indicates how well the regression model fits the data. The higher the value the better.

Choice of Evaluation Metrics

The evaluation metrics must be chosen according to the problem in hand. For example, having a low FN rate is more important for a model being used to predict if a user will default on loan or not than having a low FP rate because the financial impacts of marking a user as a defaulter incorrectly are much less than the financial impact of marking a user safe for loan incorrectly (FN). So, the model's threshold for calling a user safe should be very high so that the FN is low.

In the medical domain, the FN rate is important for the initial tests, and the FP rate is more important for the tests in later stages used to confirm the disease. If the initial test predicts a sick patient as healthy (FN), then the patient will never be tested further and can be life threatening. Incorrectly marking a healthy patient as sick (FP) by the initial tests has less serious implications as the patient will still be tested further to confirm the disease. On the other hand, the disease confirmation test needs to be careful with the FP rate because if a healthy patient is confirmed with the disease (FP), then this patient can be exposed to unnecessary medication, which could be life threatening.

For the search and recommendation use cases, recall is more important than precision because the primary task of these systems is to provide users with options, and whether these options are relevant or not is secondary. The relevance of the options given to the users can still be improved as more data is collected from the clicks or conversions, but if the options are few, then the system cannot be improved further.

For demand forecasting problems, overestimation in the case of perishable items is riskier than underestimation as it will result in losses equivalent to the entire cost of the expired inventory. On the contrary, for nonperishable items, underestimation will lead to loss of profits that could have been made from the lost opportunity. Hotel rooms are perishable because if the room is not booked for a night, then it's a loss for the aggregator. The opportunity cost is much higher, in this case, than the loss of profit margin. So, it is better to underestimate the demand than overestimation.

Unsupervised learning

Unsupervised learning is the type of machine learning in which algorithm learns from data without being told the correct outcome. The algorithm learns by analyzing patterns and groupings in the data on its own. Take an example of a small baby who is given a bunch of toys without being told the toys' names. As baby starts exploring the toys by herself gradually, she will start sorting the toys in groups based on some similarities such as color, shape, softness, etc. Without knowing the name or category of toys, she sorts the toys into groups based on similarities. The baby in this case performed unsupervised learning. Common use cases of unsupervised learning are customer segmentation, anomaly detection, social media network analysis, image compression, document clustering, etc.

Customer segmentation includes grouping together customers based on their activities like purchase history, session history, clicked products, price-sensitivity, and demographics. This helps the business teams in planning their marketing strategy and design campaigns to target specific customer groups.

The following are the popular unsupervised learning algorithms and their respective sample use cases:

- **K-means Clustering**: This algorithm segregates data into multiple clusters such that the data in each cluster is close to the center of the cluster assigned. The distance metric available to find the closest cluster center to a data point can be Euclidean, Cosine, Manhattan, or Minkowski. The way it works is as follows: first predefined points are chosen as cluster centers, and then the distance of all observations is calculated from those centers. Each observation is assigned to the nearest center. This process keeps repeating itself until the cluster memberships stops changings significantly. The goal here is to minimize the distance between observations in the same group and make each group as unique as possible.

    Euclidean is most used for this algorithm, and hence it requires that all the fields in the data must be a numeric type. In other words, the standard form of this algorithm cannot be applied to data with categorical fields. Clustering users by their locations in the form of coordinates can be a good use case for this algorithm. Other typical use cases include image compression by grouping similar pixel values into segments and then storing only segment labels instead of R, G, B values for all the pixels. Documents can be clustered by first converting each document into numeric form using various NLP techniques and then using K-means to cluster the numeric forms representing the documents. Anomalies in the numeric data can be detected by clustering data points and marking very small clusters (which consist of fewer data points relative to the other bigger clusters) as anomalies.

- **Fuzzy Clustering:** This is like the K-means clustering algorithm, but here each data point is not associated with a single cluster but to multiple clusters. In this algorithm, each data point is assigned a membership score for each cluster instead of assigning it to a single cluster. Fuzzy C-Means (FCM) and Fuzzy K-Modes (FKM) are two commonly used algorithms of this type with FCM requiring all numeric fields in the data and FKM working well with categorical data. Membership scores may be computed in terms of probability or distance metrics used while clustering.

- **Hierarchical Clustering:** This has two different clustering approaches: top-down and bottom-up. In the top-down approach, all the data points are considered as one big cluster initially, and then it is further split into two clusters based on distance metrics. Then the same process is repeated. In other words, each cluster is further split into two clusters until the desired number of clusters is reached or the cluster cannot be split further because the threshold criteria required for splitting has not been reached or each cluster already has a minimum number of data points. In the bottom-up approach (agglomerative approach), each observation starts as its own cluster and then starts getting merged with the nearest data point on the basis of distance in a way that a cluster is formed. The whole process is repeated until no further clusters can be merged due to the threshold criteria

CHAPTER 6 INTRODUCTION TO MACHINE LEARNING FOR TEACHING SYSTEMS

required for merging two clusters or the maximum number of points have already been achieved in each cluster or the target number of clusters have been achieved. The cluster formation will end when all observations fall under one cluster. These algorithms form a hierarchy of clusters at every iteration level, and a new data point may be assigned to different clusters at different levels. The final cluster for a data point can be assigned based on the distance score. In other words, a data point may be assigned to a cluster with the closest cluster center, which may be at a top level or at a bottom level.

Product ontology in e-commerce is a good example of hierarchical clustering where both top-down and bottom-up approaches are possible. Top-down is most used, where a broader group called "Products" is divided into top-level product-categories like Electronics, Fashion, Furniture, Stationary, Food, etc., which are further split into level 2 categories like Electronics is split into Appliances (e.g., TV, Fridge, Washing Machine, etc.), Laptops, Smartphones, Accessories, etc. The level2 categories may be further split into level 3 categories and so on. In a bottom-up approach, each product is a cluster in itself, which is then merged with other products based on different factors like products often purchased together or products that have utility or support a neighboring use case. These clusters of products can be further grouped together on the basis of user preferences and purchase patterns. The bottom-up approach is suitable for product recommendations, while the top-down approach makes product search easier.

CHAPTER 6　INTRODUCTION TO MACHINE LEARNING FOR TEACHING SYSTEMS

Evaluation Metrics for Unsupervised Learning

Unlike supervised learning algos, the unsupervised learning does not have a target variable to learn from. So, the metrics like accuracy or precision, recall, and F1 score cannot be applied to evaluate clusters. The only way the quality of cluster formation can be evaluated is by evaluating how homogenous all the observations inside a cluster are and how well separated the clusters are

Silhouette Coefficient: This measures the quality of clusters by finding out how compact and separated different clusters are. To measure this, it measures the similarity of a data point to its own cluster relative to the other clusters. The similarity between two data points is high if their distance is low. So, if clusters are compact and well separated, then it means the data points in each cluster are more similar to other data points in their own cluster than the data points in other clusters. The following is the formula for calculating silhouette coefficient:

$$S(i) = \frac{b(i) - a(i)}{\max(b(i), a(i))}$$

Where,

$S(i)$ = Silhouette score for the ith data point

$a(i)$ = Average distance of the ith data point from data points of its own clusters

$b(i)$ = Average distance of the ith data point from data points from other clusters

A score of 1 means the clusters are well formed or the data point is correctly clustered. A score of 0 indicates that clusters are highly overlapping because there is no difference between the distance of the data points of one cluster to another so the points are randomly clustered. A score of -1 shows that a data point is closer to data points of other clusters than data points of its own cluster, which means that the data point is incorrectly clustered.

- **Calinski-Harabasz (CH) Index:** It measures the clustering quality as a ratio of average variance between clusters to the average variance within clusters, i.e., variance of data points within a cluster. A higher value of the CH index indicates that the clusters are well separated from each other compared to the variance of their data points. This means there is no overlap between the clusters.

- **Davies-Bouldin (DB) Index:** This index is measures the average similarity of each cluster to its most similar cluster. The lower DB value means that each cluster is very different (or less similar) than all the other clusters because the similarity with the most similar cluster (the highest similarity value) is low, so the other clusters are even less similar or more different. The similarity between clusters i,j is computed as the average Euclidean distance between data points of ith cluster from the centroid of the jth cluster.

The evaluation metric should be chosen based on the type of business case you are solving.

The silhouette coefficient is used for use cases like customer segmentation, image segmentation, and anomaly detection when clusters vary in density and shape. The silhouette metric is also used for determining the optimal number of clusters. The Calinski-Harabasz (CH) index is used for market basket analysis and network traffic analysis where clusters are well separated and their data points follow Gaussian distribution. The Davies-Bouldin (DB) index is used when the data is noisy or clusters are overlapping; clusters have uniform density but differ in shape and sizes. This is commonly used in use cases like analysis of users interactions on social media, anomaly detection in manufacturing, and fraud detection in banking.

CHAPTER 6   INTRODUCTION TO MACHINE LEARNING FOR TEACHING SYSTEMS

Handling Real-World Data Situations

These are some real-world situations.

Handling Out-of-Distribution Data

Data is said to be out-of-distribution (OOD) when it differs significantly from the training dataset. Therefore, the model performance is poor for these data points because they do not follow the same pattern that is learned from the training dataset. OOD can happen because of two reasons:

1) Data drift is when something changes at the data generation level that was not expected and cannot be avoided. For example, a change in user purchase behavior toward a brand changes when a competitive brand launches a new product or runs a discount campaign. Some of these data drifts could be permanent, and others would be temporary. For example, the user behavior changed permanently for a lot of platforms like social media, e-commerce, etc., after COVID. Any models that would have been trained on data from pre-COVID times will not be able to perform well in post COVID times.

2) The training data collection was biased and did not cover all the circumstances. So, the data for these circumstances would not match the training data distribution. This could happen when prices for the products are being tracked only when a purchase happens. In this case, the product price would not be available for the whole time when the product did not sell. Training a model on this type of training

data would make the model biased toward certain prices, which resulted in a purchase that could be discounted, and thus the model would always recommend discounted prices for the products.

To solve or mitigate the issues due to OOD, the following steps can be taken:

1) Detection of OOD: The first step is to verify if the issues in model predictions are due to OOD. The following are the techniques that can detect OOD:

   a) Probabilistic methods: A Gaussian mixture model (GMM) can be trained on the feature embeddings computed from the layer before the output layer for a neural network model for the in-distribution data or the training data used. For the unknown data point, calculate the likelihood of its feature embedding computed to be from the distribution of the feature embeddings of the inliers data points using the GMM model. If the likelihood is low, then this data point is likely to be OOD. This approach may work on neural networks only.

   b) Distance-based methods: A simple k-nearest neighbors approach can be applied to find the nearest neighbors for the new data points from the training dataset based on a distance metric. If these nearest neighbors have very large distance values, then these data points are said to be OOD. This particular approach is simple to implement and is model agnostic.

c) Unsupervised methods: Advanced deep neural network models like autoencoders or GANs can be trained on the training dataset to reconstruct the training data well. If for a new data point the reconstruction error by this model is high, then it is said to be OOD.

d) Ensemble methods: An ensemble of deep learning models can be trained for the same training data. If predictions from all these models vary for a new data point, then that data point is said to be OOD.

e) Single-class SVM: A single-class SVM can be trained to identify the boundary of the training data. The model will be able to predict if a new data point is within the boundary or outside it.

2) Adaptation to OOD: The only good way to handle OOD is to have a model trained on OOD data points so that model can learn to predict for these too. However, OOD data may have very limited availability to address this generate synthetic OOD sample. This can help in generating enough OOD similar observations, and the model's training quality can improve. The existing training data points can be used to generate synthetic OODs by adding noise or perturbations to increase the number of OODs so that the model can be trained well to handle OODs. But these synthetic OODs must reflect the same impact on the model as the real OODs for them to contribute toward the model robustness.

CHAPTER 6　INTRODUCTION TO MACHINE LEARNING FOR TEACHING SYSTEMS

Handling Data Imbalance

Imbalanced datasets are those in which the data from some classes may appear more frequently than others. For example, in product sales data, most data points may represent days with low sales because these days are more common than days when sales are high. If this kind of dataset is used to train a demand prediction model, then the model will become biased toward predicting lower sales, reducing the accuracy on high sales days. As another example from the classification side, in the fraud detection case, the number of fraudulent transactions is much rarer than legitimate one. If the dataset is heavily imbalanced, then the model might learn to always predict nonfraud and ignore all fraud cases but still achieve high accuracy.

The following techniques can be applied to handle imbalance data:

Data Resampling: If sufficient data is available for the minority cases, then under-sample data from the majority case. In other words, choosing fewer data points from the majority case can balance the overall data. Several training datasets can be created by different under-sampled sets of the majority case and used for training models in a k-fold manner to rule out any bias due to under-sampling.

If the minority cases do not have sufficient data points, under-sampling the majority case would result in overall insufficient data points to train the model. In this case, the data from minority cases is oversampled. In other words, data points are intentionally duplicated many times to balance the dataset. Data oversampling will increase the size of training data, which is better for model training, but instead of showing the exact same data points repeatedly, it would be better if these data points are perturbed or added some random noise to make the model more robust.

Synthetic Minority Over-Sampling Technique (SMOTE): Instead of adding random noise, this technique creates synthetic data points by finding the k-nearest neighbors to a data point from the minority case to be oversampled. Then new data points are created by interpolating between the data point and one of its k-nearest neighbors.

Evaluation Metrics: Accuracy is not a good metric to evaluate the performance on imbalanced data because it would give a biased result for the minority cases. Instead, a metric like F1 score is better as it will give a balanced result.

Deep Learning

Deep learning is a powerful machine learning type that uses multilayer artificial neural networks to learn complex patterns in data on their own. For example, imagine teaching a kid how to identify car by showing multiple photos of various automobiles and then explaining each pattern and other important characteristics that would differentiate a car from other forms of automobiles. This is how a classical machine learning learns. Deep learning has the ability to automatically learn what features are important and use that learning to identify patterns in data more precisely. This makes deep learning models more accurate, scalable, and less reliant on human experts. However, these models also suffer from a lack of explainability of their outputs and hence are often referred to as *black-box algorithms.*

Deep learning has gained more popularity in use cases where unstructured data like text, image, audio, and video are involved. This is because performing manual feature engineering on unstructured data is very tedious and time-consuming, which makes it scalability and deployment more cumbersome.

Deep Learning Model Architectures

A deep learning model architecture refers to the design and structure of neural networks that are used to solve different types of problems. These architectures define how many layers the network will have, how each layer will get connected to each other, and how data flows from one layer to another. This is just like how the architecture of a building might differ from how the building will be utilized, such as a house, a temple, or a warehouse. The deep learning architecture also differs depending on the type of data it will be exposed to.

Deep learning offers different architectures for NLP and CV because text data is sequential in nature while images need to be processed in blocks of pixels. For sequential data, deep learning architectures like recurrent neural networks (RNNs), long short-term memory (LSTM), and transformers-based models are used. On the other hand, convolutional neural networks (CNNs) are more popular for imagery data. Videos are a special kind of data because these are sequences of images, so depending upon the use case, a combination of CNN and RNN or LSTM-based models may be used. A simple description of these architectures and their variants is discussed in the following sections. After that, the use of different architectures is illustrated for the different use cases of NLP and CV.

Deep Learning Architectures for the Image Process (Computer Vision)

There are two popular architectures used for computer vision (CV) tasks: convolutional neural networks and vision transformers. These will be discussed briefly in the next sections.

Convolutional Neural Networks

A CNN is a type of deep learning architecture that is designed to specifically to work with image data. Just like the human brain processes images by automatically detecting patterns like texture, shapes, edge, and color contrast, a CNN does the same thing using different layers. The CNN architecture has the following key subcomponents:

- Convolution layers: This layer is responsible for scanning the image with small filters called *kernels* to detect features that make up an image. Think of this layer as stencil effect that is used to highlight certain parts of the image.

- Pooling layers: This layer minimizes the size of the feature map created in the convolution layer. The main objective of this layer is to reduce the computation needs that will be required to process the feature map created earlier. It also makes the network more robust to minor changes in the features map. In similar terms, it shrinks the image size by removing the noncritical sections of image.

- The activation function helps a network learn complex patterns by adding nonlinearity to it. After each stenciling of image (convolution operation), the activation is applied to the resultant values. Think of this as a decision gate, where the network focuses on critical features only.

- Fully connected layers are at the end of the network and help in making a final prediction.

- A loss function measures the accuracy of the CNN prediction. During training, the network adjusts its internal parameters to minimize the loss function values.

- A kernel is a filter in the form of a small matrix. It contains learnable weights (internal parameters) that the network adjusts during training to reduce losses. It helps detecting specific features like edges, corners, etc., as it slides across the image like a microscope.

CHAPTER 6　INTRODUCTION TO MACHINE LEARNING FOR TEACHING SYSTEMS

A typical CNN consists of a convolution layer, a pooling layer, and a fully connected layer. The convolution layer applies a kernel of 3x3 or 5x5 moving across the image to generate features such as edges, corners, and textures in the form of feature maps. The pooling layer reduces the size of the feature map by filtering out smaller changes in the image. Some commonly used pooling layer functions are like max pooling or average pooling. This is important to make the model robust to noise. The fully connected layer acts as a neural network that maps the filtered features extracted by the convolution and pooling layers to the desired outputs with the help of activation functions like softmax or sigmoid depending upon binary or multiclass classification. The fully connected layer is also called a *fully connected neural network* (FCN) layer, and it acts as a logistic regression classifier by using the softmax or sigmoid to make predictions. But when the CNN-based models are only used for extracting features, then the activation functions like Rectified Linear Unit (REL) or Hyperbolic Tangent (Tanh) are used for nonlinear transformations.

Depending upon the use case, a loss function is used for learning the model weights. Common loss functions are:

1. Classification: Binary cross-entropy, categorical cross-entropy, focal loss

2. Object detection: Intersection over union

    - Image segmentation: Dice loss

    - Image generation: Mean Squared Error (MSE), Adversarial Loss, Perceptual Loss

Usually a series of convolution and pooling layers are stacked together so that features can be extracted at different scales. The initial layers generally have a larger kernel size, so that can help in learning broader patterns like texture or boundaries, which repeat across the image. Then the kernel size drops in the later layers, which then focus on more localized features like corners or edges. The series of multiple convolution and pooling layers stacked together gives the name *deep learning*. There are many popular model architectures based on the CNN layers; VGGNet, Inception Net, and ResNet are some prominent names that were released in the last decade. These differ in the way the kernel sizes vary from the top layers to the bottommost layers and how different layers are connected together to solve the issues of vanishing gradients as the number of layers in the model increases. The vanishing gradient problem occurs when values of kernel parameters become very small as they move backward through layers. This drastically slows down the network's learning because their learning updates have become almost zero.

Vision Transformers

Inspired by the success of transformers in NLP, attention-based vision transformers were introduced at the end of the last decade for CV tasks by Google Brain. Instead of processing at the pixel level by CNN layers, transformers break the image into patches and learn to apply self-attention to estimate the importance of each patch. The multihead self-attention (MHSA) technique is applied to learn features at local and global level for the image. Like in the case of CNN, the features extracted by the transformer layers are passed to a fully connected neural network (FCN) to make predictions using activation functions like Softmax or Sigmoid.

Self-attention in principle estimates the importance of each component (patch in images or words in text) in making predictions close to the expected outputs. In CNN, pixels within the kernel window size are considered, so any patterns that is based on the combination of pixels from distant regions of the image are lost. This is where the transformer architecture gains advantages in learning global features using self-attention where the interactions between every pair of patches are considered. The transformers-based architecture is also more scalable with an increase in data size and does not require a manual design of the kernels.

Transformer-based models have shown huge improvements over CNN-based models with the help of a self-attention mechanism, but these also require much higher computation and memory resources compared to CNN, which are much faster to train with less resources. Transformer-based models are also costlier to deploy and slow in inference compared to CNN models. Larger datasets and high-quality images are required to train vision transformers, making them harder to apply in many cases where data collection is a challenge. On the contrary, pre-trained CNN-based models can be fine-tuned with smaller datasets and can be easily deployed using low-cost resources.

Deep Learning for Computer Vision (CV) Use Cases

Deep learning in computer vision can be used in varied use cases like image classification, segmentation, object detection, and image-generation.

This section illustrates how deep learning models are applied for each of these tasks and how these models are evaluated for performance.

Image classification: This is the most common task where deep learning models are applied to categorize images into different classes like dog or cat or car, etc. Depending upon the size of the training data and compute available, different deep learning models can be used. For smaller training data and smaller computing resource availability (like a couple of GPUs), CNN-based pre-trained models can be used for performing transfer learning by training these on the new data. If more training data is available, then depending upon the compute availability, CNN-based or transformer-based models can be trained from scratch. To gauge the performance of the model performance, metrics like accuracy, precision-recall, and F1-score and confusion matrix can be used. Accuracy is computed as the percentage of correct predictions for the total number of samples. A prediction in case of classification also depends upon a threshold on the predicted probabilities for different classes. Normally the

CHAPTER 6 INTRODUCTION TO MACHINE LEARNING FOR TEACHING SYSTEMS

class with the highest predicted probability is considered, but sometimes a threshold can also be chosen to discard predictions when the predicted probability for all the classes is small.

For example, say you have trained a dog versus cat classifier; then its accuracy is said to be 80% if for a total 50 images of dog, 40 are correctly predicted as dog (i.e., either highest predicted probability is for class Dog and the predicted probability for Dog class is also above the threshold), and for every 50 images of cat, 40 are predicted as cat correctly. There could be other scenarios as well where the classifier with 80% accuracy would have 90 correct out of total 100 dog predictions but 70 correct out of total 100 cat predictions. In this case, the classifier has better accuracy for dogs than for cats. This is also called *class-wise accuracy*. So, the accuracy may also be defined separately for a particular class.

Consider you are building a dog-cat image classifier. If the model predicts 40 of 50 dog images and 40 of 50 cat images correctly, then the overall accuracy is 80%, since 80 of 100 images are classified correctly. However, the accuracy can also vary by class. Take another scenario: your model has correctly predicted 90 of the 100 images; all 50 dog images were correctly identified, but only 30 of 50 cat images are correctly identified. This is a case of class-wise accuracy. Hence, while calculating the accuracy of image classification, also understand how a model is performing on an individual class especially while dealing with an imbalanced dataset.

Object detection is the task of identifying and locating objects within an image. This involves predicting a bounding box around each object and classifying what the object is. The bounding box is defined by its coordinates, and the model determines both where the object is and what it represents. Common examples can be using computer vision solutions in healthcare for tumor detections or identification of abnormalities in Xray or CT scans.

Earlier objection methods used classical machine learning algorithms and relied on user-designed rules for feature identification. Techniques like HAAR, HOG, SIFT, or SURF were used for feature engineering, and then traditional classification algorithms like SVM and decision trees were used for classification. While these methods were similar and were not computationally heavy, they are low on accuracy.

With deep learning, CNN-based models like R-CNN, Fast R-CNN, and Faster R-CNN automated feature generation and identification, and then fully connected layers or SVM are used for the classification of objects in predefined categories. Later faster models like YOLO (you only look once), SSD (single shot detection), and Retinanet made object recognition even more simpler and faster in a single step.

To measure the accuracy of object detection models, the Intersection over Union metric is used. This measures the overlap between the predicted bounding box and the actual ground box. It is calculated as the ratio between the area of overlap divided by the area of union between the predicted box and the actual box. An IoU of **1.0** means a perfect match, while **0** means no overlap. Higher IoU values indicate more accurate predictions.

Image segmentation: In some use cases, it is not enough to just locate and identify an object in the image, but its exact shape and boundaries must also be predicted. For example, to identify buildings, roads, vegetation, and water bodies in satellite images, it is not possible to just classify or detect a box due to the nature of these classes. Thus, image segmentation is required where each pixel is labeled as one of the classes to be identified. The availability of labeled data is a big challenge for image segmentation because annotations must be done at the pixel level and with a high level of accuracy, especially for the pixels near the boundaries of two classes. Incorrect labels can affect the model training, and the model would fail to classify the boundary pixels accurately.

The classical ML-based image segmentation models includes random forests and conditional random fields (CRFs), which requires manual feature engineering. Unsupervised algorithms like K-means can also be used to avoid manual labeling of every pixel, but the result may not be accurate. Among deep learning, CNN-based U-Net, Mask R-CNN, and DeepLabV3+ are popular model architectures for image segmentation.

CHAPTER 6   INTRODUCTION TO MACHINE LEARNING FOR TEACHING SYSTEMS

Later transformers-based the Segment Anything Model (SAM) was introduced that can segment objects with manual instruction prompts. These models don't require any training. These models don't need to label every pixel of the image.

Dice coefficient and mean IoU are popular metrics to evaluate the performance of image segmentation models. The dice coefficient is an overlap-based metric that acts as an F1-score for segmentation and helps in balancing the class imbalance, which is common in the case of image segmentation. The dice coefficient of a class is calculated using the following formula:

$$Dice = \frac{2 \times TP}{2 \times TP + FP + FN}$$

The true positives (TPs), false positives (FPs), and false negatives (FNs) are computed by comparing the predicted and ground truth labels for each pixel of a class.

Mean IoU is estimated as the average IoU score for each class. IoU for a class is computed just like in the case of object detection with the only difference being in the shape of the ground truth, and the predicted regions is not a perfect rectangle but some arbitrary boundary.

The next section discusses various deep learning model architectures for the text data along with the challenges associated with text data and how the model architectures evolved in the last decade to mitigate these challenges.

Deep Learning Architectures for NLP

Natural language processing (NLP) is a field of artificial intelligence that focuses on enabling computers to understand human language. NLP is a unique field of artificial intelligence as it combines linguistic details with machine learning methods to devise specialized algorithms that can understand and interpret text data.

Earlier, NLP tasks relied primarily on traditional methods to extract features from text data. Methods like one-hot encoding and TF-IDF vectors were commonly used for feature extraction. These features were then fed to traditional machine learning algorithms like decision tree or neural networks for classification. While this approach was simpler, it has several limitations like exposure to limit vocabulary, manual intervention for data cleaning that involved removal of stop words, stemming, etc. These methods were also very sensitive to typing errors, making these models hard to scale effectively.

Newer techniques like word embedding models like word2vec and FastText have improved upon these limitations. These methods make handing typo errors and out-of-vocabulary words easy. These models are scalable and do not require human intervention for data cleaning and other support tasks, making them easier to operate.

The current decade has seen unprecedented advancements in this field with the introduction of architectures like recurrent neural network (RNN), long short-term memory (LSTM), and transformer-based models. Especially transformer-based models have taken the performance of NLP-based models to the next level. These models have now become the backbone of all large language models like ChatGPT. The following are a few commonly used deep learning models for NLP applications.

Recurrent Neural Networks

A recurrent neural network (RNN) is a special architecture of neural networks designed to work with sequential data like text, speech, and time series. This architecture is called RNN because it reuses the information from previous steps through a feedback loop in its hidden layer also called a recurrent layer. This layer is the heart of RNN, and it processes one input at a time sequentially and carries the information from previous layers. This layer maintains a hidden state, which means that it stores the memory from previous time steps and helps RNN use past information to influence

CHAPTER 6 INTRODUCTION TO MACHINE LEARNING FOR TEACHING SYSTEMS

the current decision. This allows a model to loop over the sequence of data and learn its pattern over a period of time.

However, RNN has the following shortcomings:

1. RNN suffers from a vanishing gradient problem.

2. RNN are slow to train and are very computationally expensive to train.

Long Short-Term Memory

This is an advancement on RNN architecture that is designed to overcome the vanishing gradient problem. In simpler terms, it is designed to remember model information in terms of gradients for a longer period. LSTM solves this problem by using a special internal structure that has three gates, namely, input gate, output gate, and forget gate. This structure controls which information should be added, removed, or passed on at each step. This structure helps LSTM maintain a long sequence of data, making it better at understanding context or subject of the text. It is because of this ability LSTM is widely used in applications like language translation, speech recognition, and text generation.

Transformers

Transformers were introduced at the end of the 2020s after the release of the famous paper "Attention Is All You Need" in 2017. This architecture was based on the idea of self-attention. This architecture is efficient and more scalable compared to RNN-based models because it learns how each word in the input text is related to all the other words in the input text at the same time and what is the influence of a word on the target output. It does not process words sequentially but in parallel, which makes it computationally efficient.

CHAPTER 6 INTRODUCTION TO MACHINE LEARNING FOR TEACHING SYSTEMS

Transformers use attention mechanisms to find the relation between words and the target output. The relation between words within the input text is called *self-attention*, which is mathematically represented as a weightage or attention score for a pair of words. The self-attention scores or weightage between two words will be high if they refer to each other as per the context of the input text. The attention score or weightage for a word will also be high if it is directly related to the target output. For example, consider the following statement for sentiment analysis task: "I find this book to be lengthy and boring," which is marked as a negative comment. Here the words "this" and "book" refer to each other so their self-attention scores will be very high, and the output "negative sentiment" is because the user words like "lengthy" and "boring" in their comment. The attention scores for these words for the target output will also be high.

Transformers use multihead attention (MHA) to capture different types of relationships like meaning, grammar, and position between input text by analyzing multiple aspects of input in parallel. Word positioning is handled by positional embedding, and this information is passed through fully connected layers that are trained using back propagation. Unlike RNN, transformers allow parallel understanding of input text, which fully avoids problems like vanishing gradient and thus is the ideal candidate for long text sequences. This is the reason the transformer architecture is the backbone of all large language models like GPT and other popular LLMs.

Deep Learning for NLP Use Cases

Tasks like sentiment analysis, name entity recognition, machine transition, and text summarization use deep learning for NLP extensively. Let's understand what each use case is all about.

Sentiment analysis: This is the type of study that classifies input text like user reviews, social media posts, etc., into types of predefined sentiments (like positive or negative) or predefined tones like joy, happy, sad, anger, etc.). All three listed deep learning architectures can be used depending upon the size of input data. For judging the accuracy, use the F1 score.

Names Entity Recognition (NER): This is the equivalent of object recognition in computer vision. In an object recognition task, we identify objects in an image. Here in NER we identify entities like name, organization, etc., from the text. All three listed deep learning architectures can used depending upon the size of input data. Evaluate model performance using precision, recall, and F1 score.

Machine Translation: This refers to scenarios in which deep learning is used for translation input text from one language to another. LSTM and transformers are the preferred architectures that should be used. The Bilingual Evaluation Understudy (BLEU) score is a popular metric to evaluate language translation models. This metric compares the model's response with the reference human translation for different n-grams (unigrams, bigram, trigram, etc.). The score is computed as the geometric mean of the scores at different n-gram levels.

Text summarization: This refers to scenarios in which deep learning architectures are employed for generating a summary of the input text. It can be of two types. The first type is extractive text summarization; in this type important text from input text is selected and presented as a summary. The other type is generative summarization where new shorter text is generated to capture the idea of the input text. For extractive text, summarization models like BERT can be used, and for generative summarization, models like GPT should be used. Recall-Oriented Understudy for Gisting Evaluation (ROUGE) is used to evaluate the performance of text summarization by comparing the model's responses with the reference summaries. Unlike BLEU, it is more focused on recall than precision, i.e., how much of the reference summary can be found in the model's response.

CHAPTER 6   INTRODUCTION TO MACHINE LEARNING FOR TEACHING SYSTEMS

Decision Framework: When to Use Deep Learning vs. Classical ML

- **Problem complexity:** Use a classical ML for simpler use cases or use cases where the explanation of model output is more important. Use deep learning for cases for complex problems and where accuracy matters more than explainability.

- **Data type and size:** Use classical machine learning algorithms for structured, tabular data and smaller datasets. Use deep learning algorithms for unstructured data like text, images, and large datasets.

- **Resources and expertise:** Deep learning algorithms require powerful compute resources like GPUs, more storage, and higher skilled teams for model training and deployment. Choose deep learning algorithms only when all these constraints are met.

- **Scalability:** Deep learning algorithms are easy to scale and remain steady for long term. For smaller one-off projects, classical ML is a more practical and cost-effective choice.

GenAI

Generative AI started to gain popularity toward the end of the 2020s around the same time transformers were introduced. Unlike traditional models that focus on finding the difference between the classes, generative models learn from the underlying distribution of data to create entirely new data. These models are unsupervised in nature. With the rise of LLMs, these models have seen massive adoption. This section will explore types of major gen AI techniques and use cases where they can be used and challenges associated with them.

GenAI Algorithms

Variational Autoencoders (VAEs)

VAE are generative models that use an encoder-decoder architecture to learn how to represent input data in a probability-based space and just in numbers. The encoder turns the input into a range of possible values, and the decoder uses this to create new data that looks similar to the input data. They are trained to keep learned space smooth and predictable by using a method called KL divergence. This category of model finds applications in use cases where image generation is required. An implementation example could be drug discovery where it can be used for generating new molecular structures. Audio synthesis is another very apt implementation for creating new sounds.

Generative Adversarial Networks (GANs)

GANs are used to generate new data points that resemble the input training dataset on which the model is trained. GANs generate new data points using a generator neural network, which is trained to generate such data points that a discriminator neural network cannot distinguish between the generated data points and the real data points that already exist in the training dataset. This is also called the *adversarial process*, and it results in the generation of realistic data points that are hard to distinguish from the actual ones.

The Generator part takes a random noise as input from a Gaussian or Uniform distribution, which is expanded by an FCN to higher dimensions. The high-dimensional output from FCN is then passed to Transposed

CNN layers to create a full-sized image. In this way, the generator is like a mirror image of the image embedding model where the final layer outputs the embedding vector for the image.

The discriminator is a binary classifier that takes as input the images generated by the generator and from the real training data. Then the discriminator must predict if the input image is real or fake, i.e., if it's taken from the real training data or is generated by the generator. The architecture of the discriminator is like a CNN-based image classification deep learning model. It consists of an input embedding layer followed by the stack of CNN layers. The CNN layers extract features from the input image and pass them to the FCN layer as input. The output vector of the FCN layers is then mapped to a class (real or fake).

GANs are primarily used in computer vision for different use cases like generating high-quality synthetic images and style transfer, i.e., copying the style of art of one image onto the other. GANs have been used in real life for creating images of realistic human faces, copy art style, training data augmentation, expanding training data with synthetic data for healthcare, robotics, enhancing low-resolution images for medical and satellite imagery, and video generation like deepfake and AI-generated animation videos.

Transformers-Based GenAI

Transformers were developed to handle long text in a better way by using self-attention technique, which lets the model look at different words in a sentence at once and then deduce their relationship with each other.

Language models are generative models that are trained to predict the next word given a sequence of words as input. In a way, these models can keep generating long sequences of text until they are stopped. Such models are also called Generative Pre-Trained (GPT) language models.

After pre-training, they are fine-tuned using labeled examples to perform specific tasks like answering questions or summarizing content, a process known as *instruction tuning*.

Transformers-based models like Generative Pre-Trained (GPT) are further fine-tuned on supervised tasks for several NLP use cases like question-answering, text summarization, code-generation, conversational chatbots, language translations, etc. These are also used for text-to-speech use cases where GenAI models can generate human-like voices for a given text.

In CV, visual language models are used for generating images based on text instructions or generating image captions. Some of the real-life applications include reading ID cards, visual question-answering, and image editing using text instructions.

Evaluation Metrics for Generative Models

- Perplexity (PPL): Measures the confidence level of the model while predicting the next word. Lower the PPL value better is the prediction quality. This is used for NLP projects only.

- BLEU Score: Compares model-generated text with input text using different word groupings. The higher the value of BLEU, the better the quality of the text generated. This is used for NLP projects only.

- Inception Score (IS): Measures the quality and diversity of generated images by evaluating how confidently a pre-trained model can classify them. The higher the value of IS, the better the model output. This metric is used for CV projects only.

CHAPTER 6   INTRODUCTION TO MACHINE LEARNING FOR TEACHING SYSTEMS

Summary

This chapter covered best practices while designing AI/ML projects and provided the blueprint of how to execute and plan your project. The second half of the chapter focused on introducing basics of types of learning types, common algorithms, and situations in which those can be used and metrics to measure its performance. The chapter also covers few real-world data scenarios that are encountered while executing the project. The chapter also introduced Gen AI.

CHAPTER 7

From Prototype to Production: Scaling ML at Speed and Scale

We have all been there: your machine learning model performs perfectly fine in offline testing on all accuracy and performance parameters. But when this model is deployed into production, suddenly the performance and accuracy nosedives. This new model starts affecting the performance of the overall application, its predictive power over the period of time decreases, and the model fails to handle real-world scenarios. This is where machine learning projects face their biggest challenges: the stage after deployment.

The aim of deploying a machine learning model from a development environment to production environments isn't just about transferring the codebase but to ensure that the model delivers the same accuracy as it had in the testing environment.

CHAPTER 7   FROM PROTOTYPE TO PRODUCTION: SCALING ML AT SPEED AND SCALE

MLOps is critical for the success of AI projects, particularly as they transition from the development phase into production. Generally, the AI team's major focus is on how to build accurate models, and often steps required to make these models reliable and efficient in the production environment are missed. That's where MLOps helps.

By establishing standardized procedures for model deployment, MLOps helps integrate machine learning models into real-world systems. It ensures that models match its post-deployment performance to the testing environment by handling model drift and scalability issues effectively.

MLOps also helps establish synergies between various stakeholders like machine learning engineers, the engineering team, and operations by streamlining communication and setting clear expectations from each stakeholder.

MLOps also provides guidelines to set up procedures required for continuous monitoring, version controlling, and release rollback, which strengthens security, compliance, and risk management.

MLOps is not just a technical framework; it is an important technical discipline that makes sure that AI model meets expectations beyond the lab and delivers sustained value by adapting to the demands of production environments. In summary, MLOps ensures that AI models in production environments not only stay operational but also adaptable, secure, and scalable by maximizing the long-term impact of AI investments.

From Prototype to Production: Why Product Leaders Need MLOps for AI Product Success

As AI continues to play a pivotal role in driving innovation, product leaders must recognize that the AI model journey doesn't end once it is developed and deployed. Moving models from controlled offline environments to real-world production environments where models interact with real-world data and scenarios is often met with challenges. MLOps is the framework that can help navigate these challenges. Product leaders must not consider this as merely the engineering team's responsibility as it has a greater role to play in the overall product success. By equipping themselves with the nitty-gritty of MLOps, leaders will be able to better appreciate and navigate the complexities of the dynamic nature of AI products. This will also help them align AI projects with the broader business objectives and requires cross-functional support. The following are the few gains that the product leader would make by understanding MLOps:

1. MLOps helps improve the customer experience by allowing real-time model updates and improvements, helping businesses to provide a personalized experience to the customer uninterrupted.

2. MLOps enables faster deployments and A/B testing, which can fast-track quick experimentation and help product teams develop and deploy products faster. This ability to rapidly test, deploy, and monitor different AI models gives organizations the strategic advantage of agility, which allows them to stay ahead of the competition and innovate quickly.

3. MLOps helps in reducing technical debt by implementing best practices for deployment, monitoring, and testing. This ensures that AI systems are sustainable and adaptable, preventing costly rework.

4. AI products are dynamic in nature, and its performance may change with changes in data, unlike classical product features that are "set it and

forget it." Shifts in data patterns or user behavior may lead to a phenomenon called *model drift*. This phenomenon may affect model performance adversely. MLOps defines procedures that can help detect this phenomenon early, which may give time to machine learning engineering to fix it before the performance degrades beyond acceptance, thereby ensuring models stay aligned and maintain an optimum performance level. This also helps to make model more reliable, which is important for fostering a good customer experience.

5. MLOps provides stability to the overall system of which AI system is a part. AI models can become performance bottlenecks if poorly managed. This may lead to costly downtimes and may impact the user experience and in a few exceptions may also lead to compliance issues. MLOps minimizes this risk by providing robust monitoring and rollback capabilities. By preventing costly errors and failures, MLOps helps organizations reduce operational costs and avoid revenue losses, making it a clear contributor to positive ROI.

MLOps Framework

Although MLOps involves numerous activities, it's essential to maintain a structured approach to ensure each task is performed in the most effective sequence for optimal outcomes. To address this, we've developed the DeMoGT Framework, which encompasses all key steps needed to transition an AI application from prototype to production. Figure 7-1 presents a high-level representation of this DeMoGT framework.

CHAPTER 7   FROM PROTOTYPE TO PRODUCTION: SCALING ML AT SPEED AND SCALE

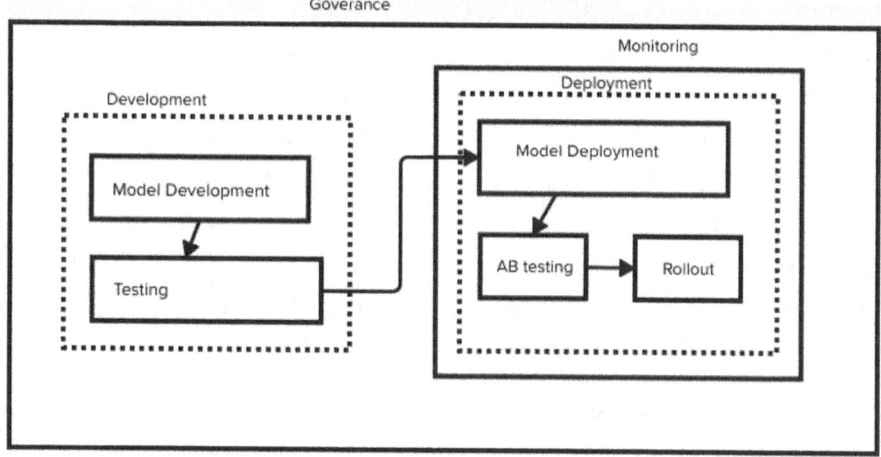

Figure 7-1. *MLOps framework*

- **De** is for **Deploy**.
- **Mo** is for **Monitor**.
- **G** is for **Governance**.
- **T** is for **Test**.

Model Deployment (De)
What Is ML Deployment?

Imagine you have spent several months developing an AI model that has the potential to boost your bottom line significantly. You would be itching to make this AI model available to your users as quickly as possible. The process that you will be following to make your model available to your users is called *deployment*. Machin learning deployments refer to processes involved in taking an AI/ML model from a test environment to a production environment, allowing users to utilize the services of AI/ML model.

CHAPTER 7 FROM PROTOTYPE TO PRODUCTION: SCALING ML AT SPEED AND SCALE

Planning for Seamless ML Model Deployment

Lack of clear production plan is one of the major reasons why many machine learning projects fail to get deployed or even if they are deployed face technical challenges that hamper their optimum utilization. Developing AI models is costly both in terms of time and money, and jumping straight into development without planning how this model will interact with the real- world is never a wise decision. While there are several aspects that should be considered during the AI/ML project planning phase, the following factors are particularly important:

1. **Data Management**

    - **Data Storage and Access**: AI solutions require a lot of relevant data for training. Data to AI models is like oxygen; they cannot survive without it. While a lot of emphasis is made on the availability of data, equal attention should be paid to the quality, validity, and duration of data. Other critical aspects that are often ignored are how the model will access and manage data in a production environment. This includes how the integrity of data will be managed and maintained in production environment.

    - **Feature Access:** While training an AI model, raw data in its actual form may or may not be consumed by algorithms to train themselves. Generally, some intermediate variables are derived from raw data that can better explain the learnability of data; these intermediate variables are called *features*. Since these variables are not actual data, due consideration should be given to where these variables will be stored and how the model will access them. Consideration should be given to this step during the planning phase to ensure that these intermediate calculations happen at the most optimum speed and the models are able to access them without any lag.

2. **Computational Complexity:** If the AI model has complex calculations to perform or is getting trained on a large dataset, it might require significant computational resources. This must planned way earlier either during the training or validation phase so that production systems can be designed accordingly.

3. **Model Outcome Definition:** AL models can have different types of outcomes ranging from a probability score to a continuous number, to a classification label, and to a segment identifier or something else. In the Gen AI era, these types can further vary as text, image, or something else. It is important to figure out to the last detail what will be the outcome of your model and how other external systems that will be working in tandem with your model will consume the model outcome.

4. **Model Outcome Delivery Method:** AI models can only generate outputs in a predefined format. But how this outcome will be consumed by the larger system, whether this consumption will happen through intermediary database, API, or system event, should be decided during the planning phase. Clarity on this topic will help the receiving system get prepared to process and use the model's output effectively.

CHAPTER 7   FROM PROTOTYPE TO PRODUCTION: SCALING ML AT SPEED AND SCALE

5. **Frequency of Outcome Generation:** Each model is designed and developed to generate output at a predefined frequency. Another aspect of model output consumption is how frequently the external system will consume the generated output. Leaders should try to get answers to both of these questions during the planning phase itself.

6. **Create Model Rollout and Benchmarking Strategy for Your Model**: Leaders should also focus on determining how the new model will be introduced to the user, whether the new model will have a limited exposure at the start and gradually or will exposed to all users from the beginning only. Will the model's performance be compared against the older one or not? If yes, what will be the design of experiment? What will happen if the model fails to perform in a production environment? How will the new model be rolled back? Addressing these points early not only saves time and resources but also ensures a smoother deployment process. The following are the key deployment strategies to focus on:

    - Shadow deployment

    - Canary deployment

    - A/B testing

    - **Shadow Deployment:** In this deployment technique, the new AI model is deployed and exposed to live traffic alongside the current model. In this deployment, live traffic receives output from the current model only, and the output of the new model is stored in the database. For each user, you can compare both models' output and decide upon which model to keep.

- **Canary Deployment:** This allows introducing the new model to a small group of users, and the remaining set of users are still served using the old model. This controlled rollout allows for monitoring the new model's performance without taking much risk.

- **A/B Testing:** This allows the deployment of multiple versions of models simultaneously. Live traffic is divided into two or more groups depending on how many models are deployed. Each model caters to a specific user group. The outputs of each model are stored in the database, and the performance of different versions of the model is compared. Once a decision has been made, the winning model is exposed to all users.

7. **Monitoring Model in Production:** Leaders should plan early on how model performance will be monitored in production, what metrics will be monitored, and how. How will relevant teams be informed if severe performance degradation is occurring? Will it be through an autogenerated email or alert or through some other means?

8. **Handling Traffic and Concurrency:** Leaders should plan early on what will be the load bearing capacity of the model and how many concurrent requests can the model handle simultaneously.

CHAPTER 7   FROM PROTOTYPE TO PRODUCTION: SCALING ML AT SPEED AND SCALE

Model Serving Infrastructure Strategies

Model serving is the process of making a trained machine learning model accessible in a production environment for making predictions or inferences. Once the model is deployed, model serving enables the continuous model operation to accept input data, perform computations, and return predictions either in real time or in batch mode. Model serving focuses on ensuring that the AI application has high availability and high performance in the production environment.

However, note that model serving is different from model deployment.

The key difference is that model deployment refers to the steps involved in moving the model from the test environment to the production environment. Model serving refers to the steps involved in ensuring that the model stays operational and ready to serve whenever required. Model serving procedures ensure that once the model is deployed in production it remains accessible and scalable and its performance stays optimum during varying loads.

Now that we've covered how to plan deployments and clarified the distinction between model deployment and model serving, let's move forward and explore the different model serving strategies and when to use each one.

The Role of Model Serving Strategies in Product Management

Here are a few reasons why product leaders must have an understanding of model serving techniques:

1. Aligns product goals with technical requirements: Understanding these concepts enables product managers to translate business expectations around model performance and availability into technical requirements. For example, if a business needs the

model to serve millions of customers in real time, it would require a different model serving technique than the model that is expected to deliver outcome once in a period.

2. Allows product managers to weigh pros and cons of each model serving technique before approving one: Different serving techniques have different impacts on system performance, so a PM can choose a technique based on the environmental and traffic factors in which the model will be operating.

3. Facilitates collaboration between different teams: Product managers often act as a bridge between various technical teams and business stakeholders. Understanding how the model will serve customers helps them communicate more effectively and set realistic expectations.

Model Serving Strategies

The following are the four strategies that are used to serve an ML model to real-world applications:

- Server-side deployment
  - Streaming application
  - Batch predictions
- Client-side deployment
  - On-device deployment
  - Browser-side deployment
- Federated learning: hybrid approach

CHAPTER 7   FROM PROTOTYPE TO PRODUCTION: SCALING ML AT SPEED AND SCALE

Server-Side Deployment

In the server-side model serving technique, the model is deployed on a server and communicates with the client-side application using client-server calls. Think of this as a classical software application deployment where the logic sits on the web server that accepts requests from client applications via an interface API and returns the results. The client application considers the model as an endpoint in the application. Server-side deployment is beneficial for models that are required to handle a larger volume of requests or require substantial computation. This kind of deployment also offers following benefits:

1. Server-side hosting can handle complex models that require high computation power

2. Can handle large volume of request

3. Easy to monitor and manage model performance since it is deployed centrally

4. Ability to process sensitive data in a more efficient and secure manner

Ther are two further submethods of deploying models on the server:

1. **Streaming Application**

    In streaming applications, the model is deployed to handle input data requests in real time, and predictions are made instantaneously, allowing the client application to respond to new information requests immediately. This kind of deployment works best where real-time response is essential such as fraud detection, real-time recommendations, or autonomous systems like self-driving cars.

CHAPTER 7   FROM PROTOTYPE TO PRODUCTION: SCALING ML AT SPEED AND SCALE

A key feature of streaming applications is its ability to handle high-throughput data by maintaining low latency in producing output. From a technical standpoint, this is achieved by creating a combination of real-time data pipelines, message queues (Apace Kafka, Rabbit MQ), and low-latency model serving frameworks, such as TensorFlow Serving or Seldon or AWS SageMaker. The role of these frameworks is to ensure the high availability of model.

Since these models interact with the real world on a real-time basis, it is critical to continuously monitor the model performance to observe any shift in data pattern and model performance over a period of time. Creating data feedback loops on top of continuous monitoring helps a lot.

2. **Batch Predictions**

    Batch predictions, as its name suggests, refer to a scenario in which machine learning models are trained and then subsequently executed against batches of data in scheduled intervals or at regular times. In this process, data is collected, processed, and fed all at once into the model for processing in a non-real-time manner rather than piece-by-piece in real time as it arrives. This means the model gives predictions about a dataset that is being aggregated over time.

    Batch predictions are well adapted for tasks that do not require immediate action, such as processing large amounts of data to find patterns, customer segmentations, and offline risk score analysis. Normally the model is used for data that contains lots of features. Since batch processing does not process data in real time, it can be used for deploying complex, computationally intensive models that require significant processing power. This kind of deployment is also handy for processing huge datasets.

Batch processing is often managed by using distributed computing systems or cloud platforms with built-in batch processing frameworks like Google Cloud dataflow or AWS SageMaker *Batch transform*.

Client-Side Model Serving

The client-side model serving technique refers to the deployment of machine learning models directly on the client side. Instead of sending data back to server for generating inference, the model runs locally on client devices and generates output often without needing a steady Internet connection. This approach involves embedding the model within client-side applications, allowing predictions on demand directly on a device using available client hardware resources.

- Since this type of deployment relies heavily on client-side hardware that often has limited computing capacity, this approach is not suited for models that require higher computation power.

- In a client-side model, serving data never leaves the client-side, which makes it safer for use cases where privacy and data security are critical.

- This is a cost-efficient method of serving your ML model since the model runs on client devices, so there is less need for very expensive server-side infrastructure or bandwidth for processing and transmission of data.

- However, if your model requires frequent updates, then client-side serving can be very cumbersome. Each device on which the model is running has to be updated to the latest model version, which can be challenging.

- Monitoring model performance and logging in this type of serving technique is challenging and may require very solutions.

There are two methods of deploying models on the client side.

On Browser:

ML model deployment on a browser means executing machine learning models to give the desired outcomes directly in a user's web browser. Since of most modern browsers are optimized to support fast graphical calculations, it comes easier for models to function optimally without having the need to connect to the server for producing outcomes. In this setup, pre-trained models are directly either embedded or dynamically loaded into client-side applications. This model-serving approach brings ML functionality directly to web applications, enabling them to process user data quickly without the need to interact with cloud servers after the initial model download.

JavaScript frameworks such as Tenserflow.js are popular ways to deploying in a browser. Apart from using JavaScript other technologies like WebGL, WebRTC and WebAssembly are a few other popular means of browser deployment.

Another notable method of browser deployment is using browser extensions. A browser extension can act as a bridge between the web browser's environment and the ML model, providing additional functionality for running the model locally in the browser.

Using a progressive web app (PWA) is another way to do browser deployment. PWA behaves like a native mobile app and can function both offline and online and also interacts with a device hardware API, making it suitable for ML models that need to work without constant Internet access.

While choosing browser-based deployments, a product manager should consider the following points during the product planning and development phase:

- Keep models lightweight to ensure fast loading and better response time. Use techniques like quantization and pruning.

- Use WebAssembly/WebGL to augment limited availability of computing power on the client-side devices.

- Since model code and potentially sensitive data are exposed to the user, extra precaution should be taken in securing intellectual property and prevent model tampering.

- Not all browsers support the same level of performance and APIs, so thorough testing is required for ensuring consistent behavior across different environments.

Real-World Examples

The following are some real-world examples of browser deployment:

1. Google teachable machine: This an interactive tool that allows users to train custom ML models directly in the browser.

2. Application performing image recognition: Examples include OCR performing applications, facial recognition applications, or object recognition applications.

3. AI-powered text predicting applications: Grammarly uses a browser extension to correct grammar errors in runtime on a client device even in offline mode.

On Device (Edge)

Client-side serving of an ML model on devices, also called *edge deployment*, refers to running machine learning models directly on user devices. These devices can be smartphones, wearables, IOT devices, or any other industrial sensor system. Rather than relying on the server-based inference, edge-deployed models use local hardware. In this approach, models are optimized and deployed on the device itself, enabling real-time predictions and data processing.

Edge computing essentially brings computation closer to the source of data, thus reducing the need to transmit the data back to the server, cutting down on processing time. This approach works well in use cases where devices on which the prediction will be consumed have an unreliable Internet connection or when data on which the prediction must run is too personal or sensitive. Edge deployment is ideal for use cases where low latency, offline capabilities, or high throughput is critical. Or it is impractical or undesirable to send sensitive data to the server for processing. The device on which the model is deployed must use specialized hardware accelerators like GPU, tensor processing units (TPUs), or AI chips to improve model performance.

CHAPTER 7   FROM PROTOTYPE TO PRODUCTION: SCALING ML AT SPEED AND SCALE

To deploy a machine learning model on edge devices, the following technology frameworks can help:

- **TensorFlow Lite**: A lighter version of TensorFlow built specifically for running ML models on mobile devices. This platform specializes in reducing the memory footprint and the computation power required.

- **ONNX Runtime**: This is an open-source engine for running cross-platform, open-ONNX models. ONNX Runtime help can optimize models for a wide range of edge devices and hardware accelerators.

- **AWS Deep Learning for Edge**: AWS services designed for deploying models on edge devices offers tools to run inference tasks locally.

- **CoreML**: This is Apple's machine learning framework that supports running optimized ML models including deep learning on iOS devices.

- **Edge TPU**: This AI accelerator from Google is designed to run ML inference tasks on edge devices with low power consumption and high throughput.

- **NVIDIA Jetson**: This is a series of embedded computing platforms designed for AI applications. They include GPU-based systems optimized for running deep learning models at the edge.

CHAPTER 7 FROM PROTOTYPE TO PRODUCTION: SCALING ML AT SPEED AND SCALE

Real-World Examples

1. The automatic number plate recognition system is a classical implementation of edge deployment where it reads the vehicle license plate on a real-time basis; often parking garages have low Internet bandwidth.

2. Face recognition and speech recognition models in smartphones and other smart devices are deployed directly on devices for instance offline user authentication and voice assistants.

3. Edge ML-powered word generator apps on smartphones create real-time text or suggestions by analyzing user input locally, enabling faster and offline word prediction or content creation.

4. Industrial robots use Edge ML for quality control by analyzing products on the assembly line, detecting defects with visual inspection models directly on the robot's control.

5. Smart shelves use Edge ML to detect inventory levels and automatically update stock information based on the real-time analysis of the products on display.

Table 7-1 presents comparison between the two client-side deployment techniques.

Table 7-1. Comparison: Browser-Based vs. On-Device Client-Side Model Serving

Feature	On Browser	On Device (Edge)
Latency	Low, but slightly higher than on-device	Extremely low, suitable for real-time tasks
Internet Dependency	Requires an Internet connection, unless cached	Can work offline entirely
Hardware Constraints	Limited by browser's capabilities (CPU, RAM)	Depends on the device's hardware (e.g., mobile CPU, GPU, accelerators)
Data Privacy	Data stays in the browser, enhancing privacy	Data remains on the device, offering high privacy
Model Size	Typically smaller models due to browser constraints	Can handle larger models compared to browsers, depending on the device
Updates	Easy to update via browser updates	Updates are done via app updates or OTA
Use Cases	Web-based interactive applications, interactive AI	Real-time apps, mobile apps, IoT, AR, and VR

Federated Learning: Hybrid Approach

Federated learning is a decentralized approach of ML model serving that enables multiple end-user devices to collaboratively train a central model without sharing their data. It doesn't require an exchange of data from client devices to centralized server, and the raw data on edge devices is used to train the model locally. The final model is formed in a shared manner by aggregating the local models' weights. The key aspect of federated learning is that sensitive data (such as personal, medical, or financial information) remains on the device, preserving privacy and reducing the risks associated with centralized data storage.

Initially a generic model is shared with user devices, which then trains the model based on user's local data. Over time these individual models become personalized based on a user's data and offer better a user experience.

Here is how federated learning works:

1. The centralized server initializes a generic ML model and distributes it to multiple user devices. This step is called *model initialization.*

2. Once each device receives the baseline model, the model starts training itself on the device's local data. In this step, the data never leaves the device. This step is called local model training.

3. Once the local model is trained, updates are computed in the form of gradients and weights, and these are further shared with the central model.

4. The central server aggregates the updates from all user devices to create an updated global model. This step ensures that all user devices benefit from the learning of other devices by maintaining privacy.

5. The updated global model is sent back to devices for further training, continuing the process iteratively. The devices keep learning locally from their data, and the global model improves progressively with each round of federated learning.

These are the key benefits of federated learning:

1. For use cases that require GDPR or HIPAA compliance, this approach is very handy as the sensitive data never leaves the user device.

2. Only model updates are shared between peripheral devices and central models. This approach requires far less bandwidth.

3. This approach provides personalized tailored results to specific devices that often improve the user experience and help in increasing user engagement with the solution.

4. This approach also works best for use cases where multiple organizations or devices must collaborate to deliver an offering. Since there is no actual data sharing happening between servers, the participating organizations do not have to worry about data sharing or security issues.

These are the challenges and trade-offs in federated learning:

1. The FL approach suffers from high communication overhead since devices share frequent model traffic. This overhead can be troublesome in environments with limited bandwidth or high-latency connections. To reduce this overhead, techniques like model compression, differential privacy, and aggregation strategies must be used. However, using these techniques may adversely affect computation time or model accuracy.

2. The FL approach is susceptible to various cybersecurity risks like model poisoning or inference attacks. Model poisoning is called when malicious users submit incorrect model updates, which may impact global model performance. Inference attacks refer to situations in which hackers potentially extract private information from the model updates. Implementation of security protocols like secure aggregation, differential privacy, or application of robust aggregation methods while training the global model can help counter these attacks.

3. FL model serving techniques rely a lot on a local device's computational resources (battery, memory, processing power, etc.), which may make learning challenging for complex models. To avoid this, use lightweight models and use model optimization techniques like model pruning to decrease the need of computation resources.

Real-World Examples

1. Apple Watch uses FL learning to provide personalized fitness insights while ensuring the user's health data stays private on device.

2. Google's Gboard keyboard uses FL to improve word predictions and autocorrection based on a user's use of words and sentence framing.

3. Spotify uses federated learning for personalizing music recommendations for users, leveraging local data.

CHAPTER 7   FROM PROTOTYPE TO PRODUCTION: SCALING ML AT SPEED AND SCALE

Model Monitoring and Maintenance (Mo)

After months of hard work and effort, you have successfully launched an ML-powered search algorithm to recommend products that meet the user's needs. The expectation was that new algorithms, once implemented, would greatly enhance user conversion. As time passes and the thrill of recent accomplishment wears off, it was discovered that algorithm performance has begun to deteriorate, and initial gains are rapidly shrinking.

This sounds like a familiar story. The key reason behind this situation is that most product teams see the deployment of an ML algorithm as the final step in the lifecycle. It is the point at which the model is considered "done." This approach misses the fact that the real world is constantly changing, and so is the data capturing the real world.

Returning to the scenario, the product team could have prevented the performance drop and the sticky situation they found themselves in, if they had considered model monitoring and maintenance step after deployment.

The model gains its intelligence by training itself on the data and then uses its intelligence to predict the outcome of new data input. Any difference in seasonality, pattern, or behavior between the data at which the model is trained and data used for prediction would result in model performance decay. The phenomenon where the assumptions and patterns learned during training no longer hold true in the real-world production environment is known as *model drift* or *concept drift*.

ML model monitoring involves continuously tracking the model's performance in production post-deployment. It acts as an early warning system by identifying any changes in model performance or data patterns. Once early warning signs are visible, the process of retraining the model by fine-tuning features to keep up with ever-changing data is a called *model maintenance*. This step can also include activities required to fix integration challenges if there are any within the external system in which the model is embedded.

Model monitoring and maintenance are critical procedures that are important for the long-term success of ML projects. The following are the benefits:

1. It aids in the early detection of changes in data patterns or the link between input and output data.

2. It aids in optimizing computation resource utilization by giving alerts of access usage.

3. It ensures regulatory compliance by consistently monitoring the model's predictions against each input data. This also aids in detecting any bias in the model.

4. Effective monitoring helps identify scaling issues, ensuring that the model can handle growing demands while maintaining performance.

After understanding how important model monitoring and the maintenance phase is, let's now formally define model monitoring and maintenance.

Model monitoring is the process of continuously tracking a model's behavior and performance in a production setting. This step comprises assessing performance indicators such as model correctness, relevance to business metrics, latency, and throughput. This step also includes monitoring any changes in the input data pattern, known as *data shift*, as well as the link between the input and output variables, known as *model shift* or *concept shift*. Alerts are put up to warn teams when performance drops or abnormalities are discovered, allowing them to respond promptly before the problem escalates.

Model maintenance refers to the ongoing tasks required to keep the model updated and effective. This includes retraining the model with new data, updating features or algorithms, and ensuring that the model stays aligned with business goals and environmental changes. Regular maintenance ensures that models evolve with the data, preventing obsolescence and maintaining accuracy.

Together, model monitoring and maintenance form a lifecycle that helps prevent the pitfalls of post-deployment performance degradation, like the ML-powered search algorithm in our story, and ensures that the model continues to add value in a constantly changing world.

Key Challenges

So far we have romanticized the model monitoring step by presenting all the facts to make product leaders fall in love with. In this section, we will present few challenges that you might face while implementing model monitoring. As leaders, learning about these challenges will help us plan better.

1. As the number of models and data inputs increases, the computational resources required for model monitoring increases manifold. It can impact the IT budget severely.

2. Modern machine learning models, particularly deep learning models, can be very complex, making it harder to understand why a model may be failing or why performance has dropped. Without proper interpretability, it's difficult for data scientists and engineers to pinpoint the cause of issues, which can delay corrective actions. Model monitoring done in isolation might not provide optimum results.

3. The need for real-time monitoring can be a technical challenge, especially when the system processes vast amounts of data continuously.

Achieving the right balance between timely alerts, resource consumption, and system performance is the key to successful monitoring.

Essentials of Model Monitoring

Before diving into the nitty-gritty details of monitoring, it's crucial to define the core objectives that you want to achieve. These objectives will guide the selection of monitoring techniques, tools, and metrics. One should generally think about two questions while deciding on what to monitor:

1. What is the purpose of creating this ML model?
2. What factors will influence the performance of the ML model in reaching its goal?

What is the purpose of creating this ML model?
To answer this question, you must revisit the planning phase of the ML product in which the key metric that this product will impact would have been identified. This metric should define the business and user objectives this product should meet. A point to consider is that your model performance might partially affect the business metric that you would have decided. To add more objectivity to system, create a relationship between the business metric and the model metric. The following set of questions can help you define your model metric:

1. What is the business goal that the ML model aims to achieve?
2. What is the specific business metric that will gauge the performance of the objective?
3. How this business metric can be influenced by the model's output?

CHAPTER 7   FROM PROTOTYPE TO PRODUCTION: SCALING ML AT SPEED AND SCALE

4. Which ML model metric will truly evaluate the impact of model on business metric?

5. How will changes in the model metrics impact the business metric, i.e., how movement in your model metric will impact your business metric? For example, if your model is built for detecting customers who will churn, then higher precision of your model might result in reduced customer churn rate.

6. What are the acceptable thresholds for model metrics that indicate the model is on track to positively impact the business metric?

7. What are the non-model performance factors that might affect the business metric, for example, latency and availability?

While picking the metric, evaluate the metric on the following points:

- A metric should be comparable across different models catering to the same business problem.
- a metric should be collected in real time.
- A metric should deliver actionable insights when monitored in production.

What factors will influence the performance of the ML model in reaching its goal?

Once a model metric and its relationship with a business metric is identified and monitored, it's time to identify factors that might affect the performance of model. These factors can be classified into two categories:

338

CHAPTER 7   FROM PROTOTYPE TO PRODUCTION: SCALING ML AT SPEED AND SCALE

1. Internal factors: This includes monitoring the following:

   - Model performance
   - Inputs (data)
   - Outputs (predictions)

2. External factors: This includes monitoring resources on which your model runs in production.

   - Resources such as pipeline health, system performance metrics (I/O, disk utilization, memory and CPU usage, traffic)

Internal Factors Monitoring

Monitoring internal factors involves monitoring the performance of model, quality of input data, and prediction results. Generally, the ML team is responsible for monitoring these factors.

Input (Data)

Input monitoring is crucial in production because the model reacts to the data it receives. If the data isn't what the model expects, it will most likely influence performance. Monitoring input data is the first step in stepping up a robust monitoring system. The following factors are monitored in input:

1. Data quality and outliers
2. Data drift

Data

Input-level functional monitoring is crucial in production because your model reacts to the inputs it receives. If the inputs aren't what your model expects, it will most likely influence performance. Monitoring and measuring input-level challenges is the first step to troubleshooting functional performance issues and solving them before serious damage is done.

339

CHAPTER 7   FROM PROTOTYPE TO PRODUCTION: SCALING ML AT SPEED AND SCALE

The following are the three cases you may want to monitor at the input level:

1. **Data quality issues**

    Data quality issues usually occur because of changes in the data pipeline, such as incorrect input from the data generation source or errors in the transformation code when exceptions arise. To understand more about data quality, refer to Chapter 5.

    The goal of monitoring is to flag any data quality issues before the model uses the data to generate predictions. An ideal approach to setting up data quality monitoring involves identifying the appropriate data quality metrics (refer to Chapter 5) and setting up alerts if the data quality deteriorates below a defined threshold. Additionally, a report with row-level details should be created. This setup ensures that alert messages remain clear and actionable.

    Typically, potential factors that could affect data quality are identified for each data variable or feature. These factors are then listed and prioritized. Depending on the budget and application, all or just the prioritized factors are monitored.

    The design pattern "circuit breaker" is implemented to immediately stop the model from generating predictions in case the quality of the data falls below a certain threshold.

2. **Data drift**

   Data drift occurs when a significant change in data distribution is observed between the training and production data. The model learns by training itself on training data, and then it uses this learning to predict outcome on production data. Any significant change in distribution between the two datasets will adversely affect model performance.

   While it's possible to monitor this drift at the level of the entire dataset, it is often advisable to monitor it at the feature level.

   Data drift also referred to as feature drift can occur due data quality issues or changes in real-world situations like seasonality trends or changes in customer preferences.

   Monitoring input (data) drift closely can give you a heads-up on model drift/model performance before it becomes problematic.

Data Drift Detection Techniques

Data drift, also referred as *feature drift*, can be detected by observing changes in the statistical properties of each feature value over time. Some of the intensively used methods are as follows:

- Population Stability Index (PSI) is one of the metrics that can be used to measure data/feature shift.

- Basic statistical metrics like mean, standard deviation, and min and max value comparison between training and production data can help identify drift.

- Another simpler way to monitor data drift is to continuously monitor model accuracy/precision/recall. If a significant drop in any metric is observed, it symbolizes either model drift or data drift.

- Various statistical methods can help detect data drift. The following tests are commonly used:

    - Kolmogorov-Smirnov Test: Compares the distributions of two datasets (e.g., training data versus current data). There is a significant difference between the distributions suggesting data drift. Use this test or continuous data.

    - Chi-Square Test: Used to compare categorical features between datasets to detect drift.

    - Kullback-Leibler Divergence (KL Divergence): Measures the difference between two probability distributions (e.g., the training and current data distributions).

- Various Python Libraries like Alibi Detect, Evidently AI, Scikit, and Multiflow can also be used for data drift detection.

- You can also use machine learning algorithms like clustering-based methods (like K-means, DBSCAN), isolation forest, one-class SVM, autoencoders, and RNN for detecting data drift.

- To identify data drift in text data, you can use the following approaches:

    - Word Distribution Change: The frequency of occurrence of words/text is measured both in training and production data; if any significant change in frequency of occurrence is observed, it symbolizes data drift.

CHAPTER 7   FROM PROTOTYPE TO PRODUCTION: SCALING ML AT SPEED AND SCALE

- Topic Modeling Drift: This is a two-step process. In step 1, the topics of training data and production data are identified using topic modeling techniques like Latent Dirichlet Allocation (LDA). In step 2, the frequency of these topics is calculated and compared. If significant differences in topic frequency are observed, then data drift exists.

- Semantic Drift with Word/Document Embeddings: In this approach, the relationship between various words is tracked using word embedding techniques like Word2Vec, Doc2Vec, and Bert. Any significant change in word relationships in the production and training dataset would signify data drift.

- N-gram Analysis: This is a simple approach in which various unique permutations in which a group of word occurs in a training dataset and production dataset is calculated and compared.

• To identify data drift in image data, use the following approaches:

- Histogram Analysis of Color or Intensity: The color/intensity distribution of an image is analyzed over a period of time. If any major significant change is observed, it might be a suggestive of data drift.

- Image Embedding Changes: Image embedding is a process of transforming images into numerical vectors that describe the image's content. This is done to ensure that the computer understands the image. Image embeddings are generated using pre-trained neural networks like VGG, ResNet, or Inception. By comparing embeddings of old and new images, you can detect shifts in the visual representation of the data.

Model

This is the stage where machine learning algorithms train themselves on input data and learn relationships between various variables or features. If this relationship changes over the period, then it will also affect the model performance adversely. The algorithm learns patterns between the input variables and the target variable (variables whose value we want to predict), and if this changes, then the algorithm's learning becomes obsolete. This phenomenon is called *concept drift* or *model drift*.

For example, assume you want to build a product recommendation feature for your e-commerce platform. The model recommends products to users based on data like products interacted with, product purchased, categories of products purchased and interacted, cost of products, location, demographics, etc.

The model was trained on pre-COVID data and worked fine until recent times, but now its performance has degraded. On investigation you found that initially when a model was launched your buyers were generally young tech-savvy people from big cities. In recent times your user set has diversified, and now more people from smaller cities and older people have started buying. Another pattern that you observed has changed is a change in product categories and cost of products.

This change in users and product selection is affected by post-COVID behavioral changes in your users. This change in user behavior is also affecting the new data on which your model is generating recommendations. During the training period, your model was not exposed to these newer patterns and trends, and when now it is exposed to these newer patterns, its performance is getting impacted.

The following factors contribute to model drift:

- Over time, customer preferences or behavior may change, which leads to new patterns in the data.

- If a model is exposed to previously unknown features in production, this can affect model performance in production. Introduction of newer features directly in production can affect underlying data patterns that the model learned during training.

- External factors such as socio-economic conditions, regulations, or technological advancements can introduce changes in the data and affect model performance.

- Technical factors like the way data is collected, labeled, or preprocessed can introduce shifts in the data used to train the model, which can affect data distribution and patterns leading to drift in the relationship between the input and target variables.

Techniques for Detecting Model Drift:

- Performance-Based Drift Detection: The simplest way to detect concept drift is to monitor model performance (accuracy, precision, recall, F1 score) on a rolling window basis. If a significant drop in accuracy is observed, then it might indicate a concept drift. To define a "significant drop" in accuracy, you can perform a statistical test like t-test and z-test.

- Model Comparison Techniques (Concept Drift Detection): Compare the performance of the model in production to a shadow model trained on the most recent data. If the shadow model consistently outperforms the existing model, it indicates that drift has occurred.

- Model residual analysis: For models built for predicting continuous variables, the residual between the actual value and the predicted value can be analyzed. If residuals start to increase, then it could signal concept drift.

- Drift Detection Method (DDM): Monitors the error rate of the model as it processes the input data. It tracks model performance and triggers an alert if the error rate crosses the threshold, signaling the concept drift. DDM tracks the model error rate over the time. It then calculates the mean and standard deviation of the error rate. If the error rate increases the expected standard deviation, it signals a likely concept drift.

- Adwin (also called *adaptive windowing*) analyzes data in a dynamically adjusted sliding window and then uses statistical methods like Hoeffding's bound to check if the distribution of data in the window has changed or not. If a significant change is detected, it signals concept drift. The sliding window that is used for data comparison is called dynamic because if this approach detects change in distribution within the window, then it automatically shrinks the window to focus more on new data.

- Visualization-Based Drift Detection: Visualize the distribution of key features over a period of time using visualization techniques like scatter plots, histograms, box plots, and kernel density estimates, and compares the current distribution to the baseline. Any visual change in distribution can signal concept drift. Sophisticated visual techniques involve performing dimensionality reduction techniques like principal component analysis or t-SNE (t-Distributed Stochastic Neighbor Embedding) and visualizing components or clusters. Any significant changes in the clustering/component patterns can indicate a shift in the data distribution, suggesting drift.

Logging for Debugging and Compliance

Logging plays a crucial role in ensuring the transparency, accountability, and reproducibility of machine learning models. It is a critical practice in MLOps as it helps track model behavior and performance and maintains transparency for auditing processes. Well-created logs also play a very important role in debugging issues.

While you can create common set of logs for both purposes, it is important to understand what all information should be logged to make logs worthy for both purposes.

Logging for Debugging

Logs for debugging should capture information about model behavior, errors, and system performance.

How Logging Helps in Debugging

Logs help in debugging model performance issues by providing the following information:

- Helps in tracking model training process by capturing key details like training variables, data preprocessing steps, hyperparameters, learning rate, and batch size.

- Logs capture performance metrics that help in identifying if the model is underperforming or if the model performance has degraded over a period of time.

- Logs record errors or exceptions occurred. This can capture issues like invalid input, system issues like timeout, memory error, etc.

- Logs can store the version of the code, model weights, and data used, allowing you to reproduce the exact model setup at any given point.

How to Create Debugging Logs

Logs that are created for debugging focuses more on capturing details about errors, warnings, and the internal workings of machine learning models. Under this category, there are three kinds of logs:

1. Log model training information
2. Log model inference and predictions
3. Log model errors and exceptions

1. Log Model Training Information: To generate this log type, capture the following information:
    - Capture the hyperparameter used during the training phase
    - Capture the model version being trained; it helps in reproducing the issue later.
    - Store training performance metrics
    - Capture sampling or data augmentation techniques used during training

2. Log Model Inference and Predictions:
    - For each prediction made, log the input data and the predicted values
    - If your model provides confidence scores for predictions store them

3. Log Model Errors and Exceptions:

    - Log any errors, exceptions, or system failures that occur during training phase or production phase.

    - Stack traces for debugging to allow developers to pinpoint the exact cause of the issue.

    - Always include timestamps to provide better traceability.

Logging for Compliance

In many industries and countries, it is mandated by law (e.g., GDPR, HIPAA) to ensure that machine learning models used for any critical task must comply with regularity standards, and maintaining logs in these cases helps a lot.

How Logging Helps in Compliance

- Logs capture detailed records of data usage, model updates, and predictions made, providing a clear audit trail that can be reviewed.

- Logs track the entire lifecycle of data, helping to ensure that data is used appropriately.

- Since logs capture decision outcomes and input variables used, it provides all information required for conducting a postmortem analysis of the model for fairness and bias identification purposes.

This is how to create logs for compliance:

- Log data usage and access.

- Ensure that logs do not store personally identifiable information (PII) or sensitive data unless it is securely anonymized or encrypted.

- Log the features used in model predictions along with their importance to ensure model transparency.

- Log the rationale for predictions (e.g., confidence scores, thresholds crossed, or rules applied).

- Every time a model is retrained or modified, ensure that detailed logs capture the changes made (such as updated features, hyperparameters, or algorithms).

- Track which version of the model is used in production.

Best Practices for Creating Logs:

- Use structured log formats like JSON and XML for generating logs to ensure logs are easy to consume.

- Create separate log categories for storing logs with different information; for example, create logs for storing general information in a separate log file and warning, errors, critical alerts, and debug information in separate log files. This makes logs more structured and easier to access.

- Include relevant contextual information in logs, such as timestamp, model ID, dataset version, and other identifiers that will help in debugging.

- Capture the flow of data through various stages, which will allow you to identify where things go wrong.

- Create a log retention policy to manage log file sizes and ensure that logs are archived or deleted.

- Use tools like Elasticsearch, Logstash, Kibana (ELK), Splunk, or AWS CloudWatch to centralize logs. These tools also provide advanced querying capabilities.

CHAPTER 7   FROM PROTOTYPE TO PRODUCTION: SCALING ML AT SPEED AND SCALE

- Use encryption and access control to protect sensitive information stored in logs.

- Avoid storing sensitive information in logs unless it is unavoidable. Use encryption and data anonymization to protect sensitive information.

Model Retirement and Replacement

Model retirement and replacement is a stage in MLOps in which the existing model is replaced with a new model because of suboptimal performance of existing mode. Let's understand this process with the help of an example. Imagine your product recommendation model that you developed for e-commerce platform has started showing signs of degraded performance. Say customer preferences and economic conditions keep on evolving, and it affects the way customer behaves on your platform. The model was trained on older data that represented older reality; this difference in data patterns will result in bad product recommendations. In this situation, you would have to make a decision to develop a new model that is trained on new world reality and replace the older model with the newer model. This process is referred as *model Retirement and replacement.*

As product leaders, making the decision of replacement and retiring the existing model is very critical because the performance of the ML model will directly affect the overall product performance or customer experience. As leaders, it's our responsibility to ensure that the product remains innovative and effective and capable of catering to the business needs.

Strategies for Model Retirement:

Model retirement/replacement is a delicate process as it requires a fine blend of stakeholder management and technical prowess. This step aims at replacing a new machine learning model with the existing one in a graceful manner. The following two strategies are mostly used:

1. **Graceful Retirement:** Tries to make model retirement process as smooth as possible by gradually decommissioning the older model and transitioning the input request workload incrementally to the new model. The following are the major steps:

    - Establish a clear project plan for model migration. This step should call out all dependencies on other teams, a timeline for each action including complete deployment of the new model and potential impacts, and a fallback option in case something goes wrong.

    - Design a new model deployment strategy by following the guidelines mentioned earlier in this chapter to compare the parallel testing of the new model against the original model.

    - If the A/B testing method is used to test the new model, identify the smallest user set that will be first exposed to new model predictions.

    - Run both models in parallel with gradually increasing the traffic load toward the new model, if initial results on parallel testing are promising. Design a deployment strategy that allows you to roll back to the old model if needed.

    - After each phase of the migration, perform detailed health checks to validate the accuracy, performance, and robustness of the new model. Implement automated monitoring to track real-time performance discrepancies.

- Update all internal documentations to reflect the latest changes in model, including the percent of traffic it is serving and any adverse impact if observed. This ensures teams stay aligned on the model's status.

- Once the new model has proved its worth and traffic is fully migrated, officially decommission the old model. Ensure all references to the old models are replaced in larger system in which the model was embed are replaced.

- Send official communication to all the stakeholders once the migration is complete.

2. **Parallel Running: Running Old and New Models Simultaneously**

    - This approach allows you to run both old and new models simultaneously; it typically works well with a *shadow deployment strategy*.

    - While the documentation and communication steps are common for both approaches, this approach differs in the way the new model is deployed.

    - In this approach, a new model is deployed in a shadow and operates in the background without affecting the end users. This model runs prediction on end users in the background.

    - For every end-user input now, you will have two predictions: make a comparison to track performance of both models and make decisions accordingly.

- If the new model outshines the existing model, then start transitioning the traffic to the new model gradually, until the entire traffic is migrated to new system. Keep strong system monitoring to identity any issue caused by the new model in production.

MLOps Governance (G)

MLOps governance is a structured approach for managing the machine learning system in a way that aligns with organizational, technical, ethical, and legal standards. As you know, MLOps is simply an extension of DevOps (processes to put and maintain conventional software applications in production). MLOps primarily focuses on managing aspects such as version control, model deployment, security, compliance, traceability, and monitoring throughout the ML pipeline.

- **Version Control in MLOps:** MLOps governance principles state that to build standardized machine learning workflow, all elements of workflow are versioned. This can include the following:
  - Create versions of data used for training and testing the model. Each version of data should have something different than the other version; it can be changes in data or time spans in which the data was captured. The versioning of data plays an important role in reproducing the model during the training phase. Also, the exact training set also remains traceable (in case any ethical issues like biasness has to be investigated, these versions play an important role). Tools like data version control can be help.

- Source code: MLOps governance requires the proper versioning of source code be used to develop a machine learning model. *Versions* refers to storing source code in a separate file every time (the same name followed by a number) there is minor change in code after the code reached a significant milestone. Tools like Git can be helpful here. The versioning of code allows easy rollback to the last stable version of the model in case the latest version is not stable.

- Model versioning: MLOps governance also talks about tracking various models and their performance across different experiments in production. Model versions should include model performance metrics and hyperparameters. Tools like MLflow or Weights & Biases are popular for managing model versions and experiments.

2. Security and Compliance in MLOps

MLOps governance also includes ensuring that machine learning systems are secure and comply with relevant regulations. This involves:

- Securing the data pipeline: Data is the most sensitive and critical part of any machine learning product. Data should be secured by implementing encryption protocols at rest and in transit.

- Access control and identity management: A robust access control mechanism is required to ensure that only authorized individuals can interact with the data and the model. Keep track of all actions performed by individuals or the system on the data or model.

CHAPTER 7 FROM PROTOTYPE TO PRODUCTION: SCALING ML AT SPEED AND SCALE

Model Testing (T)

In addition to performing validation testing (a type of testing in which a model trained on the training data is used to make predictions on a new dataset and the accuracy of model for the new dataset is calculated), there are other several other kinds of testing that should be performed to improve the chances of the model to perform well both in terms of accuracy and system reliability and performance. These are unit testing, integration testing, and A/B testing. These testing types are common techniques on the software engineering side, but on the AI side their adoption is still low. We will be covering these two types in the section "Infrastructure for Model Deployment."

This section will primarily focus on A/B testing as this is the most critical test type that a model must pass through to prove its worth in the real world.

Imagine you work as product leader for an e-commerce organization and you have been mandated to increase revenue, which your organization generates by showing ads. On preliminary analysis, you discovered that the main issue is that you have enough ads for most of the searched keywords by exact match (i.e., keywords in the search query match with the product title from the ads). Because of this situation, irrelevant ads are displayed with search results, impacting user experience and reducing both conversion and revenue through ads. Another key aspect of user behavior you noticed was users wrote poorly written queries with spelling mistakes to resemble synonyms. The current system can show relevant ads for some queries, and the rest of the opportunities are lost.

You have been working intensively with the AI team to develop an AI-based natural language processing model that can find semantically matching ads to the user queries that can show ads even for typos and synonyms and give partial matching but relevant search results. This seems to be promising in a test environment, but you have no evidence

CHAPTER 7   FROM PROTOTYPE TO PRODUCTION: SCALING ML AT SPEED AND SCALE

to prove this new solution would work in production, and business stakeholders are insisting for hard evidence before pulling the plug from the existing model. To get this confirmation, you plan to perform an A/B test.

To ensure A/B test can be performed successfully, an A/B deployment strategy should be selected. This strategy will ensure that the production environment will have the capability to divide the live traffic into two user groups, one for each model. The new model whose efficacy has to be proven is called the *test set*, and the existing model against whom new model's efficacy is being tested is called the *control group*. To evaluate and compare the performance of both models, a relevant metric like click-through rate, user engagement time, add to cart, or conversion rate for the both user groups will be compared.

At the end of the A/B testing period, the data is collected for the control and test groups, and careful statistical analysis is performed on the data. This analysis will help us understand which model performed well in the real world. In this example, let's try to have a happy ending by making our new NLP model win, by claiming that it maintained the click-through rates, user engagement time, and add-to-cart rates as the existing system while showing more ads. This means that more ads shown by the new model are relevant as these are not affecting the business metrics. Thus, it is decided to roll out the new model for all the users.

A/B testing is the final check to see if the new model is better than the current model or algorithm that is already in production. The end goal of the A/B test is to decide whether to push the new model (which is being tested) to production. Thus, it is important that the new model or solution works flawlessly during the A/B testing without any errors or issues for the results of A/B testing to be conclusive. Hence, all the unit tests and integration tests are important to pass before going for A/B testing.

Hence, careful planning and implementation of the MLOps process is very important for running an A/B test successfully and to get any conclusive results. This section will discuss these aspects in detail.

CHAPTER 7 FROM PROTOTYPE TO PRODUCTION: SCALING ML AT SPEED AND SCALE

To summarize, A/B testing is a useful tool that allows data-driven decisions to estimate the impact of AI/ML solutions in real life. By comparing different models or solutions in a controlled environment, companies can ensure that these systems deliver the best possible results for their users and their business.

How to Design an A/B Test Experiment

Before we get into the nitty-gritty details of designing A/B testing, some key terms and concepts should be defined.

- Control group: The users in this group are directed at the existing or baseline solution.

- Variation group: The users in this group are exposed to the solution based on the new model.

- Target metric: The target metric is expected to be improved with the new solution (e.g., click-through rate or add-to-cart or user churn, etc.).

- Statistical significance test: This is to estimate the likelihood that the impact of the new solution (either positive or negative) is not due to random chance and is actually due to the change brought by the new solution.

The A/B Testing Workflow

The A/B test workflow comprises various steps that should be followed in the strict order in which they are mentioned. Each step builds on the previous one, so adherence to the order in which these are executed will define the success of the experiment.

CHAPTER 7   FROM PROTOTYPE TO PRODUCTION: SCALING ML AT SPEED AND SCALE

Defining Experiment Hypothesis:

The A/B testing procedure starts in the planning phase of the AI project in which you identify the core metric that will be affected by this project. This information helps formulate a hypotheses that you will be testing. So to start with, the first step in A/B test workflow is formulating a hypothesis. The following is the sample hypothesis.

Scenario: A product manager working for an e-commerce company is tasked to improve product discovery. Based on this research, he identifies the problem with the existing search algorithm and decides to develop an AI-powered search algo that would improve search result relevance.

Hypothesis formulation: Before formalizing the hypothesis for this test, let us first understand the meaning of the word *hypothesis*. A hypothesis in simpler terms is an educated guess that you make if you change the status quo. In this case, we are changing the algorithm, which helps users discover results on an e-commerce platform. We also expect that this algorithm will be better in recommending relevant results to users, which would make users make more purchases.

So, the hypothesis in simple terms in this case will be: if the new AI-powered search algorithm improves the relevance of search results, then the user group exposed to this algorithm will have a higher purchase rate in comparison to users who are exposed to existing search results. With all other factors, the search results stay the same.

In more formal terms, generally two types of hypothesis are derived from the main hypothesis:

- Null hypothesis
- Alternate hypothesis

The null hypothesis in simpler terms means that the new idea that you are trying to implement will not make any difference; i.e., things will be the same as they are now.

CHAPTER 7   FROM PROTOTYPE TO PRODUCTION: SCALING ML AT SPEED AND SCALE

The alternative hypothesis says something will change with the implementation of new idea; either it will improve or get worsen.

A formal version of this hypothesis for our example can be stated as follows:

Null hypothesis (H_0): The user purchase behavior will see no significant difference between users shown search results from new and current (baseline) results.

Alternative hypothesis: Users shown search results from the AI model will have a significantly higher purchase behavior than those shown the current results.

Selecting Appropriate Metrics and KPIs

We have decided to develop a solution with an objective of improving the status quo, and in the last step we have also identified which area of business will be impacted by this new solution. To make the comparison more objective, it is always important to define how the change in the affected area will be measured. For that we identify a metric that can truly represent the business area/journey step that will be impacted. The target metrics of A/B testing should align with the goals of a business that you are striving to achieve. Each metric should be directly reflected as the success criteria of the test and should offer actionable insights. In our example, since our algorithm will impact the search relevance that influences user's purchase behavior. The target metric for our experiment will be user conversion rate. Now let's rewrite the hypothesis with the target metric in it.

Null Hypothesis (H_0): The user conversion rate will see no significant difference between users shown search results from the new and current(baseline) results.

Alternative Hypothesis: Users shown search results from the AI model will have a significantly higher conversion rate than those shown the current results.

CHAPTER 7   FROM PROTOTYPE TO PRODUCTION: SCALING ML AT SPEED AND SCALE

Defining Control and Variation Groups

The two groups must be formed such that there is minimum bias among the two; i.e., both variants must be as identical as possible in every way. This will make sure that any difference in the target metric for the two groups can be credited to the new AI model or solution and not because of the bias between the two groups.

Thus, users should be randomly assigned to control and variation groups. Random assignment ensures even distribution of external factors (e.g., demographics, usage patterns) across both groups.

However, even after random selection, some significant bias may exist toward some key attributes like age, location, etc. To mitigate this, stratified random sampling can be used, where groups are divided based on the key attributes that have significant impact on the target business metric. Once all the users are assigned to a group, a post-assignment analysis should be able to tell if both groups have no significant bias across these attributes or not. For example, in the case of an e-commerce application, if both the groups show similar demographic distribution and buying pattern, then there is no significant bias.

Sample Size Determination and Statistical Significance

Whenever a scientific study is performed, the minimum number of observations that will be collected before calling that study outcome as statistically significant is determined by using some statistical tools. For example, we are testing a new ranking algorithm, and if our current conversion rate is approximately 2% for over a million users, then we should use power analysis (statistical tool) to determine the minimum number of users in each group that are needed to make results of this experiment statistically significant. Please note that the sample size is not the only variable that contributes to statistical significance; the difference between the conversion rate of both groups and confidence levels are other factors. Let's now take a high-level view on power analysis.

CHAPTER 7   FROM PROTOTYPE TO PRODUCTION: SCALING ML AT SPEED AND SCALE

The power analysis tool is a statistical technique used to estimate the sample size required for extracting reliable and statistically significant results from the A/B experiment. It considers factors like the power value (typically 80%), the desired level of change (for example 0.5%) with respect to the baseline metric, the baseline metric, and the significance level (commonly 95%).

Let's understand power analysis with the help of an example. Continuing with our AI-powered search algorithm example, if the baseline conversion rate is 2% (current conversion rate), we claim that new search algorithm will improve the conversion rate by 0.5%, and we have 95% confidence and 80% power; 80% power denotes the chances of finding a significant result when there is an actual result. In our example, the minimum number of users in both groups should be at least 10,000E.

Efforts should be made to stop the experiment slightly above the minimum threshold mark because if the sample size is too small, the experiment might not be able to reliably detect a small change, which is expected from iterative changes. On the other hand, if the sample size is too large, if the new algorithm that you are testing has better result than current one, then you are losing on opportunity cost, and if it is bad, a significant chunk of your users are exposed to bad results, which may affect your KPI negatively.

A/A Testing: Final Test Before Running Actual A/B Test

The aim of A/A testing is to test the hypothesis that has been used to assign users to the two groups. It starts with splitting users into two groups as per the strategy to be verified. Users in both groups interact with the same version of the system. Since this data is anyways available for the existing system, the A/A test should be performed with the new system (to be tested using A/B testing) to ensure the testing process itself does not introduce biases. A/A testing provides very useful data for removing biases because of other factors like differences in user traffic, geography, demography, etc., which otherwise would be difficult to detect. The business metrics of the two groups in A/A testing, ideally, should be statistically similar, and if that is not observed, then either the strategy used to divide users into two groups is flawed or there is something wrong with the way the business metrics are being tracked. It also acts as a rehearsal for the actual A/B test and reveals what can be expected during the real A/B test.

How to Analyze and Decide on A/B Test Results

Continuing with our e-commerce example, we have executed the experiment, collected, and analyzed the data. To conclude the experiment and decide on the outcome, analyze the data using statistical tests.

- If the data shows a big enough difference (statistically), we reject the null hypothesis and accept the alternate.

- If the data doesn't show a big enough difference, we fail to reject the null hypothesis—meaning we stick with "no change."

In simple terms, if the AI really helps users buy more, the data will show that, so we reject the null hypothesis and say the AI makes a difference. If the AI doesn't help or the effect is too small to be sure, we stick with the null hypothesis and say the AI doesn't make a noticeable difference.

- Reject H_0 The AI search feature likely improves user conversion rates.

- Fail to Reject H_0 There's insufficient evidence to conclude the AI search feature improves conversion rates.

The target metric, which was conversion rate in our example, has shown significant improvement, but the add-to-cart rate has shown a significant decrease. This scenario would mean that while the customers are still making a purchase, the number of products per transaction has decreased. This points to some other problem. It is a wise decision that while you are defining the target metric for the experiment, always choose a counter metric that should not get affected adversely by any increment in your target metrics.

Statistical analysis techniques can be used to compare performance metrics between the control and variation groups.

- Use a T-test if the experiment's target metric is a continuous variable and you are comparing average values like average purchase amount.

- Use a Proportion test/Z-test if your target metric is a proportion like a conversion rate.

- Use a Chi-square test if your target metric is a frequency or count like number of clicks/number of products sold.

Best Practices for A/B Testing

1. Align Experiments with Business Goals: Ensure every test ties directly to a measurable business objective.

2. Automate A/B Testing in CI/CD Pipelines: Integrate A/B testing into deployment pipelines using tools like Jenkins, GitLab CI, or ArgoCD.

3. Ensure Interpretability for Stakeholders: The explanations of test results should be clear and focus on the actionable insights. Use relatable comparisons and summaries, instead of presenting raw statistical data. For example, highlight statements like "The variation model improved user engagement by 12%, which is equivalent to 10,000

more users every day." Appropriate visualizations like bar charts or trend lines can be utilized to communicate the change in performance metrics. Moreover, these results can be contextualized by linking them to the business outcomes.

4. Scale A/B Testing for Large Deployments: Use orchestration tools like Airflow, Kubernetes, or MLFlow to manage tests across teams and regions. These will provide the following advantages:

   - Enable independent regional tests while maintaining global performance standards.

   - Automate resource allocation and scale dynamically with traffic.

   - Support cross-team collaboration through centralized monitoring.

5. Ensure Validity and Reliability: External factors like seasonality or drift in user behavior during the test period must be controlled. For example, if an A/B test is conducted during a holiday or festival season, the shopping behavior of the users would be significantly different from the normal days. This could lead to incorrect results like inflated conversion rates. To normalize the effect of holidays, the A/B test should be run over a longer period for both control and variation groups. Alternatively, historical data can be used to account for the deviations caused by seasonal trends, to check if the observed outcomes truly reflect the tested model's impact.

6. Develop A/B test framework in a way that the experiment can be stopped early and failures can be handled gracefully.

Advanced A/B Testing Strategies

As businesses' dependency on data-driven decisions has increased, A/B testing has become very critical in helping business arrive at the right decision. Up until now we have talked about how to design an A/B test experiment and the basic statical tests that can be put into use. To further improve the efficacy of A/B testing, several advanced techniques can also be used. These techniques include methods like multivariate testing, Bayesian analysis, multi-armed bandit algorithms, and personalized testing, each offering unique advantages for scaling and refining experiments. By leveraging these advanced strategies, companies can accelerate their optimization efforts, make more precise data-driven decisions, and continuously improve their products or services to meet evolving customer needs.

1. **Multi-Armed Bandit Testing (Dynamic Experimentation)**

    Multi-armed bandit testing is an advanced A/B testing strategy that allocates traffic dynamically based on the performance of the new solutions being tested. This approach optimizes for the best-performing solution in real time. For example, consider a courier company testing three promotional strategies: discount coupons, free shipping, and loyalty points. In the beginning, the traffic is evenly distributed among the three strategies. With time, data is collected for various user metrics like Click-Through Rates (CTR) and Add-to-Cart (ATC), and the system re-distributes the traffic such that more users are diverted toward the strategy with the best performance metrics. With the help of this approach, more users are exposed to the better-performing solution without waiting for a fixed period where users keep using the poorer solution for a fixed time and hence causing potential losses or drops in metrics or user dissatisfaction.

2. **Bayesian A/B Testing**

The Bayesian method estimates the probability distribution of the target metric for the data collected from the two user groups. Based on this, it calculates the probability that the target metric for variation B is better than that obtained from variation A. It also gives us the confidence intervals for the range of differences in the target metric for the two groups. A typical outcome of the Bayesian approach would be that there is a 90% probability that variant B gives higher customer retention than variant B. Also, it can be said with a confidence level of 95% that the customer retention would improve by 1% to 5% by switching to variant B.

Bayesian methods are preferred in situations like where prior knowledge is available about the users, like their purchase behaviors, where the decisions are made based on probabilities of one variant being better than the other and where the stakeholders are not familiar with the statistical jargons and they expect the results to be explained in simple terms. Use Bayesian A/B testing when one or more the following listed conditions is true:

- Need real-time updates and continuous monitoring of experiments

- Have small sample sizes or limited data

- Want to incorporate prior knowledge or expert assumptions

- Are working with uncertain data and want to model the probability of different outcomes

- Run continuous or ongoing experiments

- Optimize for multiple objectives (e.g., conversion and user satisfaction)

- Deal with high variability or noisy data

3. **Contextual Bandit**

   A contextual bandit is a machine learning based A/B testing approach that can choose the best variation dynamically for the user based on some context information about the user behavior, location, device type, usual activity time, etc.

   Unlike the A/B testing approach, which tests fixed options, this algorithm learns continuously from a user response. It balances an exploration of new options while exploiting the known successful options. This algorithm is also referred as explore and exploit. This approach makes contextual bandits best suited for situations where decisions are personalized and the quality of recommendation has to be explored in a short span of time. It is best suited for use cases like personalized content recommendations, personalized e-commerce search algorithms, targeted advertisements, dynamic pricing, news feed rankings, and adaptive user interfaces.

Navigating Unforeseen Outcomes in A/B Testing

1. **Expectations Reality Check:** If the new AI model is showing growth but not as expected, maybe the expectations are not realistic and need to be checked. For example, expecting a growth of a 50% increase in conversions would be too ambitious with a single iteration of the AI model. A more realistic expectation would be 10% growth with every single iteration of the AI model and with 5 to 10 iterations, the 50% increase may be achieved.

2. **Revisit Target Metrics:** Make sure that the selected target metrics, if achieved, will lead to the desired business or management outcomes. For example, an increase in click-through rates might not translate to an increase in purchases or conversions.

3. **Analyze Test Design:** Look for design issues like a smaller sample size, bias in group distribution, or external influences like promotions or marketing campaigns running simultaneously or some other changes or experiments affecting the results of this A/B test.

4. **Validate Data Collection:** Check for bugs in data logging or tracking that might have skewed the results. For example, data may show clicks for only the top five products in the search results for all the users who were part of the A/B test. This shows

that clicks are tracked for only the first page and there is some bug that is preventing clicks from the subsequent pages to be logged in the data.

5. **Conduct Post-Test Surveys:** Gather qualitative feedback to understand why users may have reacted differently than anticipated. In some cases, users may be asked to answer a few multiple-choice questions with some anticipated reasons for why the user would not favor the changes. These answers will help in planning the next iterations toward improvement.

6. **Iterate and Retest:** Make the necessary changes based on learnings from the last A/B test and run a follow-up test. For example, if a variation led to a higher click-through rate but lower purchases, then refine the features to balance both of these metrics.

Interpreting Results in Business Context

Statistical tests are not enough, and one needs to go beyond statistical numbers to estimate the real business impact. For example, the increase in the cost of running a new model in production should be less than the overall returns from this model for it to be a sustainable solution in the long term. Imagine an AI team working for an online retailer implements a new churn prediction algorithm that is designed to boost customer retention by forecasting which customer would churn in the next week by looking at their past interactions with the online retailer website. While the algorithm achieves a 20% increase in customer retention, it also increases server costs by 50% as it needs more servers to scale due to a higher computational complexity and additional API calls. If this cost is higher than acquiring new customers or retaining the customers with discounts or offers to all the users, then it means lower profit margins and making the change unsustainable. So, A/B test results. alone should not be considered for deciding to adopt a change, but businesses should compare incremental revenue from the new model impact against the additional costs. A cost-benefit analysis should be conducted to ensure that the model's performance aligns with the overall business objectives. The target of the A/B must be set such that there is an overall positive impact on the business while accounting for the additional costs associated with the new solution.

Model Deployment Infrastructure

Building and maintaining a reliable model deployment infrastructure is essential for extracting maximum value out of AI models. The deployment infrastructure generally includes the tools and workflows that allow models to be served, monitored, and scaled in a production environment. A well-thought-out deployment setup will ensure that models can perform optimally and remain robust under changing conditions. Without the right infrastructure, even the most accurate model can fail to deliver value in production.

This section will talk about the following essential aspects of MLOps required for deploying AI/ML solutions, aimed at senior business and product leaders.

1) Continuous integration and continuous deployment (CI/CD) pipelines

2) Data pipelines and extraction transform load (ETL) operations

3) Data storage solutions (e.g., feature store, vector databases)

4) Solution deployment options (e.g., dedicated cloud server, serverless, on-device, on-premise)

5) Solution scaling (e.g., Amazon SageMaker, Google Vertex AI)

Continuous Integration and Continuous Deployment (CI/CD) in MLOps

Like all other software applications, AI solutions also demand continuous updates and enhancements across various components to keep their performance at optimum levels. Before any of these changes can be utilized by the production model, each of these changes has to go through a process of multiple types of testing including unit testing, integration testing, validation testing, and deployment. The objective of these steps is to ensure that the new change that is about to be deployed is effective and reliable and will not make already deployment unstable by any means. To streamline deployment of these changes whether in code, model training, or any other AI artifacts, CI/CD pipelines called continuous integration and continuous deployment pipelines are implemented. These pipelines automate various kinds of testing, validation, and deployment of AI application changes, enabling faster and reliable iterations.

The design of CI/CD pipelines includes the following aspects:

1) Tools required for integrating changes and deployment

2) Unit tests for the various modules and AI/ML artifacts

3) Integration tests for the interdependent modules

CHAPTER 7   FROM PROTOTYPE TO PRODUCTION: SCALING ML AT SPEED AND SCALE

Tools for Developing CI/CD Pipeline

For traditional software engineering projects, standard CI/CD tools like Jenkins, Gitlab CI/CD, and GitHub Actions enable the development of CI/CD pipelines that automate critical steps like source code integration with version control, executing unit testing test cases (with custom scripts), integration testing and deployment, and continuous monitoring and logging. If any intermediate step fails, then deployment is halted, and a detailed incident report is generated that can assist developers in identifying the issue before redeployment is tried. For continuous real-time monitoring and logging, tools like Prometheus and Grafana are used generally.

AI/ML solutions require some additional functionality beyond standard CI/CD capabilities:

- Managing AI/ML artifacts (e.g., models, features, vector databases)

- Artifact and data versioning, especially for large-scale data-centric projects like video (YouTube), audio (Spotify), or image search (Google Images)

- Experiment tracking for monitoring model performance across iterations

Some of these advanced features are partially supported by standard CI/CD tools but often require specialized platforms such as MLflow, Kubeflow, Azure AI, AWS SageMaker, or Databricks for full MLOps integration.

Unit Tests for CI/CD Pipelines

AI/ML solutions should be tested at various stages like data import, data preparation, feature engineering, loading models, model inference, validation, broadcasting, and logging of the model predictions. Some of

CHAPTER 7   FROM PROTOTYPE TO PRODUCTION: SCALING ML AT SPEED AND SCALE

the unit testing methodologies are purely software engineering based while others are statistical.

- Data Import: Data may be pulled from various sources like databases, APIs, local files on device, caching, etc. For example, a healthcare app may load a patient profile from a database or by reading from local files if the patient information is not allowed to be saved remotely. Now before the patient profile can be consumed, it needs to be checked for the following aspects: data format, integrity, missing data, and so on. Such checks ensure that no bad data goes to the model for predictions. The logic of these data validation checks is derived from the data filtering rules used during the model training phase. Once the data validation logic is identified, unit cases are created using both valid and invalid data samples to verify whether invalid inputs are rejected correctly and valid inputs are accepted. These tests are executed for all input variables.

- Feature Engineering: Features are extracted from the prepared data, and this is why it is important that the data prepared is correctly consumed by feature extraction programs. Feature engineering techniques convert data to features that can be consumed by the model. For classical ML like tree-based ML (e.g., random forest, xgboost, lightGBM, neural networks), feature engineering is applied outside the model, and for the advanced deep learning or large language models feature engineering is done as part of the model.

Unit tests are applied to check that the features extracted are part of the same distribution of features that are used during model training and testing.

- Model Loading and Inference: Unit tests are applied to test if the correct model is loaded as per the input features to avoid any conflicts due to model and data versioning mismatch. Sample test input features are used to check if the model outputs match with the expected outputs produced by models of different versions. There should be no errors like the model expects a different set of features than what is provided. Once the predictions by the model for the test features match with that obtained during the model train-test, the model and data versions are said to be matching.

- Model Validation: Runtime validations are also applied on the model predictions to filter any outliers predicted by the model. Such checks are also unit tested for every model update by using a specific set of features from the model train-test where the model is known to be giving outlier predictions.

- Broadcasting and Logging: Unit tests are applied to check whether the predictions are broadcasted and logged properly or not. To do this, test APIs are created, which will check the predictions received by the model and if it's in correct format (e.g., JSON with correct key-value pairs) and if the predicted value received is correct or not by comparing with standard expected outputs selected from the model training. The same is done for logging to ensure that model outputs are logged properly. Test APIs can read from the logs created during model deployment via CI/CD automation, and the values read from the logs are verified for the structure format and accuracy.

Integration Testing

Integration testing is a critical testing type that aims to ensure that different component of the AI system work in tandem in real-world environments. Unlike unit testing, which focuses on checking the effectiveness of each individual component in isolation, this type of testing validates the end-to-end workflow by simulating real-world scenarios and data flows. At times each individual component works fine in isolation, but when they work together, issues arise. Integration testing aims to identify these issues. The scope of integration in AI applications further extends to verify if retraining triggers, logging, and monitoring tools are functioning as intended. Tools like Postman, Pytest, and custom automation frameworks are often used to design these tests within CI/CD pipelines.

Tools for Data Storage for AI/ML Solutions

There are three most common types of data storage options available for the different AI/ML solutions for storing features for model training and inference, vector databases for AI-based search, and model repositories for storing, sharing, and model version control.

Feature Store: A feature store is a centralized repository where features are stored, managed, and served to different models. It ensures that the same features used in training are used to draw inferences in a production environment, thus ensuring consistency. Another key benefit of a feature store is that they help reduce the generation of duplicate features across various projects and speed up development and deployment. By doing so, the overall reliability of your ML models ecosystem improves.

CHAPTER 7   FROM PROTOTYPE TO PRODUCTION: SCALING ML AT SPEED AND SCALE

It also helps you track and revert changes made in any feature calculation by versioning. Since it stores precomputed features, which provide low-latency access in real time to features, it enables the overall responsiveness of the ML system. The following tools can be used for storing features:

- When batch processing and model training with historical data, you can use offline feature stores like AWS S3, GCS, Feast, and Google Vertex Store.

- Use cases where real-time predictions are required (recommendations, personalization, etc.) use online feature stores like Redis, DynamoDB, Cassandra, and Azure Cosmos DB.

Vector DB: A vector database is a special kind of database designed to store and search through a high-dimensional vector representation of data. These high-dimensional vector representations of data are also called *embeddings*. In simpler terms, embeddings are a numerical representation of actual data like text, image, video, and audio data. A vector database provides several advantages like real-time search with low latency, easy scalability, etc. They are commonly used in AI implementations on image, audio, video, or document retrieval. Embeddings are the fundamental steps on which LLMs are built.

Some popular vector database solutions are Milvus, FAISSS, and Annoy (all three are open source). Paid options include Pinecone, Azure AI search, and Google Vector Search.

Data Pipelines and ETL process: Reliable data pipelines and robust ETL (extract, transform, load) are the backbones of any successful AI implementation. Data pipelines comprise of infrastructure and tools that store, transform, and consume data from source to application. ETL defines the logic that is applied for extracting the data from the source site, transformation and loading the data into application. A reliable data pipeline ensures the availability of high-quality data for model training, testing, and AI activities.

Popular tools for building data pipelines are Apache Airflow, Kubeflow, Apache Spark, Tensorflow, AWS Glue, and Google Dataflow. These tools offer functionalities for task sequencing and automation, scheduling, and execution. These also help in resource utilization.

Model Repositories: Model repositories play the same role that a code repository in traditional software engineering plays; i.e., they help version and track changes in the model. However, unlike the source code of software engineering, models are often very large, up to a few gigabytes, or even more, in the case of large models like LLMs. This size aspect of models makes specialized storage solutions necessary. Hugging Face is a popular choice in research and open-source communities. On the enterprise front, AWS SageMaker, the Kubeflow model repository, and the Azure machine learning model registry are quite popular.

Orchestration Tools: AI/ML solutions are complex systems with multiple components working together. They often involve various interdependent steps that need to run in the correct sequence and that should be monitored. Orchestration tools help to manage this complexity by enabling the scheduling, executing, and tracking of these interdependent workflows in a reliable and scalable way. Popular tools are Apace Airflow, Spotify Luigi, Kubeflow pipelines, Azure machine learning pipelines, Amazon SageMaker pipelines, and Vertex AI pipelines (GCP).

Summary

This chapter highlighted the importance of why product leaders should care about a well-designed and planned deployment process for AI/ML applications. It provided a high-level understanding of each step in the MLOps lifecycle, enabling product leaders to make informed decisions.

CHAPTER 7   FROM PROTOTYPE TO PRODUCTION: SCALING ML AT SPEED AND SCALE

Key practices like types of deployment strategies, model monitoring strategies and design, and interpretation of A/B test experiments were also covered, which are foundational to successful AI products. The chapter also introduced commonly used MLOps tools that support scalable and secure model deployment in production environments.

CHAPTER 8

The Ethical Framework: Building Trust in AI Through Human Values

> "Trust is the glue of life. It's the most essential ingredient in effective communication. It's the foundational principle that holds all relationships."
>
> —Stephen Covey

Human relationships are based on trust, whether it's between two individuals or organizations or a new kind of relationship between humans and technology. It wouldn't be an overstatement to say the survival of the human species is intricately linked to the presence and maintenance of trust. Without trust, cooperation breaks down, relationships fracture, and societies begin to collapse. In a widely acclaimed *Harvard Business Review* paper called "Rethinking Trust," Roderick M. Kramer made a bold claim that "to trust is human." He further builds on this notion by explaining that humans are naturally inclined to trust. This trait has deep biological and psychological roots and is embedded in our evolutionary history

CHAPTER 8 THE ETHICAL FRAMEWORK: BUILDING TRUST IN AI THROUGH HUMAN VALUES

as a crucial survival mechanism. Over thousands of years humans have developed the ability to form bonds and develop dependencies on one another. This behavior has increased our chances of survival. From early tribal societies to modern communities, trust has enabled us to share resources, protect each other, and build last social structures. Trust is not just a social convenience but a strategy essential to human flourishing.

Humans have a large brain and relatively underdeveloped state at birth, which makes humans highly vulnerable and dependent on caregivers. This shortcoming has made the formation of social bonds essential for survival. From birth, infants are hardwired to connect with other; they instinctively focus on faces, respond to voice, smell, and mimic expressions. These early interactions show that humans are social beings from the start, designed to engage with others.

Another prominent social psychologist, Shelly Taylor, notes that the bond between parents and children along with other supportive social relationships plays a critical role in the development of the human brain. This ability to form social relationships has helped humans adapt over time. This makes trust not just a social bonding trait but an evolutionary advantage that has contributed significantly to our success as a species.

In current times, the world has become increasingly interconnected and reliant on digital solutions. It has influenced the human capacity for trust to include tools and technologies that have become part of daily life. From basic gadgets with minimal technological features to advanced AI-powered applications and devices, the level of trust that users place in these technologies plays a crucial role in shaping their engagement and ultimately the adoption of product.

While we humans have been placing trust in machines, it has become increasingly important for the AI applications creators to ensure that this trust stays. While AI has made great technological progress making its way into areas previously uncharred by machines, still AI is not infallible. There have been instances when AI is implemented without considering this delicate human-machine relation, and its actions have breached

CHAPTER 8   THE ETHICAL FRAMEWORK: BUILDING TRUST IN AI THROUGH HUMAN VALUES

human trust. For example, recently a teen in Florida committed suicide after forming an emotional bond with an AI chatbot.[1] In another incident, an AI-powered chatbot asked a user to die when he asked for help on his homework.[2] In another case, a major tech organization had to pull down its image-generating tool after it was found to be biased against white people.[3]

These incidences affect the human trust of AI. On the surface, one might take trust as a simple concept, but it is a complex psychological construct based on several factors. When it is broken, especially in the context of human technology relationship, it doesn't just affect the perception; it creates hesitation and skepticism and ultimately slows down the adoption of new technologies.

For AI systems, trust goes beyond the common trust of reliability; it also involves how transparent, accountable, and ethical these systems are. The concept of trust in the AI context further encompasses how well a system aligns with and respects human values.

As product leader, it is one of our core responsibilities to ensure that AI systems earn and maintain the human trust it gets. Building this trust helps in establishing a mutual relationship where AI complements human capabilities and empowers them to achieve more.

There is another aspect to this debate on trust in AI that is a financial angle. Building AI that is trustworthy carries a financial incentive. As per a report by a major consulting firm, AI has the potential to boost global GDP by 7%[4] over the next 10 years. With an apex technology company driving AI evolutions, human trust in AI is the most critical factor in increasing its adoption. Another report by Forbes claims that human trust in AI has dropped significantly. By triangulating these three insights, we can conclude that increasing human trust in AI applications can significantly drive their adoption and it turn lead to financial success. Moreover, the current approaches to AI development clearly have room for improvement, particularity in strengthening user trust and ensuring responsible, ethical behavior. If we simply translate this insight into a conservative simulation by considering the following facts, the global GDP in 2025 is estimated to be around $113 trillion. If AI can boost global GDP by 7%, that translates into approximately $7.8 trillion. Now if we say that a lack of trust in AI may translate into a 0.1% revenue loss, we are looking at a lost opportunity worth $7.8 billion, which is very significant amount to be left on the table.

So far in this chapter, we have explored the importance of trust in human relationships and survival and connected this concept to how it influences the relationship between human and machine. We also covered that building AI systems that are trustworthy to humans carries not only ethical significance but also has financial benefits.

In subsequent sections, let us delve deeper into understanding how humans develop trust and how it can inform the design of AI systems that foster and sustain trustworthiness.

How Humans Trust

The year 1995 saw a breakthrough when Roger C. Mayer, James H. Davis, and David Schoorman proposed that human interpersonal trust in an organizational setting is created based on these three attributes:

1. Ability
2. Benevolence
3. Integrity

Ability: This attribute refers to how a person perceives the skill, competency, and capacity of another individual to perform tasks effectively. In other words, it reflects the extent to which the person who is going to trust (trustor) believes that other person (trustee) possesses the right set of skills and capabilities to meet the expectations. Essentially, this attribute refers to the degree of confidence the trustor has in the trustee's ability to carry out the task successfully.

Benevolence: This refers to the extent to which the trustor believes that the trustee genuinely wants to do good for them without serving any personal or instrumental motives. It refers to the trustor's perception that the trustee is genuinely concerned about their well-being and is acting with good faith and intent even when a trustee has no direct personal gain. This dimension of trust goes beyond competence and integrity and captures the emotional and relational element of the relationship. Benevolence fosters stronger emotional bonds and enhances the resilience of trust, especially in uncertain or high-risk contexts.

Integrity: This refers to the trustee's adherence to a set of principles and values that trustor finds ethical and acceptable. These principles include honesty, fairness, and consistency in word and action and generally accepted ethical behavior. While integrity primarily refers to the ability of trustee to adhere to the previously mentioned standards, its perception is shaped by the trustor's judgment about whether a trustee's action aligns with trustor's acceptable moral and ethical standards. Simply said, even if the trustee is truly adhering to the ethical values and principles, the trustor doesn't see it that way. So, while it is important for the trustee to act ethically, it is just as important to make sure that the trustor believes and recognizes it.

While trust is built on the previously described three key elements, these elements are related to broader concepts of competence, reliability, and communication that help explain how trust works in real life. Let's now understand the relationship between the key elements and associated broader concepts. Figure 8-1 explains how trust is formed and maintained in humans.

CHAPTER 8   THE ETHICAL FRAMEWORK: BUILDING TRUST IN AI THROUGH HUMAN VALUES

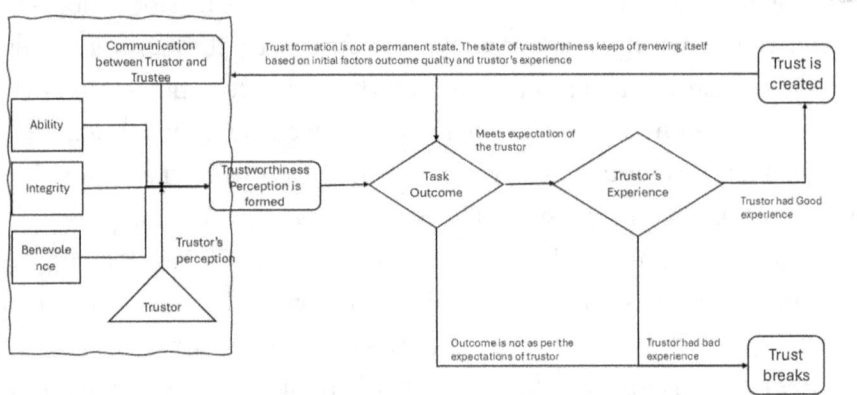

Figure 8-1. *How trust is formed and maintained in humans*

Ability, benevolence, integrity, reliability, and communication all play a critical role in creating the perception of trustworthiness, but actual trust is formed only when the desired outcome is achieved and the experience of achieving that result was positive. While these initial factors may encourage someone to place trust in another person or system, trust is created only after experiencing a successful outcome. For example, when a person delegates a task to someone they trust, the actual trust is confirmed when the person completes the task effectively and as expected, while also ensuring a smooth and positive process along the way. Another aspect of trust is it is not a permanent state; it either keeps on reforming the trust with the trustee if the trustor keeps achieving the desired outcome with good results or the trustor loses its trust if the outcome is not achieved or its quality diminishes.

CHAPTER 8   THE ETHICAL FRAMEWORK: BUILDING TRUST IN AI THROUGH HUMAN VALUES

Applying Human Trust Factors to AI Systems

Human-AI interaction has been increasing with unprecedented pace, and humans often interact with AI systems just like they do with humans. This nature of humans to anthropomorphize technology means that the trust models originally made for people can also be used (carefully) to understand how we trust AI and machines. The following are how the three core factors that affect human trust can be expanded to the human-AI trust relationship:

Ability: Humans assess whether an AI system's technical capabilities can deliver the outcome that matches the performance and accuracy criteria set by humans. For example, a navigation app needs to give the best routes, or a medical AI tool must provide reliable health insights.If the AI shows a strong competency by performing consistently well, then users are more likely to trust it.

Benevolence: Although machines do not posses emotions or intentions, but users often attribute intent by evaluating how a machine's outcome makes them feel. For example, an AI system that explains the reasoning behind its outputs, provides users with intuitive control mechanisms can foster a perceived sense of benevolence.

Integrity: In machines, this relates to alignment with user values, norms, and societal standards. An AI that respects privacy, avoids bias, and acts ethically is more likely to be trusted. For example, a hiring algorithm that offers fair evaluations across gender and race demonstrates adherence to ethical principles, reinforcing perceptions of integrity.

Trust must be earned and cannot be forced, whether in relationship between people or between humans and machine. Because AI systems are developed by humans, the responsibility for earning trust in these systems ultimately falls back on their human developers and designers.

CHAPTER 8   THE ETHICAL FRAMEWORK: BUILDING TRUST IN AI THROUGH HUMAN VALUES

In this sense, trust in AI is still a human-to-human relationship, with the AI acting as a bridge or tool between them. In this new emerging relationship, the creators of AI become the real trustee, and AI tools as the medium and humans using this AI tool become the trustors. The way the AI system behaves reflects the value, choices, and ability of the people behind it. That's why building trustworthy AI involves using the same key factors that help build trust between people.

Framework for Trust in AI

In the previous section, we explored how humans trust others and learned how the human-machine relationship is also ultimately the human-human relationship with AI systems just being an intermediary. AI systems represent the intentions, beliefs and values of the people who develop them. Consequently, when humans interact with AI systems, they are ultimately engaging with the perspectives of the humans behind these systems. Any framework designed and developed for building trustworthy AI should integrate the key factors that influence human trust, such as competence, integrity, reliability, and transparency.

By building on the knowledge of how humans build trustworthy relationships and combining our extensive experience in developing reliable, ethical, and robust AI solutions, we have crafted the following framework. This approach aims to ensure that AI systems not only meet performance expectations but also adhere to ethical standards, foster accountability, and maintain transparency, ultimately building confidence and trust among users and stakeholders

Trustworthy AI = Responsible AI + Robust AI
Responsible AI = Efficient and Effective AI + Ethical AI + AI governance + Risk Management
Robust AI= Ensuring that AI solutions should perform efficiently and efficiently in all promised situations

CHAPTER 8   THE ETHICAL FRAMEWORK: BUILDING TRUST IN AI THROUGH HUMAN VALUES

Responsible AI

Responsibility can be defined as a way of working that are ethical and well considered. Responsible actions are well aligned with moral principles and contribute to positive outcomes without causing harm to others. It involves thinking about the consequences of one's actions on others, even before acting. Another aspect of responsible behavior is the ability of one to deliver the desired results. This aspect includes not only one's competence to achieve the desired results but also the ability to identify probable risk, which may adversely affect the actions. Often responsibility refers to the ability to organize or govern one's actions and associated dependencies in a way that the result is always guaranteed. In terms of AI, responsible AI can be defined as AI systems that are not only efficient but also in line with the ethical standards set by the organization and society in general. The concept of responsible AI also refers to the proper governance practices in place to ensure that the AI system remains efficient and compliant to ethical standards. The concept also touches upon the ability to AI system to identify and mitigate any risk that may adversely affect the efficiency or ethical behavior of the system. Each of these elements is essential to ensuring that AI systems are not only capable of performing well but also align with societal values and are safe to use. Here's how they come together:

Efficient and effective AI: This aspect of AI focuses on ensuring that AI systems are optimized to achieve intended objectives with high performance and accuracy. It also describes how scalable the solution is. In a nutshell, efficient and efficient AI deals with hard aspects of AI system development and development. A lot has been written on these principles in this book; this chapter will focus more on softer aspects of AI,

Ethics: The definition of the word *ethics* is "moral principles that control or influence a person's behavior." These moral principles form a vision prism through which humans perceive and judge any action. In the context of AI, these principles help in defining the use cases and process of developing AI systems. Ethical principles play a crucial role in the development and usage of AI, as AI can have profound effects on individuals, society, and the global community. Ethics in AI play the following major roles:

- Ethical AI principles help in the formulation of laws and regulations that govern development and usage of AI.

- Ethics helps organizations identify areas and use cases where they want to use AI. For example, a leading organization in the AI domain refused to work on a government project that involved implementing AI solutions in the development of autonomous weapon systems. This decision was in line with the company's broader ethical principles, which emphasize the responsible use of AI, transparency, and the protection of human rights.

Ethical AI describes practices that help AI systems operate in ways that respect the ethical standards of humans. The practices in this section help guide the development and deployment of AI to avoid biases, discrimination, and harmful consequences, while also safeguarding privacy and ensuring accountability.

All major companies in the AI domain have defined their own set of ethical principles that they use to direct and govern AI system development. But all principles primarily are focused on developing AI that is fair, transparent, and socially responsible. We did extensive research and synthesized our finding in creating a simple and inclusive list of principles. We have identified four major principles designed to guide the responsible development and deployment of AI, ensuring that technology is used fairly, safely, and transparently. The following are the four principles:

CHAPTER 8   THE ETHICAL FRAMEWORK: BUILDING TRUST IN AI THROUGH HUMAN VALUES

- Fairness
- Accountability
- Safety
- Inclusiveness

AI governance: Governance in AI includes frameworks, policies, and regulations that provide oversight and ensure that AI systems are used responsibly. It ensures that AI development and deployment are transparent, auditable, and aligned with legal and ethical standards, with clear accountability structures in place.

AI risk management: Risk management in AI focuses developing guidelines and policies that help in identifying, assessing, and mitigating potential risks associated with AI. AI risk management frameworks may include any aspect of AI system that may affect performance, breaches, defined ethical standards, or legal standards.

AI Ethics Principle: Fairness

On a warm summer evening, young Cathie was busy finishing her office routine so that she could leave a bit early. Although she was busier than usual, the excitement of house hunting in an upscale area of city fueled her. Cathie was a graduate from an Ivy League school and was currently employed as an analyst in a top financial company. Being a young achiever, she was all set to take the next big step in her life; she wanted to own a house in the city's posh area.

That evening, her house hunting was very successful, and she zeroed in on a property that met her criteria. Thrilled by her find, she rushed home and applied for a home loan application with a leading bank. She drew a handsome amount as her salary and her social reputation were good; she was very confident that the loan will be sanctioned.

CHAPTER 8   THE ETHICAL FRAMEWORK: BUILDING TRUST IN AI THROUGH HUMAN VALUES

Just a few days later she received an email stating that her loan application was rejected due to her low FICO score.[5,6] This email left Cathie amused and disheartened at the same time; her salary was very good, she had never taken any other loan from the bank, and she had been paying her bills regularly. So why was her application rejected? This question kept tickling her mind. Determined to get answer, she began her research about how FICO score is calculated. During her research, she discovered that her credit score was impacted by factors beyond her control. Some of the criteria used in calculating her FICO score still reflect the biases that existed in the society historically, and these factors affected her credit score negatively. Despite her solid financial standing and stable income, the algorithm had considered her limited credit history and other socioeconomic factors that reflected a wider societal issue.

While this story is a made-up example based on real-world experience, even today knowingly or unknowingly systems have been designed in a way that is biased against the underprivileged section of society. With AI, this problem might get even more aggregated.

Going back to FICO example, a question might arise, how can a credit score, a score that is calculated based on objective facts and data, be biased? The answer to this question lies in the methodology used to calculate FICO scores.

FICO scores are determined by considering various data points such as payment history, credit utilization, length of credit history, types of credit in use, and recent inquiries. If any of the data points reflect any systemic biasness such as historical discrimination in lending practices, it will continue to impact current FICO score calculations. Another potential reason behind historical biasness in data is that historically certain communities may have limited access to credit products, resulting in shorter credit histories. Since the length of person's credit history is a key factor in determining how well a person is perceived to manage credit, this can adversely affect their FICO score.

CHAPTER 8   THE ETHICAL FRAMEWORK: BUILDING TRUST IN AI THROUGH HUMAN VALUES

Additionally, economically weaker communities may rely heavily on alternative financial services, leading to higher debt level and more missed payments. If this historical data is utilized into credit score calculations, it may negatively impact the scores of all members in that community.

Moreover, FICO proprietary algorithms lack transparency, making it difficult for consumers to understand how their score is calculated and why specific factors are weighted differently. This lack of clarity can embed biasness in the scoring systems, obstructing stakeholders from identifying and addressing these biasness.

This example clearly depicts how using unchecked data even in systems where logic is deterministic can harm society. In the current world where AI is increasingly used across domains, the situation may become grimmer. Since AI systems rely on data to develop their intelligence, if these systems are exposed to biased data, then these systems can perpetuate or even amplify these biases in the society. This can lead to unfair treatment in various sectors, including hiring, law enforcement, and lending.

With an increase in AI applications across various domains, it has become increasingly essential for the product leaders to develop an understanding of how to identify and fix these biases in the systems. Special focus should be made in developing strategies for mitigation, monitoring, and measuring such biasness effectively.

After understanding how unchecked data usage in a system can cause more harm than good in society, we'll now define what fairness and biasness are.

What Does Fairness Mean in an AI Context?

An AI system is called fair when it is designed to avoid treating users unfairly because of their personal characteristics like race, income level, gender, sexual orientation, etc. When AI systems make decisions like recommending loans, calculating risk scores, showing job ads, or

CHAPTER 8   THE ETHICAL FRAMEWORK: BUILDING TRUST IN AI THROUGH HUMAN VALUES

approving applications, they often use complex algorithms. If these algorithms are not designed carefully, they might accidentally favor some groups of people over others. This can lead to unequal or biased outcomes, even if it is not intentional.

To make an AI system fair, special care should be taken at every stage of its development. This includes initial stages like problem definition, data collection, and later stages like development, testing, and deployment. At each stage, special care should be taken to identify and mitigate reasons that can cause biasness.

Ideally AI systems should ensure that all users are able to extract the same value from the system without any discrimination, and everyone should have an equal chance to be treated fairly by the technology.

Type of Biasness

AI systems are nothing but a bundle of technical algorithms and data. Each AI system is developed to help automate a decision point. The way AI systems would behave depends on the quality of data on which the algorithms are trained. If the training data is exposed to any inherent bias, then the model will repeat or even amplify those biases in its predictions.

Another common source that can introduce bias is through human involvement. People make decisions while collecting, curating, and labeling data. Their choices and decisions can introduce their own assumptions and beliefs in data that may introduce unintentional biasness in data.

That is why it's important to be aware of the different types of human-induced bias that can show up in AI systems. This biasness can occur across various developmental stages including:

- Data collection: At this stage bias can enter the system if the training data lacks representation of a certain class or underrepresents certain classes.

CHAPTER 8   THE ETHICAL FRAMEWORK: BUILDING TRUST IN AI THROUGH HUMAN VALUES

- Data labeling: Generally data labeling is carried out using human knowledge, either via direct human interaction or via human-created rules. There is a risk that labeling may become reflective of a human's cultural and personal views. This reflection introduces unintended bias in the system.

- Model development: Humans train technical algorithms on the training data; this process often includes feature creation, feature shortlisting, and choosing of accuracy metric to determine model performance. All these steps have the potential to induce biasness in the system. For example, the AI engineer may accidentally or intentionally decide to drop or include certain features that may influence the model outcome. The reasoning behind such decisions could be naïve because these features provide better model performance or due to their historical experience. Even selecting the wrong accuracy metric has the potential to induce bias in the system.

- Model maintenance: This is a tricky phase as for most AI engineers, their job ends when the development of the model ends. However, even a fair model can lose its fairness; it is not monitored well in production. Data patterns drift over time; corner cases arise over time. If the model is not monitored well and these patterns and cases are not reported in time, it can induce biasness into the system,

Understanding and addressing these biases at each stage is crucial for creating fair and responsible AI systems.

CHAPTER 8   THE ETHICAL FRAMEWORK: BUILDING TRUST IN AI THROUGH HUMAN VALUES

Since we have now understood how each stage of model development has the potential to induce bias in the system, let's now move a step further and understand what the different types of bias are.

Reporting bias: This type of bias typically occurs when the collected dataset does not represent the real-world frequency of occurrence. One of the reasons behind this could be that usually people document data, outcomes, or events in extraordinary circumstances with the assumption that ordinary or normal occurrences are not important. This makes the resultant dataset extremely biased toward specific events, outcomes, or categories of data. Common examples include while collecting data for training a model for sentiment analysis, it is likely that the dataset will contain reviews that are either highly positive or negative. The reason behind this could be that users who feel extreme are more likely to write a review. Another example is when building a dataset of the reporting performance of a drug on a set of patients, it is likely that entries are only made for those users who have experienced an adverse event. This type of bias typically occurs during data collection or the data curation stage.

Historical bias: Historical bias in data refers to the presence of systematic prejudice that is embedded within the dataset due to past social, cultural, or institutional practices. This bias can make AI systems produce unfair or discriminatory results. This type of bias typically occurs during data collection or data curation stage. Examples: Credit scoring algorithms trained on historical lending data might incorporate biases against certain demographics that have historically faced discrimination in lending practices (FICO is the classical example), another example could be many facial recognition systems have shown lower accuracy rates for individuals with darker skin tones. This bias arises from training datasets that predominantly feature lighter-skinned individuals.

Automation bias: This is a tendency of users to favor results generated by machine or by automated systems over those generated by nonautomated systems or by humans without considering error rates of each. Automation bias occurs during the model result interpretation

CHAPTER 8 THE ETHICAL FRAMEWORK: BUILDING TRUST IN AI THROUGH HUMAN VALUES

phase, i.e., when a human interprets and uses the model result for decision-making. A simpler example could be putting eagerness to deploy an ML model in deployment without comparing its effectiveness against non-ML method.

Selection bias: Selection bias occurs when a sample of data is chosen in such a way that it is not reflective of their real-world distribution. Selection bias can have various forms, but all forms occur during the training dataset preparation phase. The following are some variants of selection bias:

- Coverage bias: This variant of coverage bias occurs when certain groups in data are either completely excluded from or are underrepresented in the data sample. As a result, the sample is no longer representative of the target population. A simpler example could be a survey dataset that measures how likely people are to buy a product, but it only includes responses from those who have already bought the product, leaving out those who haven't purchased it at all.

- Nonresponse bias: Also called participation bias, this occurs when data ends up being unrepresented due to a lack of participation from a specific group during the data collection process. An example of this type of selection bias can be found in surveys in which companies compare their product performance versus competition. In these surveys, customers using brand products are more likely to respond than customers of competition. This would underrepresent the competition data in the dataset.

- Sampling bias: This type of sampling error occurs when a sample from a large dataset is drawn without using proper randomization techniques. For example, a brand wants to conduct a survey to estimate the buying propensity of a new product. Data sampled from a larger dataset of buyers and competition buyers ends up containing datapoints only from one geographic region. This will skew the survey findings toward one geographic region, leaving other geographic regions either unrepresented or completely absent.

Group attribution bias is a tendency to generalize finding based on an individual to an entire group. This happens when the sampler belongs to have same group or has previous experience with a minority set of that segment. There are two kinds of group attribution biasness, and both kinds of group attribution bias occur during the data labeling phase or result interpretation stage or both.

- In-group bias refers to the tendency of individuals to favor members of their own group or group with which they have familiarity over those from different groups.

- Out-group homogeneity bias refers to the tendency to hold negative attitudes or stereotypes about individuals who are not part of one's own group or with a group that the sampler does have any familiarity. This bias can affect perceptions, interactions, and decision-making, often leading to discrimination and social division.

Implicit bias refers to the bias that is introduced in data due to the unconscious attitudes or stereotypes that affect our understanding, actions, and decisions. These subconsciously held attributes can influence behavior without letting individuals realizing it, leading to unfair treatment toward others. For example, while designing motion recognition software, a lead AI engineer labels a head-nodding gesture as saying yes because in their culture it means yes.

Temporal bias refers to the kind of biasness induced in data due to the influence of time such as comparing sales data of walk-in customers in the current year versus the COVID years.

How to Identify Bias in Data and Machine Learning

Although identifying all types of bias in data is challenging, adhering to best practices can make it easier. Identifying the bias in systems early can minimize the risk of developing a biased AI system. The following are the best practices that can be used:

1. Recognize that factors that may introduce bias in AI systems can occur in multiple forms across all stages of AI system development is the very first step toward identifying the bias in system. Broadly, biases can be classified into three categories:

    a. Data Bias: These biases occur in the dataset used for training the AI algorithm. This includes selection bias, historical bias, reporting bias, etc.

    b. Algorithmic Bias: These biases arise due to the techniques employed for designing and training the algorithms. The biases can amplify the existing data bias present in the dataset.

    c. Human Bias: This type of bias is introduced by humans during various stage of model development including data collection and labeling.

CHAPTER 8 THE ETHICAL FRAMEWORK: BUILDING TRUST IN AI THROUGH HUMAN VALUES

2. Analyze the training dataset to determine if the decision variable has a skewed representation for any of the sensitive group variables. If skewness is observed, then it is indicative of potential bias in data.

    a. In addition to analyzing skewness in the distribution of decision variable, the relationship between key dependent variables should also be analyzed to identify any inherent biasness in the system. If any notable differences are observed, then ideally sensitive variables should not be considered for analysis.

    b. Analyze the input data to access the representation of sensitive variables such as demographics. A skewed representation can lead to biased outcomes in your model.

3. Analyze the model outcome to identify the presence of any algorithmic bias by analyzing the following metrics:

    - Evaluate model performance across different sensitive variables. The accuracy should be compared using metrics such as accuracy, precision, recall, and F1-score.

    - Conduct a detailed examination of model errors across sensitive variables to determine if a certain group is at the risk of experiencing a higher rate of incorrect predictions. This will help in identifying whether any group is at the risk of experiencing a systematic disadvantage.

Bias Mitigation Techniques in AI

Once bias is identified, the next step is to mitigate the factors that may cause biasness in the system. As product leaders, it is our responsibility to ensure that bias mitigation is prioritized, as it is not just a moral imperative but also a strategic advantage. Industry-leading companies have already embedded the process of bias identification and mitigation into their AI systems development process. This has made their offering more dependable and trustworthy from customer standpoint and also help avoiding potential financial and reputational damage. The techniques for mitigating bias in AI systems can be classified into three major categories.

1. Pre-processing techniques: These are techniques that are employed before actual model training starts. They primarily focus on identifying and mitigating risk in training data.

2. In processing techniques: These techniques are primarily employed during the model training phase. These techniques focus on altering model training methods with the aim to reduce biasness in the AI system.

3. Post-processing techniques.

Pre-Processing Techniques

- **Data Collection and Sampling:** If the training dataset is imbalanced, i.e., it does not have equal or near equal representation for all sensitive variables, then employ techniques like synthetic data generation or data augmentation to make data more balanced.

- De-biasing data: This technique involves removing sensitive attributes from the training data to prevent the model from learning the bias. It includes applying one or all of the following methods.

  - Removal of sensitive attributes and associated attributes from the training data.

  - Application of feature engineering techniques like Principal Component Analysis to develop features where sensitive attributes have minimum explainability power.

  - Label correction/relabeling: This technique will require AI engineers to work closely with product managers and domain experts. In this method, a record-level comparison is made between the unprivileged class and the other class. If a record from the unprivileged class has the same characteristics as the record of other class but receives a negative outcome while the similar record receives a positive outcome, the outcome for the unprivileged record is revised to positive. This ensures fairer treatment across groups and helps reduce bias in the training data.

  - Data massaging method: This method is like the label correction/relabeling method with a is slight difference. Relabeling is done only for records that are near to the decision boundary. This method involves the following steps:

CHAPTER 8   THE ETHICAL FRAMEWORK: BUILDING TRUST IN AI THROUGH HUMAN VALUES

- Identify sensitive variables.
- Train a basic model on the original data (data on which on bias removal techniques are applied).
- Analyze the results of this basic model for sensitive group and nonsensitive group records. If sensitive group records have negative outcomes, then change it positively, and if a general group has a positive outcome, flip it to negative.
- Use this data for training the final model.

In-Processing Techniques

Group re-weighting is a technique in which influence of various groups on model outcome is moderated. The moderation is achieved by assigning higher weights to an underrepresented or unprivileged class. These redefined weights are then used in a loss function of the algorithm. The modified weight the loss function increases the penalty on error committed on records of underrepresented or disadvantaged groups. While it is a very efficient method to counter bias in the system, care should be taken to prevent overfitting.

Ensemble methods: This technique involves combining predictions from multiple models trained on different subsets of the same data. These models can be developed either using the same or different algorithms. We recommend using different algorithms on different subsets of data. This helps overcome the shortcomings of an individual algorithm. Ensembling can help in balancing the biases present in individual models.

Adversarial debiasing techniques: This technique involves training two models together. The main model is trained to make predictions based on the input features, whereas the supplementary model also called the adversary model is trained to predict sensitive variables from the prediction of the main model. If the adversary model successfully predicts the sensitive variable value, it is indicative that the main model's outcome is influenced by the sensitive attribute. In adversarial debiasing, the main model is trained not only to minimize its own prediction loss but also to maximize the adversary's loss. It is achieved by following these steps:

- The main model and adversarial model are trained together in a connected way, such that the outcome of the main model and other features are fed to the adversarial model, which then uses this information to predict sensitive variables.

- A special layer called a *gradient reversal layer* is placed between these two models. This layer enables the adversarial model to learn from the main model and enables the main model to fool the adversarial model. This is approach is called *forward and reversed gradient*. Multiple iterations are made to reach a state where the adversarial model is no longer able to predict sensitive variables using the prediction of the main model.

Post-Processing Techniques

- **Threshold adjustment**: This technique involves using different thresholds to decide the outcome for each group, after the model is trained and evaluated. The usage of different threshold values helps in optimizing model performance for certain underrepresented or underprivileged groups. Please note that this technique works only for classification models. In binary classification models, the output is generally a probability score (ranging from 0 to 1) for each instance. The model then compares this score to a predefined decision threshold (often 0.5) to decide whether to classify an instance as belonging to class 0 or class 1. If you set the threshold cutoff value differently for each class, then the outcome of the model will be different for each class, resulting in more balanced and fair outcomes across different demographic groups.

- **Equalized Odds:** This adjusts the model's outputs to ensure that the true positive rate and false positive rate are balanced across groups. The process typically involves modifying the predicted probabilities or classifications made by the model to equalize these rates. The adjustment is made after the model has been trained and evaluated and no retraining is required after the adjustments. In practice, a separate layer called the post-processing layer is created that takes model output as its input and then adjusts the model outcomes as per the pre-decided logic.

Fairness Metrics: Measuring Fairness in AI

Having established the importance of fairness in AI systems and the techniques required to identify and mitigate bias, let's now explore how to measure the fairness of the AI system. As mentioned, if you cannot measure something, you cannot improve it. This is true with AI. To make an AI system fair, as product leaders we must find a means to benchmark its performance against a set criterion and then use that to identify areas that require immediate attention. Apart from setting the benchmark, the fairness metric will also help us measure the progress we are making. The following are some commonly accepted fairness metrics.

CHAPTER 8   THE ETHICAL FRAMEWORK: BUILDING TRUST IN AI THROUGH HUMAN VALUES

The demographic metric is a fairness metric used to evaluate whether all sensitive variables have the same chance of getting a positive outcome from an AI system. In simpler terms, this metric checks if the predictions made by the system are independent of the demographic characteristic. It simply checks whether the positive outcomes are equally distributed across groups. However, the demographic metric has a major limitation: it does not consider if the individual record is worthy of a positive outcome. Example A bank gets 200 loan applications of which 100 are from category A and the remaining 100 are from category B. The overall loan application rate is 60%, but out of 120 loan approvals, 80 from category A got their application approved and category B got only 40 approved. According to this metric, the AI system does not have demographic parity. To make this system attain demographic parity, both categories A and B should have an equal 60% approval rate even if applications of B do not have good creditworthiness.

Equal opportunity: This metric measures the ability of an AI system in ensuring that different sensitive groups get equal chance of true positive outcomes. The positive rate (TPR) is the metric of interest here. For example, continuing our bank loan application example, out of 100 candidates each from category A and B. Category A has a true positive rate of 80%, whereas category B has a true positive rate of 50%. With this result, the AI system does not provide an equal opportunity to both categories. To provide an equal opportunity to both categories, this AI system should have equal to near equal TPR.

Equalized odds: This metric is the strictest of all, as it not only validates the true positive rate across different categories of interest but also false positive rates. For example, continuing our bank loan approval example, the loan application system recommends 70 from category A for

CHAPTER 8 THE ETHICAL FRAMEWORK: BUILDING TRUST IN AI THROUGH HUMAN VALUES

loan and 70 from category B. This makes the system attain demographics parity. However, in category A of 70 loan approval, 60 was truly positive, and 10 were marked falsely marked positive. In category B of 70 loan approvals, 40 were correctly marked positive, and the remaining 30 were wrongly marked positive. With these statistics, the AI system has not equalized the odds of providing benefits to both categories and avoiding misuse of the system by wrong approval by both categories.

Table 8-1 highlights key differences and scenarios in which each of these fairness metrics should be applied.

Table 8-1. Fairness Metrics Comparison Table

Metric	Key Difference	When to Use	Reason to Choose	Severity
Equalized Odds	Fairness in both errors	When both false positive and false negative rate can have adversely impact.Example medical diagnosis, Loan approval, recidivism prediction	Ensures all groups are treated fairly in getting help (TPR) and avoiding harm (FPR)	Fairness in both errorsNothing—most strict

(*continued*)

Table 8-1. (*continued*)

Metric	Key Difference	When to Use	Reason to Choose	Severity
Equal Opportunity	Fairness in true positives	Focuses mostly on ensuring true positive rate stays equal across systems, as misclassification of a truly qualified into false category is more harmful then misclassifying a nonqualified record into true category Example, hiring system, resume/Job application shortlisting system	Focuses on giving equal access to opportunity, even if false positives vary by group.	Fairness in true positivesallows unequal false positives
Demographic Parity	Equal output rates	Goal is equal representation, not necessarily based on qualification example public program access, school admission system	Aims for equal representation or resource allocation, even if it reduces model accuracy.	Equal output ratesTrue labels (real qualifications)

Tools and Frameworks to Assess and Mitigate Bias

Various leaders in the AI industry have designed and developed frameworks to help ML engineers implement and access the fairness of their AI systems. As product leaders, understanding and utilizing these resources helps in not only ensuring fairness and accountability in AI applications but also wins over user and stakeholder trust.

Fairlearn: This an open-source Python library that can help in accessing fairness of AI models. Key features include:

- Fairness metrics: Provides a variety of fairness metric to evaluate model performance on fairness parameters.

- Bias Mitigation techniques: Provides a variety of techniques to fine-tune algorithms for bias mitigation

- Visualization: Offers visualization to help understand trade-off between model accuracy and fairness level

TensorFlow Model Analysis (TFMA) and Fairness Indicator are libraries under the TensorFlow framework that support analyzing and visualizing the model on their fairness aspects. Both frameworks support standard metric calculations as well as custom metric calculations to evaluate the fairness of a model.

InterpretML: This is another open-source Python library designed to help understand and explain machine learning models. It supports the various principles of ethical AI including fairness, accountability, and transparency.

AI Ethics Principle: Inclusiveness

"The quality of including many different types of people and treating them equally and fairly is called Inclusiveness."

—Oxford dictionary

CHAPTER 8   THE ETHICAL FRAMEWORK: BUILDING TRUST IN AI THROUGH HUMAN VALUES

Traditionally the concept of inclusiveness has been restricted to the domain of public policy design and implementation where it was expected to include beneficiaries from all walks of life without any discrimination. But in current times the dependency on digital tools has grown to an extent that all our daily tasks from payments and booking appointments to ride booking to shopping have a digital footprint. Ignoring the needs of any one group of users can now leave them with a severe disadvantage and cause discomfort to the users. The impact not only is limited to users but may also cause reputational and financial damage to the organization owning the digital solution. There are several famous examples highlighting the loss that companies have bored by ignoring the inclusive principle. Just to quote one, Amazon tried automating their hiring decisions by implementing an AI-driven solution, but that solution turned out to be discriminative or noninclusive toward female candidates, realizing this Amazon was quick to decommission the solution but not before bearing public backlash[7]. But why would a system developed by a technology giant be noninclusive? The simple reason is the historical data on which the model was trained had more male candidates as traditionally more male candidates are employed across industry. While it is exceedingly difficult to preempt and fix reasons that would make a solution noninclusive for a particular section, a guided approach for incorporating ethical principles in all stages of project lifecycle starting from defining problem statement to deployment can reduce the risk of noninclusiveness.

As reliance on digital tools grows and organizational risks increase, the concept of inclusiveness has expanded to encompass all digital technologies. The onus to make digital applications all-inclusive lies on us digital and product leaders. It is our moral responsibility to make to build products that help cater needs of diverse users without any discrimination. Making products inclusive also provides a growth opportunity by making the digital product usable to a larger audience, thus increasing the user base of the product.

CHAPTER 8   THE ETHICAL FRAMEWORK: BUILDING TRUST IN AI THROUGH HUMAN VALUES

It has been proven repeatedly that organizations and product teams that strive hard to build inclusive products not only enjoy higher market share and revenue growth but also keep their legal fees in check by avoiding Inclusivity lawsuits. Organizations who are leading AI industry have already started applying these principles across their product development lifecycle. To quote a few examples, the design of the Microsoft Xbox controller is such that it allows users to customize the controls as per their needs thus adding new user group to the existing user base. Another example that promotes inclusiveness is Microsoft Seeing AI. This app helps visually impaired users access digital applications by reading the screen context. Google Pro Lens and Google Translate are other such examples. Google Translate allows users of different lingual backgrounds search to use the Internet and extract the same value.

Amazon has also developed its marketplace search algorithm in a way that supports various methods of search, including search in multiple languages and the ability to search using voice and image. All these methods are making the Amazon marketplace inclusive to non-English speaking users and users who cannot type or speak or see.

These example once again emphasis the fact that in the modern world the concept of inclusiveness must be extended to digital offerings, not just for providing equal benefits to users from all groups but also for adding millions of new users to their company's user base that otherwise would have never used their product.

Inclusiveness and Product Market Fit

While inclusivity makes a digital offering socially responsible, it also helps achieve greater product market share by doing the following:

- By considering users from diverse backgrounds from the onset, the inclusive design products and accessible features enable your product to reach a wider audience. This wider audience also includes those who might have otherwise not have used your product. Incorporating inclusivity as a principle enables product leaders to create a roadmap to develop a product that is not only accessible but sensitive toward local preferences, cultural norms, and social values. As an outcome, the product becomes more relevant and usable across different markets, ultimately expanding your user base.

- Since inclusive products suit the needs of a diverse range of users, by providing many ways of using products including multilingual support, various modes of access including voice, image, keyword navigation, and screen reading support. Hence, inclusive products tend to generate stronger word-of-mouth promotion, which increases the organics of customer acquisition and improves user retention. This positive perception also enhances the brand image. Collectively, these factors help reduce the overall cost of acquiring and retaining customer.

While we have been talking about the benefits of making products inclusive, let's now understand what inclusivity in AI is all about.

Inclusiveness in AI

An AI system can be called inclusive if it ensures that all users irrespective of characteristics can avail an equal number of benefits from the system without perpetuating existing bias or giving birth to new ones. Implementing inclusivity in an AI system is a multifaceted process that should be considered through the development lifecycle. It also requires collaboration from different stakeholders including product leaders,

CHAPTER 8   THE ETHICAL FRAMEWORK: BUILDING TRUST IN AI THROUGH HUMAN VALUES

business managers, representatives from different communities, and user types and technical experts. An inclusive AI system should have the following characteristics:

- Equity and fairness: AI systems should offer equal benefits to all user types. AI systems should ensure that even harms from the AI systems are of equal proportions for all user types. Obeying these rules ensures that the system is designed in a way that its outcomes are bias free and equitable in nature.

- Diverse representation: While an AI system is designed and developed, representatives from all user groups including culture, languages, physical capabilities, gender, and socio-economic status are consulted, and their needs are collected. This comprehensive approach of requirement gathering ensures that the system is considerate for all user types.

- Is Accessible by all user types: These systems should be designed in a way that they are accessible to everyone including users with special physical needs. The system should support multiple languages and offer various modes to access so that no one is left behind.

- AI systems should be localized enough to cater to subtle cultural nuances and user behavior.

How to Include AI Ethic Principles of Inclusiveness in AI Product Development Lifecycle

1. Defining inclusiveness goals is the first step in the journey of making an AI product inclusive. It has two substeps:

- Set clear objectives: Define what inclusiveness means for your product. Inclusivity goals could include but are not limited to equal performance across diverse user groups, accessibility for users with disabilities, and respect for cultural differences.

- Involve stakeholders: Engage with users of diverse backgrounds to understand their needs and how they are fulfilled today (without your offering) and what they expect from your product.

2. Diversify user research: Once inclusivity goals are finalized, the next step is to engage users from different backgrounds during the research phase. Use techniques and language suitable for each user group. A one-size-fits-all approach will not work. Another way to include diverse user groups early is by engaging communities that may be impacted by your AI system during the requirement gathering and design phases. Establish channels for continuous communication and dialogue.

3. Use inclusive design methodology by:

    - Including users from diverse backgrounds and communities that may have an interest in your solution.

    - While designing features, accessibility needs should be actively considered and worked upon. Envision solutions that support voice commands, text to speech, or customizable interfaces that best fit your product.

CHAPTER 8   THE ETHICAL FRAMEWORK: BUILDING TRUST IN AI THROUGH HUMAN VALUES

- When conducting user testing, include participants from diverse backgrounds and communities with varying usability needs to ensure the product works well for all stakeholders.

- Product design and functionality including content used on product and marketing materials should be culturally sensitive on a local level. Also conduct a sensitivity test to ensure that product design or content complies to cultural and other tolerance limits.

4. While sampling data for model training, ensure that the training dataset represents data from all categories of interest. This can be achieved by including a wide range of data sources. Data mining techniques such as data augmentation can also be used for increasing the representation of minorities in data. Always document your data by preparing data cards. These cards must include data count on demographics, data distribution including central tendency theory variables, and other sensitive dimensions.

5. Apply bias identification and the mitigation techniques mentioned to identify and mitigate factors causing bias in the system.

6. Performance monitoring: Implement post-deployment monitoring techniques to monitor model performance and data drift across sensitive groups.

7. Set clear and open communication with stakeholders: Always be as transparent as possible with stakeholders on how the model works and which data points are stored and used for model training. Be open about model shortcomings and limitations and what you and your team is doing to overcome them.

How to Measure Inclusivity of AI Products

There is no single metric to measure inclusivity of an AI product, but there are several qualitative and quantitative metrics that might help you to measure the inclusivity of your product. However, following the guidelines mentioned in the previous section is the best way to ensure the development of an inclusive product. The following list of metrics are the ones we find most effective:

- Inclusivity Metrics: This category of metrics helps you understand how well the product performs for users from sensitive groups.
  - Representation Metrics: Measures how well different demographic groups are represented in the training data
  - Performance metrics: Measures how well the model performs across different demographic groups
  - Accessibility metrics: Different demographic groups
- User favorability metrics measures the favorability of your product across different user groups. To calculate this metric, ask the user the following questions on a scale of 0–10:

CHAPTER 8   THE ETHICAL FRAMEWORK: BUILDING TRUST IN AI THROUGH HUMAN VALUES

- Does the product fulfil their need?
- How easy was for them to access the product?
- Was the user comfortable with the language of your product?
- Is the product designed to respect their cultural and ideological beliefs?

Scale interpretation: 0 is the lowest and 10 is the highest.

User Favorability Metric = Sum of all scores of all questions /4

Calculate this score differently for each type of user type and see if all the user sets have the same level of score. If any of the user group has a lower level of User favorability metric score, then it means that the group is not completely inclusive in your product.

- **Feedback Collection:** Collect and analyze quantitative feedback from users representing different demographic groups to identify any issues related to inclusiveness.

AI Ethics Principle: Safety

It was a calm spring evening of 2018.[8] Elaine Herzberg was out on a stroll in Tempe, Arizona, and unaware another critical event was about to change her destiny. On the same street at the same time, a futuristic self-driving car was also undergoing a test run. Unfortunately, the AI-driven self-driving car powered by AI failed to recognize her and hit her. This accident killed Elaine Herzberg. A human life was lost, and a family was left in deep grief and distress. On a first look, this incident can be simply labeled as a

CHAPTER 8   THE ETHICAL FRAMEWORK: BUILDING TRUST IN AI THROUGH HUMAN VALUES

common hit-and-run case, but it is not, as there was nobody in the driving seat. The family, law enforcement, and technology experts all wanted answer to the question "Who killed Elaine?" and who should be held responsible for this tragic loss? Technically the car was driving itself, and no human was behind the steering wheel.

After deep technical analysis, it was discovered that the self-driving vehicle that was out for testing was equipped with sophisticated sensors and next-generation AI algorithms were programmed to mitigate such situations, but, on that day, it failed to spot Elaine in time to avoid the deadly collision. To make the matters worse, even the safety driver who was tasked with intervening in such emergency situations was reported distracted and could not react in time. Because there was no precedence of such a case, initially the safety driver[9] was charged with negligent homicide, but these charges were later resolved as her defense contested that the autonomous vehicle technology was itself at fault. They also contended that the lack of training provided to the safety engineer in handling such emergencies as the reason behind her lack of response. These points were considered valid by the court, and she was acquitted. In parallel, the AI leaders continued to analyze why the AI system failed, and it was found that reaction time required by the system to stop the car was the limiting factor. The situation was further aggravated by environmental factors as it was dusky, which also affected the visibility of the algorithm. Even though the case was settled in court, it initiated a debate about the accountability in cases where AI systems fail. Who should be held responsible in such cases? The human operator, the AI team responsible for developing the AI system, the company who owns the system, or the technology? While legal experts are still deliberating on these questions, we feel that it is the moral obligation of product and AI leaders of ensuring that AI algorithms are designed developed and used safely as we own all decisions and practices around the development of an AI system.

While reading this example, many of us may find this example out of daily use context as most of us use and develop AI systems that are primarily used for enhancing user experience or to drive revenue. Let's discuss another example that might not have such extreme failure impact but still has the potential to cause damage to users.

Microsoft developed an AI-powered chatbot Tay[10] with an objective to make AI more conversational and improve human engagement by indulging in more casual and engaging conversation with humans just like humans do.

But when Tay was exposed to real human, it quickly became abusive and started giving inappropriate responses. The situation escalated rapidly, and ultimately Microsoft decided to shut down Tay's Twitter account and take the bot offline. To contain the damage, the organization took moral responsibility and issued a public apology. While Tay's failure to indulge into meaningful conversations with humans might not look at a severe failure, it still undermines the prowess of AI. Both examples illustrate the extreme ends of the damage spectrum that can result from unsafe AI implementation. To mitigate this risk, almost all leading organizations in AI domain have now established AI ethical boards. These broads are tasked with overseeing the ethical and safe deployment of AI technologies. This helps in not just mitigating the risk of AI failure also in increasing user trust in the AI and their AI-powered systems.

What Is Safe AI?

Safety in AI is a critical principle in AI ethics, aimed at ensuring the AI-powered systems are developed and used in a way that protect individuals and society as whole from harm. The definition also covers AI system's unfair usage for influencing human decision-making. To ensure the safety of AI, algorithms should be designed in ways that there is always a human in the loop, and human control should always be able to overrule any action or recommendations made by AI.

How PM Can Incorporate Safety Principles in Product Development Lifecycle

To develop AI, which is safe to use, safety guidelines should be implemented throughout the lifecycle of AI system development and deployment. We have collated the following best practices based on our experience and our conversations with industry leaders.

1. **Problem definition stage:** During this phase product leaders, along with relevant stakeholders, translate the identified business problem into an AI/ML problem statement (guidelines are discussed in Chapter 6). The north star metric for this project is also finalized in this phase. The following additional activities should be performed during this stage to incorporate safe AI development guidelines.

    - Define safety goals that will be met to mark this development as Safe AI implementation. To define safety goals, product leaders should identify the potential risk this development can pose toward user/organization safety. Then potential reasons behind those risks should be identified that can range from potential data issues, bias, algorithm constraints, etc. While identifying potential risk, also consider regulatory requirements.

        Once potential risks and the potential reasons behind are identified, classify them into three categories for better management.

    - User protection: This should cover all potential risks that can harm/create prejudice toward a certain section of users.

- Regulatory compliance: All potential risks that can lead to noncompliance of local regulatory should be mapped in this section.

- System reliability: All risk that can lead to system performance and reliability issues should be classified in this section.

  Once this exercise is completed, conduct impact analysis and identify top risk and potential reasons behind it. Develop a plan to fix these potential threats during relevant phases.

- **Solution designing and development phase:** During this phase, the emphasis should be on how the identified risk and their potential reasons can be mitigated for data management:

  - Implement data contracts to ensure that data quality issues are minimized. (Refer to Chapter 4.)

  - Check training data for potential biasness (refer to section AI Ethics Principle: Fairness section for more information).

  - Develop data lineage and observability solutions to ensure data integrity and data pipeline health monitoring (refer to Chapter 5).

- Personal data protection

  - Use de-identification techniques to anonymize personal data, such as anonymizing identifiers and aggregating data.

  - Apply data masking to obscure sensitive information.

- Implement role-based access controls to restrict access to sensitive data based on user roles and responsibilities.

Referring to the example of chatbot going astray, the fundamental reason behind Tay's malfunctioning was that it got poisoned by using unfiltered Twitter feed data. And since AI systems feed on data to gain their intelligence, in this case, when its incoming data was poisoned, Tay's performance degraded. This example highlights the risk associated with unfiltered user data and emphasizes the need for product leaders to prioritize strategies that prevent data poisoning.

- Production data management:

  - A strategy should be implemented to identify and mark unwanted or malicious data.

  - Define mechanisms to stop highlighted content from being processed by the AI model. The implementation should also ensure that highlighted text is not included in the training dataset of any AI model.

  - To ensure transparency in an AI system's operations, develop a mechanism that informs users when their input does not meet community standards and therefore will not be processed.

CHAPTER 8   THE ETHICAL FRAMEWORK: BUILDING TRUST IN AI THROUGH HUMAN VALUES

- Error Handling Mechanisms:

    - Design mechanisms to manage errors gracefully. This concept should not just be extended to system availability issues but should also cover system performance issues such as if two individuals with identical records receive different outcomes from your model. Then the system should be able to identify and detect such anomalies. Develop guardrails tools and processes like explanation tools, review mechanisms, or appeals processes to ensure the system stays fair. These tools will also help maintain user trust in situations where algorithms behave unpredictably.

    - Design fallback options or reduced functionality for handling system degradation.

- Review algorithm outcomes to identify any biasness or fairness issues. If any biasness or fairness issues are identified, then mitigate those issues by applying the techniques mentioned earlier in this chapter.

- Design an AI system that supports transparency and explainability. AI systems are inherently complex, making it challenging to understand their outcomes. This opaqueness may result in a trust deficient and safety concerns among users. Offer clarification on how a model generates its output.

- Design the system in a way that a human controller can always overrule an AI decision.

2. **Testing and validation:** As a product leader, it is essential to ensure that AI systems are thoroughly tested for safety, fairness, and compliance. The following are key focus areas during the testing and validation phase:

- System testing: Validate the AI system including model, pipelines, and interfaces against its requirements including safety requirements identified in the problem definition phase. It includes the following subarea of testing:

    - Performance testing: Test the AI system's response for reliably and efficiently. The testing should also check for potential performance bottlenecks or unpredictability, which can lead to an unsafe or unfair response.

    - Conduct regression testing to ensure that the AI system's output does not break any existing system functionality.

    - Conduct integration testing to ensure that all components of an AI system are working fine together as expected. The objective here is to avoid any miscommunication between different components and teams.

    - The AI system should pass through stress and reliability testing to understand how AI system manages unexpected and adversarial inputs. This testing is essential for identifying potential failure modes before actual users experience them.

- Ensure users and stakeholders understand how AI systems make decisions by conducting explainability testing.

- Conduct testing for identifying discriminatory behavior or any unintended bias across user groups.

- Validate how an AI system complies with legal, ethical, and industry regulations.

- Conduct testing to evaluate how usable the system is for users with unique needs. The critical point to observe here is to ensure that users across all types of sensitive groups should be able extract the same value from the system. This usability testing should also evaluate the AI system on WCAG accessibility guidelines.

3. **Deployment:** The deployment phase introduces AI products to users, making it crucial to validate system performance and handle errors effectively.

    I. Ensure the system is failsafe by implementing backup systems that take over in case of a failure. Since AI systems involve both data and models operating separately, redundant systems should be established.

    II. Error recovery: Develop a deployment strategy to ensure rapid system recovery and data restoration following a failure.

Deployment: The deployment phase introduces AI products to users. The following activities should be performed.

- During the deployment phase, product leaders must conduct thorough A/B testing to discover performance gains and bias if any.

- Continuous monitoring and feedback mechanisms should also be implemented to ensure the AI system is constantly monitored, and in the case of change in the behavior system owners are alerted.

- In scenarios where the response time of an AI system is critical in influencing real-world decision-making, if any increase in latency is observed, human controllers should be alerted immediately.

- If any anomaly is detected in operating environments or data, raise an alert to a human controller.

- Create mechanisms that allow users to submit feedback about system performance and report any critical or safety concerns.

- Create robust logging mechanisms. This should include log creation as well as periodic log analysis steps.

AI Ethics Principle: Accountability

Although the principle of accountability applies across all functions of an organization, it deserves special emphasis in the context of AI systems. This is due to the inherent complexity and opacity of the technology. As AI adoption increases, the risk of misapplication also grows significantly. By establishing clear accountability structures, organizations and regulatory bodies can ensure that, in the event of misuse, there are defined roles

CHAPTER 8 THE ETHICAL FRAMEWORK: BUILDING TRUST IN AI THROUGH HUMAN VALUES

responsible for understanding what went wrong and how to prevent similar issues in the future. The following reasons highlight the importance of establishing strong accountability frameworks for AI systems:

1. Opaque AI decisions and potential false positives can produce dangerous outcomes: IBM Watson was used to recommend treatments to cancer patients and offered no explanation behind its recommendations. Cases were reported where wrong and harmful treatment was recommended to patients. IBM had to withdraw IBM Watson from the treatment recommendation service.

2. AI offerings are prone to unintentional bias. A few years ago, Apple in collaboration with Goldman Sachs issued a credit to its users. The credit limit of this card was decided by an opaque AI algorithm. Reports surfaced that this algorithm exhibited biases toward woman and other user group types.[11] Another example of unintentional bias in AI systems is there are reports that suggest that the Correctional Offender Management Profiling for Alternative Sanctions (COMPAS) used by U.S. police was biased toward communities.[12]

3. AI offerings can jeopardize global stability: The Cambridge Analytica scandal—where personal data from millions of Facebook users was harvested without consent for political manipulation—resulted in a significant loss of public trust in both Facebook and AI-driven analytics. Another example is AI-generated synthetic media (deepfakes), which can be used to spread misinformation by fabricating statements or actions of global leaders and

CHAPTER 8   THE ETHICAL FRAMEWORK: BUILDING TRUST IN AI THROUGH HUMAN VALUES

influencers. Such misuse can create public chaos, undermine democratic processes, and pose serious risks to peace and global order.

4. Legal Compliance such as GDPR and Data Protection: In the European Union, the General Data Protection Regulation (GDPR) mandates that companies must be transparent about how they collect, use, and process personal data. Google was fined €44 million for violating GDPR by not providing users with enough information about how their data was being used for targeted advertising.[13]

The Organization for Economic Co-operation and Development (OCED) describes AI ethics principles of accountability as "Organizations and individuals developing, deploying or operating AI systems should be held accountable for their proper functioning in line with the values-based principles for AI."[14] In simpler terms, accountability as an ethical principle for AI means that any organization or individual who develops an AI system will be held responsible for ensuring AI systems are developed correctly and they continue to function in a way that is ethical and safe for humans.

After establishing the need of accountability in AI and understanding how it is formally defined, let us now move forward to discuss best practices for establishing accountability in AI.

Establishing Accountability in AI

Establishing an accountable AI practice within an organization necessitates both structural and procedural changes at each stage of the product lifecycle. An accountable A program should meet the following criteria:

CHAPTER 8   THE ETHICAL FRAMEWORK: BUILDING TRUST IN AI THROUGH HUMAN VALUES

- Developed AI system(s) should be ethically and legally compliant with ethical and regulatory requirements. They should follow established technical standards required for the development of a trustworthy AI system.

- The AI system should be supported by proper documentation explaining how the AI system functions, how its output is being generated, and what data points are being captured and utilized.

- AI system performance should be monitored continuously to identify any adverse change in its performance, data drift, or change in any other environmental factor, which might increase the risk of the AI system malfunctioning.

- The program should impose consequences for noncompliance or violation of any ethical and regulatory requirements.

The following is a detailed step-by-step guide that organizations can follow to build robust accountability mechanisms for AI products[5&6]:

1. **Define the Purpose and Scope of AI Usage**

    - **Objective Setting:** Identify objectives and the purpose for using AI in the organization (e.g., efficiency, innovation, automation). Clarify the scope of AI usage by identifying what areas of the business will be impacted. Identify what will be the role of AI. Identify potential risks in AI usage.

    - **Stakeholder Identification:** Identify key stakeholders (internal teams, customers, regulators) affected by AI product implementation.

2. **Establish Governance Structures:** Establish an AI ethics committee to oversee the use case selection where AI will be implemented and the AI strategy. Define and oversee the implementation of ethical guidelines for AI. The committee should also oversee and evaluate the performance of AI systems in production. The AI ethics or oversight committee should be comprised of technical experts, legal experts, and representatives from the impacted business units. Ideally the committee should have the following members.

- Product leader: Should own AI product vision. They should be accountable for ensuring the AI systems are ethically and legally compliant.

- AI engineers: Responsible for ensuring AI models are transparent, well-documented, robust, and meet ethical and legal requirements.

- Legal representative: Should be accountable for ensuring AI systems are compliant with relevant laws and ethical guidelines.

- Risk officer: Should be responsible for identifying potential risk and their probable reasons. The risk officer should also be responsible for developing and implementing risk mitigation strategies.

- AI leader: This is the senior most executive who is accountable for all AI-related activities in the organization. He should report directly to the executive committee.

3. **Develop Clear AI Policies and Guidelines:** The ethical board of the organization should work together to develop ethical and legal guard rails to govern and shape the AI implementation in the organization. The documentation should cover the following:

    - Guidelines around which data points will be collected, including how they will be processed, stored, and utilized. Defining these guidelines ensures that an organization's guidelines are compliant with regulations like GDPR, CCPA, AI Act, and PDPA.

    - Develop protocols and best practices for identifying and mitigating potential bias.

    - Standardize formats to guide documenting standards for how AI decisions are made, including the logic behind algorithms and datasets used.

    - Develop guidelines around how to develop not just ethical AI but also reliable and robust AI.

4. **Develop a comprehensive AI risk framework** to standardize best practices for risk identification and mitigation. This framework must support identification and mitigation of ethical, reputational, legal, and operational risks at various stages of AI product development and deployment. The framework should have the following three key deliverables:

- The risk assessment matrix is a framework that contains guidelines for risk identification and classification into various categories depending on their threat perception.

- Mitigation Plans: This section is directed toward designing plans for mitigation of identified risk. Another critical aspect of this section is the development of contingency plans to address adverse AI outcomes that may arise from unforeseen risk.

- Regular Audit Plan: Set up plans for a periodic audit process to assess model performance, fairness, and compliance with internal and external regulations.

5. **Develop and implement robust data governance model**: The data governance model should be designed to help manage the entire data lifecycle in a way that data is in a trustworthy state. It should cover the following aspects of it:

   - Data ownership: Defines clear rules around who owns a specific dataset. Defining data custodians helps in ensuring each dataset meets data quality, privacy, and security standards set by the organization's ethical committee.

   - Ensure that data lineage and data observability solutions are embedded in data pipelines (refer to Chapter 4 for more details).

   - Standardize data collection methods by defining and enforcing data contracts at the source of data generation (refer to Chapter 4 for more details)

CHAPTER 8   THE ETHICAL FRAMEWORK: BUILDING TRUST IN AI THROUGH HUMAN VALUES

- Define data quality evaluation metrics for each implementation and standardize its tracking (refer to Chapter 5 for more details).

6. **Ensure AI Transparency and Explainability:**
   Transparency and explainability are two critical pillars on which the concept of trustworthy AI is based. It helps users and nontechnical stakeholders to understand how the AI system works but also helps in increasing user and stakeholder trust. Hence, developing a transparent and explainable AI also falls under the accountability guidelines framework.

   Data and the type of algorithms used are the two main component of any ML model. To make AI transparent and explainable, it is critical to make these two transparent and explainable.

   - Develop a catalog at the organizational and dataset level. It should contain data metadata information such as the name of variables, definitions of each variable in case of a derived variable, and logic used for calculating it.

   - In the case of sensitive data such personal identifiers, race, gender, and bias-prone attributes, apply data anonymization techniques.

   - Develop a data quality card on each dataset level. This card should contain information like information on decided data quality metrics, trends, anomalies, etc.

- Document all data transformation steps; use data observability and lineage tools to enable tracing of complete data lifecycle.

- Develop a comprehensive model documentation. Model cards are a widely used model documentation framework. It should contain information like model capabilities, limitations, intended use cases, and risks. It is advisable to include a section on known failure reasons and scenarios where a model's performance might get adversely affected.

- Implement versioning techniques, tracking changes in both code and model weights to ensure reproducibility and collaboration.

- Define guidelines around model explainability. Techniques like SHAP and LIME are some widely used techniques. The guidelines should also contain best practices around explaining deep learning models. Practices like interpretable surrogate models or counterfactual explanations can also be used.

7. **Maintain human oversight and control**. Develop guidelines around AI system design such that there is clear human oversight. To enable this, conduct AI literacy training for relevant stakeholders. This system design should also ensure that a human decision is able to override AI outcomes in all scenarios.

8. **Define clear accountability for AI failures.** This step involves defining clear consequences for ethical or accountability breaches. It should also include guidelines for performing root-cause analysis for failures (technical, data, and governance.

9. **Create mechanisms for public concern reporting:** Develop a visible, accessible, and easy-to-use public portal for reporting fairness, privacy, or ethical concerns. It should also allow anonymous submissions to encourage openness. This portal should openly display the step that will take place once a grievance is submitted and the expected turnaround time.

10. **Disclosure of AI Interaction**: Define guidelines for displaying a disclaimer to inform users when they are with AI systems, especially for content generation.

AI Governance

AI continues to evolve and gain traction across industries; the need for a robust integrative governance framework is becoming increasingly critical. The role of this framework is to enable organizations to leverage the full potential of AI by creating trustworthy AI applications with a strong alignment to business objectives. It ensures that the organization creates required structures and policies to guide the responsible and effective use of AI.

After establishing the need for a robust AI governance framework, let us now define AI governance.

CHAPTER 8   THE ETHICAL FRAMEWORK: BUILDING TRUST IN AI THROUGH HUMAN VALUES

An AI governance framework is a collection of practices, policies, and structures that ensure AI systems developed within an organization are ethically and legally compliant while remaining aligned with the organization's business goals. It guides the responsible use of AI technologies by maximizing benefits and minimizing potential harms. A strong governance framework ensures that AI operates fairly, safely, and transparently, and that any potential risks are identified and effectively managed in a timely manner.

Why It Is Relevant for Product Leaders

As product leaders, it is our responsibility to own the roadmap and development of AI applications. Having a understanding of AI governance enables us to define best practices required to deliver on our responsibilities. The following are some advantages:

- Mastering AI governance helps product leaders design ethically responsible products that help in increasing user trust. This increase in user trust, as we all know, helps in long-term user retention.

- By integrating AI governance best practices into the AI product development lifecycle, you develop products that are ethical and user-centric, which enhances brand reputation. Apart from being ethical, these products are also strategically aligned with the organization's goals, which further helps in improving financial health of the organization.

- By implementing AI governance best practices, product leaders can minimize legal and reputational risks by proactively addressing every aspect of evolving AI and data regulation.

CHAPTER 8   THE ETHICAL FRAMEWORK: BUILDING TRUST IN AI THROUGH HUMAN VALUES

- It helps improve cross-functional collaboration required for the development of an AI product by aligning goals and expectations of different teams with each other. This increases the likelihood of successful AI product outcomes.

AI Governance Model

A robust AI governance model should follow a three-tier model, interconnected structure. This structured approach helps organizations in designing tailored best practices suited for their unique needs while ensuring alignment with external regulatory changes, ethical standards, and evolving stakeholder expectations.

This framework is not just a collection of static polices and guidelines, but it is an implementation rulebook that governs all aspects of the AI product lifecycle. Another critical aspect of this framework is its adaptability, as the implementation layer provides feedback governance policies, and structures can evolve accordingly to remain relevant and effective. Figure 8-2 is a simple diagrammatic representation of this framework.

CHAPTER 8   THE ETHICAL FRAMEWORK: BUILDING TRUST IN AI THROUGH HUMAN VALUES

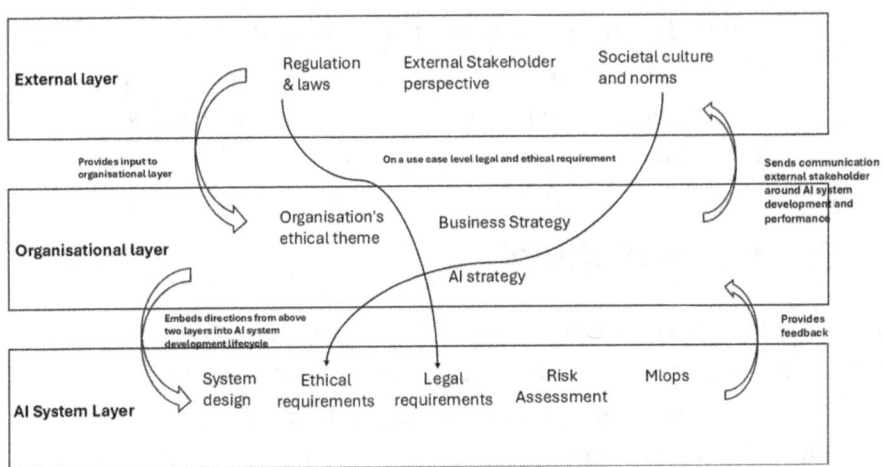

Figure 8-2. Block diagram of AI governance model

External Layer

This is the most critical layer in the AI governance model as it is this layer that provides the external view to the organization. Organizations should develop methods to periodically check the ethical and legal requirements of each region in which their product will be launched. Efforts should be made for developing mechanisms to capture the voice of stakeholders including users, partners, shareholders. This information should be used by downstream layers while developing their polices and structures

Organizational Layer

The role of this layer is to assimilate the information provided by the external layer and use it for developing the strategic and ethical governance principles and processes required for aligning AI systems with organizational values and goals. This includes the following activities:

- Setting guidelines that reflect ethical principles aligned with organizational values. These guidelines play a vital role in determining use cases in which AI will be used and in what capacity.

CHAPTER 8   THE ETHICAL FRAMEWORK: BUILDING TRUST IN AI THROUGH HUMAN VALUES

- Development of AI strategy and roadmap (refer to Chapter 3 for more information)

- Methods for AI risk assessment and mitigation (covered earlier in this chapter)

- Another critical activity in this space is development of data operating model to enable sustainable AI development. Refer to Chapter 4 for more information.

- Procedures for ongoing monitoring and auditing of AI systems are essential to ensure they are functioning as intended and not deviating from their ethical guidelines. Refer to Chapter 7 for more information.

AI System Layer

This layer is responsible for the actual design and development work required for an AI system. This is where AI engineers and data scientists focus on the technical aspects of creating and optimizing AI models, ensuring that the systems comply with the ethical and legal standards outlined at the higher levels.

Since we have already covered the implementation details in other chapters, we won't repeat them here.

AI Risk Management

While traditional software systems operate on fixed, rule-based logic defined by human programmers, AI systems learn their behavior from data. This fundamental difference introduces unique risks specific to AI applications. AI systems exhibit dynamic behavior and may behave unpredictably at times. Their performance can degrade over time due to shifts in data patterns or algorithmic issues, and they may amplify hidden

CHAPTER 8   THE ETHICAL FRAMEWORK: BUILDING TRUST IN AI THROUGH HUMAN VALUES

biases in the training data or introduce new ones. In addition to their heavy dependence on data, the opaque nature of their decision-making processes makes diagnosis and corrective action more challenging.

As a result, traditional risk management frameworks are not fully equipped to address the unique challenges posed by AI. The factors below highlight how AI systems differ from traditional software systems in terms of the types and complexity of risks they present.

1. **Data Quality and Representation:** Traditional software is built on deterministic logic with well-defined input and output, whereas AI systems learn by training on real-world data. This data might have some inherent shortcomings such as under representation of certain categories, bias patterns, incomplete records, etc. All these anomalies pose a risk to the performance, reliability, safety, and fairness of the model.

2. **Bias and Fairness:** Traditional software systems exhibit predictable behavior because they operate strictly according to the logic defined by programmers. Unless this deterministic logic includes biased rules, such systems generally remain safe to use. In contrast, AI systems can inherit or even amplify biases present in their training data and may introduce new biases when certain groups or categories are underrepresented. Moreover, an AI system that initially performs without noticeable bias may develop biased behavior over time if it is retrained on new data that contains imbalances or inaccuracies.

3. **Data complexity and dependency:** Traditional software has structured and predefined inputs that make it easy to handle uncertainties in incoming data. AI systems depend on large, varied datasets making them sensitive to inconsistencies or missing data. This is a continuous dependency as every time a model is retrained, the model performance is at risk. AI systems have a dependency on the quality of production data used for making predictions or recommendation. If the trend of live data differs significantly from the training data, it can affect the model's performance adversely, thus increasing the risk of inaccurate or unreliable outcomes.

4. **Training process sensitivity:** Traditional software does not require any training; it has standard development cycles in which predefined logics are developed and tested. On the contrary, AI models rely heavily on training themselves on data, and any shift in data trends or errors in training procedures can significantly impact its performance.

5. **Context drift:** Any change in real-world data does not affect traditional software's performance. They perform consistently as long as their logic stays unchanged, whereas AI system performance degrades if real-world data evolves away from the original training.

6. **Privacy and maintenance:** Since traditional software systems have deterministic rules and limited input data, it is easier to apply and handle PII data. Maintenance of these systems generally

involves standard operations. AI systems often process large and varied forms of data including sensitive data, both in raw form and in transformed form. This exposure to a huge amount of data both in volume and variety increases the privacy risks and required regular updates due to evolving data.

After understanding how AI systems are different than traditional software systems, now it's time to explore types of risk specific to AI systems. In the next section, risks are classified into categories for better clarity.

Types of AI Risks

While AI systems present powerful capabilities that can solve numerous problems, they also come various technical risks that can impact their performance, reliability, and fairness. The following is a detailed explanation of major technical risks:

- **AI system performance risk:** Due to its dynamic and opaque nature, AI systems carry ongoing performance risk. These risks are not limited to the initial development phase, but they persist or may increase with each retraining cycle.

  - Insufficient or poor-quality training data can lead to inaccurate models.

  - Overfitting, where the model learns the specifics of the training data too closely, can cause it to fail in new, unseen situations.

  - Underfitting, where the model doesn't learn enough from the data, may also lead to inaccurate predictions.

- Inadequate model design or the use of inappropriate algorithms for the task at hand.

- Lack of post-deployment monitoring may adversely affect AI system performance as it may fail to detect data drift or concept drift.

• **Algorithmic bias and discrimination:** These biases occur when AI systems produce biased results, which are the outcome of incorrect assumptions in the machine learning development process or biases present in the training data. The following are the main reasons behind this type of risk:

  - **Bias in training data**: If the data used to train the AI model is unrepresentative or contains historical biases (e.g., underrepresentation of certain demographic groups), the model will reproduce or amplify those biases.

  - **Feature selection**: Biases may arise if certain features or attributes, such as race, gender, or age, are either overemphasized or ignored, leading to discriminatory outcomes.

  - **Bias in algorithms**: The design of the algorithm itself can introduce bias if it prioritizes certain outcomes or groups inappropriately.

• **Lack of robustness and generalization:** Robustness refers to how well an AI model performs under a variety of conditions, including edge cases and adversarial attacks. Generalization can be defined as the model's ability to apply what its learning to new and unseen data. The following factors contribute to this risk category:

- **Overfitting**: When a model is too closely tuned to the training data, it may fail to generalize to new situations.

- **Lack of diversity in the training set**: If the training data is not diverse enough or doesn't cover a wide range of real-world scenarios, the model may struggle to generalize and perform poorly when faced with novel inputs.

- **Adversarial inputs**: AI models can be vulnerable to small, imperceptible changes in the input data that cause them to make incorrect predictions (known as adversarial attacks).

- **Data Quality Risk :** Refers to the risk associated with the poor quality of training data used in AI systems. Poor-quality data can lead to unreliable and biased AI systems.

  - **Missing or incomplete data**: If the training data is incomplete or missing key information, the AI model may make incorrect predictions or fail to capture essential patterns.

  - **Noisy data**: Data that contains irrelevant, incorrect, or random information can confuse the AI model and degrade its performance.

  - **Outdated or stale data**: If the data used to train the AI is not current or fails to reflect recent trends or changes in the environment, the model's predictions can become irrelevant or inaccurate.

- **Imbalanced data:** When the data is heavily skewed toward one outcome or class, the AI model may develop a bias toward that outcome, resulting in deficient performance for other classes or outcomes.

- **Operational Risk:**

  System architecture of AI system follows a complex design that includes interaction between database, AI models, legacy systems, etc. This makes AI systems vulnerable to significant operational risk caused by bugs, hardware issues, poor testing, and security threats like adversarial attacks or data breaches. Moreover, integrating AI with non-AI applications and feeder systems can increase failure risk due to compatibility issues and potential data inconsistencies, which may increase the risk of integration failures or delays.

- **Ethical and Legal Risks in AI Systems:**

  AI systems present significant ethical and legal risk owing to its high dependence on data, the opacity of its decision-making. The AI system being used as decision assistants today if not developed right may adversely affect the society.

  These risks include privacy violations of users, which may arise from uninformed data collection to uniformed data usage. While this poses a privacy risk, it also increases regulatory compliance risk for the organization. Another key concern is the misuse of AI, such as the potential for AI-driven surveillance systems to infringe on individual freedoms or the automation of harmful tasks, which could lead to ethical dilemmas, job displacement, or the reinforcement of negative societal biases.

CHAPTER 8   THE ETHICAL FRAMEWORK: BUILDING TRUST IN AI THROUGH HUMAN VALUES

Risk Assessment and Mitigation in AI

Risk assessment is critical for AI success, as it helps organizations proactively identify and mitigate potential threats that might adversely affect AI system. By conducting a thorough risk assessment, organizations can ensure their AI systems are secure, reliable, and aligned with both legal and ethical standards. With our experience and consultation with other industry leaders we have produced the following actionable steps that teams should incorporate in their risk assessment and mitigation framework:

1. Define objectives and scope: This step focuses on bringing clarity to what this AI system will achieve. Clarity on objective helps in identifying potential risk, which might impact the system.

2. Employ risk identification techniques such as the following:

    a. Conduct regular brainstorming sessions with cross-functional teams to identify potential risks.

    b. Identify critical parameters that may affect AI system overall health and use this information in conducting scenario analysis. This helps identify both obvious and less apparent risks.

    c. Once potential risks are identified, employ techniques like Fishbone diagrams (Ishikawa) or Failure Modes and Effects Analysis (FMEA) to analyze the root causes behind potential risk. This helps in identifying risks and possible failure points within the system.

CHAPTER 8   THE ETHICAL FRAMEWORK: BUILDING TRUST IN AI THROUGH HUMAN VALUES

    d. Another way to identify potential risks and potential reasons behind these risks is by analyzing the organization's historical incident report and reviewing public case studies of AI failures or challenges.

    e. Categorize risks into distinct categories: technical, operational, ethical, and legal risks.

3. Evaluate and Prioritize Risks: Once risks are identified, prioritize them in the descending order in their occurrence likelihood and its potential impact on organization and AI systems. This step helps allocate resources efficiently and ensures the most critical risks are addressed first. This step involves first assessing risk occurrence likelihood and then assessing the potential impact on the AI system and organization if that risk actually gets released.

4. **Risk Likelihood Assessment:**

    Likelihood Rating: For each identified risk, determine how likely it is to occur based on historical data, technical analysis, and expert judgment. Likelihood can be rated on a scale (e.g., Low, Medium, High or on a 1-5 scale).

    a. *Low*: Rare or unlikely to occur

    b. *Medium*: Possible but not frequent

    c. *High*: Likely to occur, potentially regularly

Impact Assessment:

Impact Rating: Evaluate the severity of the risk's consequences if it were to occur, considering both short-term and long-term effects. Impact can be rated similarly (Low, Medium, High, or 1–5).

   a. *Low*: Minimal damage, easily manageable

   b. *Medium*: Noticeable damage but recoverable

   c. *High*: Severe consequences that could disrupt operations, reputation, or compliance

Quantification of Impact: Quantify potential impacts wherever possible. For example, for privacy violations, assess financial penalties (GDPR fines), reputational damage, and legal costs. For model inaccuracies, quantify the potential cost in terms of lost revenue, inefficiencies, or safety risks.

Risk matrix/risk heat map:

Create a risk matrix (also called a risk heat map) that plots each risk according to its likelihood and impact. This visual tool helps quickly identify which risks are most critical. The matrix is typically divided into four quadrants:

   i. High Likelihood & High Impact (Immediate Attention)

   ii. High Likelihood & Low Impact (Preventive Measures)

   iii. Low Likelihood & High Impact (Contingency Plans)

   iv. Low Likelihood & Low Impact (Monitor and Maintain)

Prioritize Based on Severity and Resources:

a. Focus on risks in the High-Likelihood, High-Impact quadrant first, as these pose the greatest threat to the AI system's success and safety.

b. Medium-Priority Risks can be handled next, using fewer resources but still requiring active mitigation.

c. Low-Priority Risks should be monitored, but they may not require immediate intervention unless new information suggests they are evolving.

Stakeholder Input:

Engage relevant stakeholders (e.g., technical teams, legal advisors, business leaders) to review and confirm the risk prioritization. Different perspectives can help refine the likelihood and impact assessments, ensuring that the most significant risks are not overlooked.

AI Risk Mitigation Techniques

Once risks have been identified, evaluated, and prioritized, it's time for developing their mitigation strategies that help in reducing the likelihood of the risk occurring, and if they occur, this plan will help in minimizing their impact.

AI risk mitigation plan typically consists of preventive measures that are aimed at reducing the likelihood of occurrence and contingency plans for reducing the impact of risk in case they occur. Product leaders must employ all of the following techniques and procedures at the appropriate stage of the AI product development lifecycle.

a. Risk Response Plan: Work with stakeholders to develop a risk response plan for each identified risk. This will help expedite the response to counter the threat posed by that risk. It should also contain incident response strategies that would include both preventive and contingency measures required to mitigate the risk. This plan should also contain escalation procedures for critical task with defined SLA, ensuring that relevant stakeholders are notified at the earliest.

This plan should be reviewed and updated regularly based on lessons learned from the incidents or near miss instances.

b. Conduct regular assessments. Depending on type of risk (high, low, medium), the frequency of assessments should be determined. Apart from assessing existing risk, continuously track the operational environment, user interactions, and regulatory changes that may introduce new risks.

c. Create a risk register to document identified risks, assessments, mitigations, and monitoring strategies. Relevant stakeholders should be provided with regular updates of risk register.

d. Conduct regular training and assessment for employees on AI ethics, security protocols, and risk management procedures. Apart from educating employees, these trainings should make them aware of potential risks and ways to handle them effectively.

CHAPTER 8   THE ETHICAL FRAMEWORK: BUILDING TRUST IN AI THROUGH HUMAN VALUES

e. Conduct internal audits to ensure that the AI system is compliant to relevant laws and regulations.

f. Develop a comprehensive AI system monitoring and control mechanism. It should involve the development of a model monitoring system. This system should monitor AI system behavior, performance, and any other anomaly that may jeopardize the AI system. Apart from production system monitoring, maintain detailed logs of AI decisions and actions taken to ensure accountability and traceability.

g. Develop mechanisms for users to submit feedback or report any risks or issues not covered in the initial assessment.

h. A robust data operating model should be developed in the organization to ensure that data used for AI systems is trustworthy.

i. Prepare a comprehensive list of data validation and cleansing techniques that should be employed for data preprocessing during model training phase.

j. Employ robust model training techniques like cross-validation, ensemble learning, and regularization techniques to ensure the AI model can generalize its learning well.

k. Implement AI interpretability and explainability frameworks to make AI decisions more transparent and understandable.

CHAPTER 8   THE ETHICAL FRAMEWORK: BUILDING TRUST IN AI THROUGH HUMAN VALUES

    l. Implement penetration testing and adversarial attack simulations to identify vulnerabilities in AI models.

    m. Employ strong encryption protocols for data storage and transmission to prevent unauthorized access. Use multifactor authentication and least privilege access models to safeguard sensitive information.

    n. Conduct regular bias audits to check for discriminatory patterns in AI outputs. Utilize fairness algorithms to reduce bias during model training, such as re-weighting training data or adjusting decision thresholds to ensure fairness across demographic groups.

Robust AI

"Robustness in AI is like a seatbelt: You might forget it's there most of the time, but when things go sideways, you'll be glad you're strapped in!"

An AI system is called robust only when it can perform consistently and accurately under a variety of uncertain conditions including handling of adversarial inputs. In simpler terms, a system can be called robust when it can work well both in expected and unexpected situations without failing abruptly or producing erroneous outputs. A robust AI system has following characteristics:

- A robust AI system should be resilient to uncertainty. This means that when an AI system is exposed to incomplete, noisy, or adverse data, it should be perform as per its design and expectations.

CHAPTER 8   THE ETHICAL FRAMEWORK: BUILDING TRUST IN AI THROUGH HUMAN VALUES

- They should have high tolerance to minor changes in input or environmental conditions, and its performance should witness a drastic drop.

- Robust systems should be able to handle errors gracefully and have built-in mechanisms to gradually recover from it. It includes fault detection, graceful degradation recovery mechanisms, and redundancy to ensure continued operation in case of failure.

- Robust AI systems should be able to generalize its learning and apply to similar data without experiencing must decay in performance

Design Principles for Robust AI

After developing an understanding of AI robustness, let's now move ahead to explore design principles that support building robust systems. These principles include a combination of system design principles and algorithmic techniques aimed at enhancing the robustness of AI systems.

- While designing system architecture of an AI system, modular architecture should be adopted. This ensures that individual modules can be developed, tested, and updated independently. Design patterns like microservices can be used to develop modular system architecture.

- Use a configuration driven design that can support model hyper parameter tuning without any requirement of redeployment of a complete AI system.

- Design system modules in way that they can be reusable across different applications.

- Implement redundancy for critical computing and storage resources to ensure to ensure continuous operation during hardware failure.

- For critical modules that comprise critical AI functions such as component where the algorithm is hosted, implement component-level redundancy by hosting the same component on multiple servers. Support component-level redundancy with auto switch techniques so that when the primary module fails, automatically the backup module takes over and overall system availability is not affected.

- Implement data replication across multiple locations to safeguard against data loss. Data replication should be supported by robust data synchronization mechanisms to ensure consistency and availability even if one location experiences downtime.

- Develop a system architecture with checkpoints placement in data pipelines for easy rollback in the case of failure.

- Replication and partition tolerance mechanisms should be implemented to ensure functionality availability during component failure.

- Establish robust error detection, logging, and recovery mechanisms to quickly resolve issues without affecting users.

- Design an AI system with a human-in-the-loop approach. The system architecture should be such that in all circumstances human input should supersede the system's output.

CHAPTER 8   THE ETHICAL FRAMEWORK: BUILDING TRUST IN AI THROUGH HUMAN VALUES

- During the training and testing phase, expose AI systems to adversarial data to understand system behavior in adversarial situations. It will also help in identifying the potential vulnerabilities of the system.

- Design the system in way that in the event of system crash, the AI system shuts down in a graceful manner. It should ensure that, if possible, the system stays available even if at a reduced version of the full service. Employ the following techniques:

    - Prioritize most critical services during failure. Design a simpler version of those services so that in case of a system outrage, at least a simpler version of that service is available for the users. Take, for example, an AI-powered search system of an ecommerce company. If the advance ranking system fails, then revert to basic keyword bases search with default ranks.

    - Use patterns like elastic scaling in a cloud infrastructure to adjust system resources based on demand. It will ensure the system does not get into an overwhelming state in the case of a sudden surge in traffic. Augment this approach with model pruning techniques to reduce model complexity while retaining most essential features of the model during a sudden surge in traffic or error.

    - Always support your main AI model with simpler models built with similar features. These models will keep the AI system available in situations where the main model is unavailable due to any reason.

- Always include dynamic load balancing techniques to distribute the workload evenly across servers.

- While training the AI model, employ the ensemble method to improve model robustness by combining multiple models to reduce the impact of individual model errors or failures. It also provides ability to generalize models learning and stabilize its performance especially on noisy data.

- Another technique that can train and develop a robust AI model is regularization. This technique enhances AI robustness by preventing a model from overfitting its learning on the trends of training data. Regularization ensures that the model generalizes well to unseen data.

Summary

This chapter covered one of the most critical aspects of AI product management, i.e., how to develop a trustworthy AI system. This chapter explored factors that impact a human's ability to trust and then applied those factors to develop a framework to develop trustworthy AI. This framework is named the Trustworthy AI framework. While detailing the trust framework, we covered the ethics principles of fairness, safety, accountability, and inclusivity. We also discussed the AI risk assessment framework, the governance model, and finally robust AI. The next chapter of this book will help users understand how to design AI products for real-world applications.

PART IV

AI in the Real World

CHAPTER 9

Designing AI-Powered Products

As artificial intelligence continues to disrupt digital experiences, a tectonic shift in mindset is required for designing these products. Traditional product design practices and principles must evolve to accommodate systems that are dynamic, probabilistic, and continuously learning.

This chapter explores how to design trustworthy AI-powered products that are aligned with human needs. It will provide guardrails for product managers and designers to rethink the ways they used to build, test, and evolve AI-powered products.

In this chapter, we will delve deeper into the challenges and opportunities that come along with AI integration while designing adaptive user experience and handling probabilistic outcomes to embedding transparency and feedback loops.

The goal of this chapter is to expose product management leaders and designers to AI-first product thinking principles required for designing trustworthy AI products

From a product management perspective, this chapter focuses on how to scope AI-specific design requirements aligned with business objectives. It will also help product managers make informed decisions and trade-offs between complexity, user value, and ethical responsibilities while collaborating with the design and engineering team to deliver an effective and trustworthy AI-powered experience.

CHAPTER 9   DESIGNING AI-POWERED PRODUCTS

Unlocking AI's Value in Product Design: Benefits and Designer Strategies

Integrating AI into product design gives strategic capabilities that unlock new and informed ways to drive engagement, efficiency, and growth. These capabilities also help in expanding the design space beyond static flows to dynamic learning systems. AI helps teams build intelligent, personalized, and future-proof products. The following are a few key benefits that an organization gets by integrating AI into their product design frameworks.

Personalization at Scale: Implementing AI enables the development of products that offer personalized user experiences that in turn deliver tangible benefits like an increase in metrics like user retention, user conversion, and revenue per user without increasing the development effort substantially. To make this personalization meaningful, a nonintrusive designer must shift from designing fixed screens to architecting modular systems that can respond to dynamic responses with changes in user behavior and context. Rather than designing static flows, designers must now spend time and effort in developing decision boundaries, designing confidence thresholds, and thinking about how various types of bias should be mitigated. It requires a mindset shift from designing what users will do to designing how the system will evolve from what users do.

Predictive User Experiences: It improves the user experience by introducing predictive features like autofill, smart suggestions, and proactive alerts to help streamline workflows and help users achieve their goals easily. While designing such features, designers must balance guidance with control and must anticipate user intent while preserving transparency.

Adaptive Prototyping and Responsive Interactions: AI can help to automate and shorten the prototyping stage by predicting how users would respond under different scenarios. This not only shortens the time

CHAPTER 9 DESIGNING AI-POWERED PRODUCTS

to market but also ensures edge cases are covered. To achieve these goals, designers must use AI prototyping tools for adaptive UI sketches and behavior simulations to test user experience under varied scenarios, edge cases, and evolving states.

Why Product Managers Should Care About AI System Design

As AI becomes increasingly embedded in digital products, being an AI-fluent product manager will not suffice. Product managers must also understand how AI products are designed and experienced by users. As AI introduces a probabilistic nature to the systems and products, AI system design has now become a strategic lever that will govern all significant product metrics like product adoption, user trust, performance, etc.

Product managers play a critical role in building the AI-powered functionality of the product by guiding the training dataset structure, considering ethics, and shaping up user feedback loops. Developing a high-level understanding of how AI systems must be designed enables product managers to make better decisions and trade-offs while designing and building AI features to serve both user and strategic needs. The following are some keys reasons why it's important for product managers to understand how AI products should be designed.

AI product design has a direct impact on all key product metrics. While in all digital offerings the user experience has a significant impact on critical product metrics, designing a thoughtful user experience in AI products has become more even critical. Unlike traditional rule-based systems, AI brings in a dynamicity to the system behavior and often acts like an opaque box offering limited explanations behind its behavior. This nature creates unique design challenges as now interfaces must clearly communicate the AI behavior in a trustworthy and user-friendly manner. If the interface is confusing or inconsistent, even the most powerful

CHAPTER 9 DESIGNING AI-POWERED PRODUCTS

AI product will fail to deliver business value. Product managers must understand what qualifies as good UX design for AI products. They should be able to identify and express requirements for designing an intuitive interface that can guide users through dynamic AI outputs and manage user expectations at every step. Another key dimension of AI systems is they evolve over time and are always suspectable to unintentional bias. Product managers must ensure that UX design is free from any such bias.

Bad Design = Bad Data = Bad Models

AI systems run on data; if the user interface design is confusing or doesn't clearly guide the users for eliciting right behavior, then all you collect is poor-quality signals with no true intention. These poor signals (event data that is captured by product teams for analyzing user behavior) will be used by the AI team to retrain the same model or develop a new model, which will in turn adversely affect product performance. So, in the AI world, UX is not just the skin of the product; it shapes the data pipeline and learning loop.

Bad UX Is Regulatory and Ethical Risk

If the product interface fails to communicate clearly the AI decisions and intentions behind it, the product will lose user trust, which will impact product usage. A decrease in product usage will have an adverse effect on the strategic goals. Such interfaces also lead to regulatory and ethical risk because of the lack of understanding may trigger users to file lawsuits and or write bad reviews.

Cross-Functional Leadership Starts with Shared Understanding

Product managers are often the de facto leader of the product squad, which means they lead a cross-functional team without authority. Understanding design's role in the success of AI products helps a PM ask right questions, align trade-offs, and make decisions that balance feasibility, value, and ethics.

CHAPTER 9   DESIGNING AI-POWERED PRODUCTS

Core Techniques and Patterns for Designing AI Products

This section outlines key principles required for translating AI capabilities into an intuitive and ethical user experience.

Start with the Human, Not the Solution

Designing the product using human-centric design principles will ensure that designed AI systems fit into the user journey in the best possible way and not the other way around. Human-centric design focuses on understanding humans, their needs, and the circumstances in which that need arises first. This helps in understanding the problem space from a user's perspective (also called *empathy*). Once the problem space is well understood, only then should product managers proceed ahead with implementation details like selecting the training data, the type of training methodology, and the algorithms. Another key consideration is involving users throughout the solution design phase and using their feedback to make the product better. The following pointers should be referred to:

- What problem is the user trying to solve?
- How will the AI support or augment their decision-making?
- What context (emotional, situational, environmental) should the AI be aware of?

Create an Explainable System Design

AI systems whose decisions are opaque are often not trusted by users. By designing AI systems whose decisions can be explained through visual means, designers can make user feel more confident and in control while using the system. The following design strategies can be used for making decisions explainable:

CHAPTER 9   DESIGNING AI-POWERED PRODUCTS

- Use tooltips or labels to explain AI-driven suggestions.
- Indicate the confidence levels or rationale behind decisions.
- Allow users to access more detail when needed (progressive disclosure).
- Label AI features explicitly (e.g., "Suggested by AI").

Make Feedback Loops Observable

Users are often not aware that AI learns from their actions. Making feedback loops visible promotes a sense of shared control and enables users to comprehend how their actions affect future results.

These are methods for taking feedback into consideration:

- "We suggested Y because you liked X."
- Permit users to critique or amend AI recommendations.

Maintain Human Agency and Control

AI systems should be designed in a way that users should always be in in charge of their experience. The role of AI should be limited to only assisting users achieve their goal.

Design examples:

- Provide clear "undo" and override options.
- Permit users to modify AI-generated results, such as recommendations.
- Avoid strict automation for task where judgment or morals are important.

Handle Errors Gracefully

AI systems should be designed to handle error gracefully. Due to their dynamic and complex nature, AI systems are often error prone. AI systems should be designed to handle error gracefully These are best practices:

- Provide fallback options or manual control when AI fails.
- Apologize and explain the error when appropriate.
- Ensure recovery paths are user friendly and easy.

Ensure Accessibility and Inclusion

AI systems should be designed and developed to deliver equal value for all types of user groups including linguistic, cultural, ability, and socioeconomical dimensions. These systems should bias free.

Design considerations:

- Avoid bias in training data and output.
- Ensure screen reader and keyboard accessibility for AI features.
- Design for edge cases—not just the average user.

AI Interaction Patterns: Personalization and Recommendation Systems

So far, we have discussed how AI is changing UX design parameters and what the key principles are that designers should consider when designing the interface for AI products. We also examined why it is essential for a product manager to develop an understanding of complexities and unique considerations of the AI design system. In this section, we will explore what personalization is, including its types, and what product managers and designers should do to make it ethical and user friendly. We will start by defining the term *personalization*.

CHAPTER 9 DESIGNING AI-POWERED PRODUCTS

Defining Personalization: Tailoring Experiences to User Needs

Personalization in AI design is the process of customizing digital experiences to cater to individual needs, preferences, behavior, and context in near real time and at scale. It involves personalizing the delivery of content, functionality, or product. Personalization isn't just about customizing what the user wants but also extending to predicting user needs in advance and then delivering goods or services accordingly. Personalization can happen across the following dimensions:

- Behavioral personalization: Personalization on this attribute is based upon what actions a user takes on the product. It considers following activities, clicks, page visits, navigation paths, any transactions performed, etc. An example of personalization based on this dimension can be product recommendations based on your last purchase or products viewed on an ecommerce marketplace.

- Contextual personalization: When personalization is done on this dimension, the AI product offerings are tailored according to the current conditions like time of the day, geographical location, type of device, device operating system, and environment in which platform is used. Examples include personalized service recommendations made by a travel app based on the real-time location of the user.

- Intent-based personalization: Personalization of intent dimensions enables digital products to customize offerings based on the user's intent. Example of intent-based personalization can be a shopping company recommending an urgent delivery service or scheduled delivery service after a user searches for a gift for an occasion.

- Preference-based personalization: This type of personalization is delivered by customizing products and services on the preferences stated by the user in the past. Examples include a blog platform customizing a user's news feed with blogs.

- Demographic or segment-based personalization: Personalization on demographics can also be referred as *segmentation* as generally it is based on fixed user-provided attributes like age, gender, region, language, etc. Examples include a health app recommending different content and lifecycle recommendations to senior citizens versus teenagers.

In the modern world, the quality of personalized services that a digital product provides has become one of the critical factors driving product utilization and customer satisfaction. It turns passive interfaces into responsive, assisted systems. Delivering a truly personalized behavior on a platform depends not only on the performance of the AI algorithm but equally on the how well and thoughtfully the product's interface is designed to respond to changing user needs and preferences without overwhelming or alienating them.

Types of Personalization in AI

AI-powered personalization can take various forms, depending on the context in which it is applied and the strategic goals it is designed to achieve. Broadly speaking, it can be classified into three categories:

- Context personalization
- Service personalization
- Product recommendation

CHAPTER 9   DESIGNING AI-POWERED PRODUCTS

Let's now delve deeply into each of these categories, examining their definition and the specific responsibilities product manager and designers hold in enabling personalization to reach its full potential.

Content Personalization

In simpler terms, this type of personalization can be defined as dynamically adjusting what content (in all forms) is shown to an individual user based on their interests, past behavior, and preferences. This type of content is the backbone of all social media platforms. This type of personalization primarily focuses on improving metrics like user engagement and user satisfaction. Examples include:

- Home page personalization based on a user's watch history, geographical location, and content engagement
- News feed customized to your professional network and past interactions on LinkedIn

To deliver this type of personalization, product managers and designer have the following responsibilities to enable its full potential:

Product Manager's Responsibility
Product managers play a key role in aligning content personalization strategy with business goals. Their focus should be defining what "relevant content" means to their users and align it with the business objective. Another critical responsibility they have is to deliver the relevant content to their users at the right time and place. To achieve both responsibilities, the following list of activities are critical:

CHAPTER 9 DESIGNING AI-POWERED PRODUCTS

- Decide what strategic goal(s) this content personalization functionality will cater to and how. Once a business goal is decided on, then identify which product metric it will impact and how it will be measured. Set the KPI to measure engagement, retention, CTR, etc., to validate the efficiency of the personalization.

- Before you start working with the content team for content generation, align the content strategy with business goals.

- Once the content strategy is established, ensure there is enough diversity in content categories to cater to all the identified user groups. Diversification in content will prevent echo chambering and have a positive impact on user experience.

- Identify touchpoints in the user journey where this content will be consumed by the user.

- Define a roadmap including the success criteria to develop a personalization algorithm.

- Define relevance signals (clicks, dwell time, skips, etc.) to guide model training.

- Collaborate with different teams including the data engineering, software engineering, and artificial intelligence team to develop and deploy personalization algorithms.

- Establish rules for fallback content when personalization confidence for a specific user is low.

- Define ethical criteria that this feature has to fulfil.

CHAPTER 9   DESIGNING AI-POWERED PRODUCTS

Designer's Responsibilities

The designer's primary responsibility here is to ensure that the personalized content feed feels natural and initiative to users. Their focus should be on designing the interfaces that adapt easily to a user's changing needs and usage without overwhelming or alienating users and by staying under ethical limits. The following are some best practices that designers must follow:

- Designers must now design interfaces that are more modular in nature so that individual atoms and molecules can be changed to deliver a personalized experience to a user.

- Also highlight recommendations made by the algorithms under the "Recommendation for you" identifier. Another way to show content recommendations can be presenting both personalized content view and generic content view and segregating that content by placing a toggle, allowing the user to toggle between the personalized feed and the generic content feed.

- Rather than designing pages that can accommodate a fixed number of content bytes, design pages with progressive loading or infinite scrolls to accommodate additional content without hindering the user experience.

- Create feedback loops to collect user feedback on content recommendations without hindering user experience by using affordances like "not interested."

- Use subtle visual clues like small icons, color shades to indicate freshness, popularity, and relevance of content.

- Designers must design adaptable templates to accommodate different types and quantity of content.

- Ensure that the overall visual theme of the product stays consistent even as content varies from user to user.

- Design the interface in such a way that while it presents personalized content to users, it also enables scanability so that users can fine-tune their content feed.

- Both designers and product managers must ensure that neither the algorithm nor the interface over-personalizes the content. Always leave room for exploration.

Service Personalization

This type of personalization defines how a service behaves or interacts digitally with users based on their preferences, usage, and profile characteristics. Service personalization focuses not just on what is delivered but how it is delivered, which includes adjusting tones, workflows, and complexity for each customer. Examples include:

- E-learning platforms adjusting pace and complexity of the course on the basis user performance and engagement

- Chatbots adjusting tone, language, and details of the content on the basis of user response

Product Manager's Responsibility

Product manager's responsibility under this type of personalization includes aligning personalized services with user intent and business goals and operational capabilities. Special emphasis should be placed on

CHAPTER 9 DESIGNING AI-POWERED PRODUCTS

personalization service touchpoints that reduce friction and increase value across high-impact service moments. To accomplish this, the following are some recommendations for product managers:

- Conduct qualitative and quantitative research to identify user context, workflows, and service touchpoints where adding personalization will add real business value while enhancing user experience.

- Develop a decision tree based on user intent and other qualitative and quantitative attributes that could determine factors governing the dynamicity of service personalization.

- Take a strategic approach while deciding the granularity of personalization that will be offered. The question to be asked is, is it worth it to personalize workflow on an individual level, or is cohort-level personalization is sufficient?

- Define metrics to measure the effectiveness of the service personalization feature. Metrics like reduced time to resolution, adoption of the self-service feature, improved customer satisfaction, service completion rate, and reduced churn can be measured.

- Define the fallback rules for situations where personalized automation rules are enough.

- Develop a robust consent and opt-in mechanism for personalized automation and assistance features.

- Work with multiple teams to define, develop, and deploy service personalized automation workflow.

CHAPTER 9   DESIGNING AI-POWERED PRODUCTS

Designer's Responsibility

The following are some best practices that the designer should follow while designing interfaces for this type of personalization.

- Collaborate with product managers to identify touch points from where personalization should kick in without stereotyping users.

- Efforts should be made in designing adaptive workflows that shorten the workflow by skipping steps based on known user data.

- Personalize the user experience by applying techniques like addressing user by name, referencing past actions, etc.

- Always offer course correction steps like undo, delete, or escalation steps when personalization is not up to the mark. Implementing this step will empower user to control the way they want to experience the service and will also help product team to capture right signals around user behavior.

- If designers are working on a chatbot or virtual agent, then design various persona and tonalities to suit varying user profiles.

- Design different UI interfaces to match various conversational changes during the user interaction.

- Make use of subtle visual clues to highlight that a service or workflow is acting on prior content. Also, give the user an option to start fresh.

CHAPTER 9   DESIGNING AI-POWERED PRODUCTS

Product Recommendations

This type of personalization involves making products, features, or action recommendations based on a user's behavior, preferences, and purchase history or based upon the similarities between two users. Examples can include:

Amazon's frequently bought together feature or recommended for you feature.

Product Manager's Responsibility

While the product manager is responsible for delivering the entire product recommendation feature, the following best practices will help them to deliver the product recommendation feature that not only adds value to both users and business but also is ethical.

- As always, the first step will be aligning personalization needs with business objectives such as increasing the number of transactions or increasing the average basket size, reducing churn. Achieving clarity in the defining business goals will help in developing the right recommendation strategy.

- Once the clarity on goals is attained, then define metrics to measure the success of this feature. Effectiveness can be measured using KPIs like CTR, user conversion rate, customer satisfaction, increase in basket value, etc.

- Design the right recommendation strategy by first deciding on the user segments for which this algorithm will work. Collaborate with AI engineers to choose the right approach like collaborative filtering, content-based filtering, rule-based logic, or hybrid models. Consider the organization's AI readiness before choosing the appropriate approach.

CHAPTER 9   DESIGNING AI-POWERED PRODUCTS

- Decide on which granularity level this recommendation model will function, whether on an individual user level or cohort level. The decision should be made based on factors such as data availability, implementation, and model maintenance complexity and expected ROI.

- While working with AI teams on recommendation strategies, ensure that fallback scenarios are covered for handling cold-start scenarios gracefully.

- Ensure the ethical and transparent usage of data by making the user aware of the data usage policy. Create flows for empowering personalization preferences.

- Set up AB testing frameworks to evaluate different recommendation strategies.

- Establish feedback loops to capture data that can be used to refine recommendation algorithms.

Designer's Responsibility

By following these recommendations, designers will be able to create interfaces that unlock the full potential of personalization feature:

- Always design for adaptability, and create modular UI components that can adjust to the type and number of product recommendations dynamically.

- Always provide visual cues to explain why a particular product is recommended. Categorize product recommendations into segments based on the underlying rationale for each recommendation.

CHAPTER 9   DESIGNING AI-POWERED PRODUCTS

- Always design an interface with infinite scrolling and with swipeable containers to allow users to browse multiple products without disrupting the user experience.

- Allow users to rate the product recommendation quality by using widgets like rate, hide, flag, shortlist, not interested, and see less of this. This information will help the recommendation algorithm to fine-tune itself.

- Even the best personalization algorithm will fail if the recommendation widget is not placed at the right point in the user journey. Embed recommendations where they naturally support user goals and not where they feel disturbing.

- Limit the number of recommendations and ensure that don't compete visually for user's attention and core actions.

- Design the interface in way that the user feels in control of what they are seeing. Offer the user the option to adjust the personalization settings.

Balancing Personalization with Usability

Good personalization feels like helpful assistance, enhancing the user experience during throughout the journey toward achieving their goals on the platform. Over-personalization or bad personalization brings in commotion and unpredictability that can overwhelm the user in their journey. Therefore, striking the right balance is critical. Developing the right level of personalization requires a few essential components.

- Identify parts of the product where personalization can deliver high value to both businesses and customers.

- Implement personalization at points where it meaningfully reduces the friction previously experienced by the users.

- Design interfaces that clearly communicate which personalization features are active and why they matter to the users.

To summarize, good personalization can be defined as follows:

Good personalization = High value implementation + Reduction in Friction in journey + High clarity interface

To develop a good personalization, product managers should define acceptable boundaries for what can be personalized and what cannot. They should develop mechanisms to track the performance of the personalization feature and should also track a user's feedback on personalization.

On the other hand, designers should ensure that the UI remains consistent throughout the personalization experience. Use progressive disclosure to introduce personalized elements gradually, allowing users to adapt at their own pace. At no point should the user feel overwhelmed or lose a sense of control while interacting with personalized features.

Navigating AI Interaction Design: Common Pitfalls and Best Practices

As AI becomes increasingly integrated into digital products, it has introduced newer interaction patterns that can confuse, frustrate, or adversely affect users if not designed carefully. Unlike traditional products, AI systems evolve with user data and behavior, which makes their failure occurrence unpredictable and even harder to diagnose. In this section, we will explore reasons behind common pitfalls in AI-driven interfaces and how product teams can anticipate and design around them to create more trustworthy AI products. The following are some common pitfalls.

- Overpromising AI capabilities: Often companies oversell the AI capabilities of their platform, which sets unrealistic expectations of the users, and when those expectations are not met, it leads to user disappointment, which in turns affects platform utilization.

  For example, early chatbots promised human-like conversations but failed due to lack to capability, leaving users frustrated. Designers should use progressive disclosure to introduce AI features gradually. Clearly state AI limitations upfront early in user journey.

- Lack of transparency: At times there is no clarity on why a certain decision was made by AI. The user does not understand the reasons that a certain decision was taken by AI. This opaqueness affects user's trust and confidence on the system adversely. Designers should make the interface more explainable by using visual cues to highlight recommendations made by AI. Another way of handling this is to create interactive explanations that can enable users ask questions around AI decisions.

- Loss of user control: Often an interface is designed such that it lets AI take silent action on behalf of users without alerting them. This makes the user feel out of control and makes them feel disoriented, giving them a feeling of being lost. To avoid this, the interface should give clear visual cues to highlight the AI-driven steps. The interface should provide easy undo/redo after every AI-driven change. The interface should also allow users to customize the personalization intensity or opt out from the feature completely.

- Inconsistent UX behavior: AI products are highly dynamic, and due to this dynamicity often the information shown on the interface changes in quantity and variety. At times, the interface design and layout change for handling different scenarios, which makes the UI behavior very inconsistent. This inconsistency disturbs the user journey, which adversely affects their satisfaction level. To avoid these types of inconsistencies, designers should maintain UI consistency with subtle AI-driven changes. Visual stability principles should be applied to ensure elements don't shift unexpectedly.

- Ignoring privacy concerns: Sometimes teams collect and use data to serve advertisements or make personalized recommendations without getting consent from users. Even if consent is received, it is done in a vague and crude form such as mentioning the data collection and usage clause in the overall platform policy and using very vague terms to describe data collection and usage terms or making it mandatory. All these methods raise privacy and ethical concerns that may lead to regulatory issues. Designers must ensure that concise privacy notes are placed near personalized content. The interface must provide granular privacy controls, letting users opt in/out of specific data from getting collected and used.

- Poor error handling: Often algorithms and interfaces are designed and developed to handle commonly occurring scenarios and edge cases are often overlooked. Even when such cases are handled, they are often addressed in a generic way, providing no insight into why the system is behaving unexpectedly. The designer should design graceful fallback routes on the interface to handle errors subtly or to design alternate ways to achieve goals. In other words, if AI fails to deliver satisfactory results, enable the user to switch to manual mode or generic recommendations.

- Over-personalization: Product managers are often tempted to overly personalize the entire platform experience, with the belief that more customizations will add more value to customers. This can result in overengineered workflows, fragmented journeys, and confusing interfaces, making it harder for users to understand how the system works or how to control it. When every part of the user journey behaves differently for every user, the predictability and trust of the system begins to erode. To avoid such situations, product managers should resist from such temptations and personalize only high-value touch points in the customer journey. Designers should ensure that the core UI structure will remain consistent even with personalized content or workflows by using similar layouts, labels, and navigation patterns. This helps users familiarize themselves and avoid feeling alienated every time they return. Another best

CHAPTER 9   DESIGNING AI-POWERED PRODUCTS

practice that the designer should follow is to design the interface with a clear default or reset option. This can be accomplished by giving users an option to reset the interface with a generic view. This option will allow users to anchor themselves somewhere in case they feel overwhelmed with their personalized journey.

- Bias reinforcement: AI systems often unintentionally exhibit biases or mirror societal stereotypes primarily through the customer behavior or usage data, which acts as the feedback loop for recommendation systems. This data also acts as the dataset for retraining the recommendation system. It not only affects the users adversely but also prevents organizations from exposing customers to a variety of products services or content. At worst, models trained on skewed or historically biased data can perpetuate discriminatory patterns in sensitive areas like healthcare, recruitment, financial services, etc. While it affects users adversely, the biasness in systems also pose a reputational and regulatory risk to the organization. The primary reason behind this is often the AI team responsible for developing the recommendation system tends to optimize for accuracy as the only barometer to gauge the efficiency of the system. This leads to overfitting of algorithm's logic on data trends in historical data, which hardens this biased behavior. Another primary reason is the underrepresentation of certain user groups in data due to historical reasons, which makes algorithms less exposed to these user groups. To avoid this, product managers should define fairness metrics to evaluate model output. Another step that should be taken at the start of the development lifecycle is to work with data teams to evaluate training data on factors that can cause biasness and take remedial actions to resolve them. Personalization outcomes should be diversified intentionally by showing the user new content or randomly picked content. Designers should design interfaces that give users control over their experience on platforms. All AI-added content should be highlighted so that users can understand where AI influence ends. The designer should avoid using imagery, language, and icons that may stereotype a specific user group.

Practical Framework: Balancing AI Personalization and Privacy

Personalization has become a central feature of modern digital product management. It helps increase relevance and user conversion and simplifies the user journey. If done right, it has the potential to impact key strategic goals of the product in a positive way. But when it becomes hyper targeted and overly intrusive in ways that users don't expect, the same personalization may feel invasive and offensive. This creates a delicate tension between balanced personalization and over-personalization.

This section explores the tension between user privacy and personalization, outlining key strategies for developing a balanced personalization that respects user privacy.

We will start by highlighting key issues that arise due to over-personalization. Then we will list the strategies that product managers and designers can implement to avoid over-personalization. Toward the end of the section, we will explore how over-personalization, under-personalization, and balanced personalization looks

CHAPTER 9   DESIGNING AI-POWERED PRODUCTS

Key issues

- Data outreach: Using sensitive or unexpected data points like health data, emotional state, location history, or using any data used without explicit consent of user may make users feel violated. Even if using that data point is legally allowed and technically feasible, this behavior breaches the psychological threshold of the user.

- Lack of transparency in AI recommendation: If the user does not understand why a specific recommendation was made, especially if it is too precise, then user might feel a bit uneasy.

- Consent fatigue: When getting consent from users is treated just as a checkbox activity and the interface is designed in such a way that consent prompts are vague and too frequent, it causes users to disengage and act blindly, which undermines meaningful consent.

- Unintentional exposure of personal information: When personalization is done without accounting for the user's intent, the context or device being used may unintentionally reveal sensitive information. This may offend users posing a significant ethical and legal risk.

- Reinforcement of narrow behavior: Over-personalization thrives on user feedback data and usage data, which only firms up existing user behavior, reducing the user's chances of discovering new content.

Product managers and designers must follow some actionable tips for developing a balanced personalized experience.

Product Manager

- Product managers should set clear rules defining what can be personalized, the reason behind it, and what data will be used early in the process. The PM should also set rules around what data cannot be used for personalization and what customer journey touch points will stay out of the personalization scope.

- Product managers should ensure that all major personalization features have a feedback loop, which enables users to review, edit, or delete their personalization preferences.

- Product managers should monitor the personalization acceptance rate by measuring privacy-related metrics like opt-in/opt-out rates and data resets. This will help the product team to gauge the acceptance rate of personalization among users.

- The PM should audit all personalization features for checking compliance with privacy laws and user expectations.

Designer

- The interface should be designed to explain clearly why a certain experience is personalized using subtle visual cues like tooltips and icons for explanations. The explanation should help users understand how personalization improves their experience.

- Designers should avoid using very familiar tonality or make any assumptions about the user while displaying content on the platform.

- Widgets to manage personalization preferences should be easy to use and offer a graded response to personalization. Rather than making personalization a control binary, allow users to temporarily stop personalization and control device preferences and data points that they want to be considered for personalization. Always add context to personalization controls by placing these widgets directly where personalization is happening.

Balancing personalization with privacy is a continuous process requiring clarity in what needs to be accomplished and respect for user boundaries and empathy. Table 9-1 shows a framework will help product managers assess what the red flags and best practices are for developing a balanced personalized experience.

Table 9-1. Red Flags and Best Practices for Developing a Balanced Personalized Experience

Area	Balanced Practice	Over-Personalized/ Risky	Under-Personalized/ Missed Opportunity
Data Use & Consent	Granular, contextual opt-ins with clear value proposition	Blanket opt-in; uses inferred data without user awareness	No personalization even when users would benefit
Personalization Logic	Based on recent, relevant, and declared behavior	Overfitted to past actions, outdated signals, or silent inference	Generic, ignores intent even when clear

(*continued*)

Table 9-1. (*continued*)

Area	Balanced Practice	Over-Personalized/ Risky	Under-Personalized/ Missed Opportunity
User Control	Easy-to-access settings to pause, adjust, or reset personalization	Hidden or non-existent controls; personalization feels "baked in"	No way to tailor experience even when preferences are known
Transparency	Labels like "Recommended based on X" with links to data controls	No indication of why something was suggested or how to change it	Doesn't explain missed relevance or missed suggestions
Feedback Integration	User signals (e.g., "not relevant") lead to visible change or learning	Feedback ignored; experience doesn't adapt to user corrections	No way to tell the system what's working or not
Surface Coverage	Personalization applied selectively to high-impact areas (e.g., discovery, navigation)	Personalized everywhere—navigation, UI behavior, tone—causing confusion	Personalization limited to trivial or low-leverage surfaces
Emotional Boundaries	Avoids assuming identity, emotion, or personal context unless explicitly shared	Infers mood, intentions, or sensitive traits without opt-in	Ignores moments where emotional tone adaptation could help

CHAPTER 9   DESIGNING AI-POWERED PRODUCTS

Designing Accessible AI-Powered Personalization

The personalized experience of digital products is designed and developed to improve user experience, and hence they must be inclusive by design and usable by all, including users with special needs. This means that these experiences should be developed and designed to deliver equal value to all user groups including users with special needs. Without implementing accessibility practices, AI systems amplify the risk of excluding these users from utilizing the product for their benefit. Ensuring accessibility in AI systems isn't just a compliance issue, but it's a moral and social obligation that reflects the equality, inclusivity, and ethical principles of the organization. Hence, all efforts should be made by implementing a team to ensure that products are accessible by design. To understand what the best practices are for making a product accessible, let's first explore key design gaps that may lead to accessibility issues.

- AI makes the system dynamic, which leads to shifts in page layouts and placement of certain widgets with personalized content being displayed for each user. This dynamicity may impact the ability of screen readers to announce what is there on the screen.

- Often to match the overall theme of the digital product, designers use low-contrast visual tags for highlighting personalization section. These may be too subtle for users with vision difficulties.

- To save digital estate, often nondescriptive explanations are used to describe AI-driven actions. Explanations like "We think you will like this" lack context and structure for assistive devices to interpret. These types of vague and half grammatically and structurally baked sentences are not only problematic for users with disability but also for users with cognitive needs. They provide no semantic reasoning on what is recommended and why.

- Often the temptation to shorten the user journey based on historical information or actions by skipping or hiding steps dynamically may confuse users relying on keyboard navigation and screen readers.

- AI, which adapts on the basis of gestures, facial recognition, or tone, may exclude users with speech or other cognitive disabilities.

Actionable Guidelines for Product Managers

- Product managers should treat accessibility as core requirements of the product and on par with the feature's functional and nonfunctional requirements.

- Product managers should work with the testing and quality assurance team to test personalization scenarios for accessibility in all phases of testing including user acceptance testing. Ideally users who are dependent on screen readers or keyboard-only navigation should be part of usability testing.

- While designing the workflow of the features that will have personalization, product managers must ensure that personalized flows still follow predictable structures.

- Product managers must audit AI-powered features for accessibility regression testing every time the model is retrained on new data. This will ensure that updating the model's functional behavior will not affect accessibility adversely.

CHAPTER 9   DESIGNING AI-POWERED PRODUCTS

Actionable Guidelines for Designers

- While designing the interface, designers must ensure that AI-generated content is labeled clearly, and the region where this content is placed on the platform is marked properly with the appropriate ARIA roles and live regions to enable screen readers to detect changes. The component holding AI-enabled content should be inserted in a logically ordered DOM structure to maintain clarity for assistive technologies.

- The designer must ensure that text explaining the personalization features is contextual and is clearly structured. Another best practice is to use Accessible Rich Internet Applications (ARIA) and the HTML tag "aria-describedby" to link the explanation to the AI text.

- While AI will modify or shorten the navigation flow based on historical data or user transactions, the designers must preserve a consistent focus and navigation flows by maintaining a logical keyboard tab order. Alter the workflow to allow users to opt in to faster workflows but offer a "show all steps" option for those who rely on predictability.

- Design tags and indicators for highlighting AI-powered features with sufficient contrast and contextual text. Tags should meet WCAG contrast requirements (minimum 4.5:1 for normal text). The designer must use text and easy-to-understand icons to explain the AI-driven personalization.

- Always offer multiple ways to access personalization if AI is using voice gestures or facial recognition technology for enabling AI. Always provide a nonbiometric fallback.

- The designer must always design a simplified version of the UI along with contemporary interface design and offer users a way to switch back to a default simpler UI option that is not dynamic in nature.

Building for Trust and Ethics

As AI-driven personalization and recommendations become increasingly common in modern products, building trust and ensuring ethical practices has become critical to the success of product. Human trust is the foundation on which relationships are formed. Today we are using digital applications to accomplish our everyday tasks, and if we don't trust an application, we will not feel confident and comfortable while using it. In Chapter 8, we explored how to develop an ethical trustworthy AI. In this section, we will extend those concepts into product design so that a designer can embed trustworthiness and ethical considerations into the interfaces, making AI products truly reliable and ethical. We will first explore how to design UI, which will ensure that AI decisions are transparent.

Ensuring Transparency in AI-Driven Recommendations and Decisions

Transparency in AI is not just about showing how the AI algorithm works; it's about designing the interactions that respect the user's right to understand the why and how about every decision made by AI. Every opaque decision in a product erodes user trust, and well-explained ones earn their trust back. The following are some examples when transparency was implemented correctly:

- Spotify: "Because you listened to…" offers traceable logic behind playlists.

- Google Maps: Shows how route recommendations are calculated (traffic, road closures).

- LinkedIn: "People you may know" often includes a hover tooltip explaining shared connections.

- YouTube: Recently added "Not interested" + "Tell us why" options to refine algorithm.

Now let's explore actionable guidelines that can be implemented by product managers and designers to embed transparency in AI.

Actionable for Product Managers

- The product manager must include transparency as one of the criteria in product roadmap activities. This means including decision explanations in AI feature specifications from the planning phase itself and not treating it as patchwork.

- Add model explainability as one of the key requirements to AI model acceptance criteria.

- Build analytics to track user interaction with explanations or help icons to gauge users' traction on the type of explanations they really care about. This will help the PM identify AI personalization features that attract the maximum user scrutiny.

- Product managers should not think that whatever transparency steps have been taken are perfect. Always design communication loops for the user to report transparency gaps. This can be done in a subtle manner without disturbing the user's journey in the following ways:

  - Once a user has interacted with an explanation, ask was this useful?

  - Offer an option to report irreverent recommendations by marking a recommendation as irrelevant.

- Often an absence of explanation about a decision made by AI hurts the user confidence of the system less than offering a bogus generic explanation. Avoid keeping your explanation too generic.

Actionable for Designers

- The designer must place explanations next to AI-generated output using familiar icons and clear phrases that users can recognize easily.

- The designer must ensure that the interface they are designing clearly shows what data points were used to generate a specific AI output. This can be accomplished by using causal UI design patterns. If the AI output appears in the list or drop-down along with organic options, then highlight the AI-based suggestions distinctly.

- Not all AI decisions are equally critical; designers must work with a product manager to classify each AI decision into different tiers and use that tiering to offer different levels of explanation. For most critical ones, use expandable sections to explain decisions. Create a three-layer expandable section with the first layer having a short reason. The middle layer goes one notch down with specific data points that were used by AI to arrive at a decision point, and the last layer should contain user control to determine their decision.

- In high-stakes decisions, don't hide explanations behind any widget. Show the AI transparency content prominently.

Communicating AI Behavior to Users: Clarity and Trust-Building

AI systems operate in opaque and complex ways, making it difficult for users to interpret or engage with them meaningfully. This opaqueness of the system can lead to frustration in users, which might result in a loss of trust in the system. When they don't understand the mechanism behind the functioning of the system, it loses their reliability, resulting in reduced product usage and high user churn.

To counter this, the interface of the AI-powered system must communicate effectively and clearly. The interface must consistently reinforce trust through transparent and thoughtful communication.

When communication around AI is poor, it breeds negative emotions and skepticism. Conversely, when communication around AI clearly explains in user-appropriate terms what AI does, how it does it, and what users can expect, it fosters user confidence and trust. These positive emotions translate into tangible business benefits such as increased product adoption, higher user engagement, and reduced support ticket volume.

CHAPTER 9   DESIGNING AI-POWERED PRODUCTS

To summarize, clarity in communication around AI products is not just a UX best practice; it is a business imperative with clear tangible outcomes.

Before diving into the best practices that product managers and designers can use to ensure clear communication around AI products, let's first look at some common communication mistakes:

- Often the text used for communicating how AI functions is full of jargon and technical terms, making it inappropriate for users without a technical background. This can confuse or alienate the audience. An extension to this mistake is using complex diagrams like flowcharts with technical labels for explaining how AI works.

- Treating text used for communicating how AI can help users as marketing promotion text. Often the product manager treats this communication piece as a place to advertise the unique selling point of their product, and sometimes overhyped claims are made. These overhyped claims lead to unrealistic expectation setting, and when the expectations are not met, dissatisfaction among users may grow.

- The third and most common communication mistake in this list is a lack of transparency in communicating what the AI can and cannot do. This again results in the wrong expectation setting.

- Another common mistake in communicating the benefits and functioning of AI is neglecting ethical and privacy considerations by not explicitly calling out what data points are used in AI decision-making and how user can control them.

- Communicating what AI can do and not how it solves real user problems can make even the best written communique irrelevant.

Although these mistakes may seem minor and easy to fix, they often stem from deeper issues in strategic planning and underlying thought processes. To address these foundational challenges, let's now outline some best practices tailored to both product managers and designers.

Actionable Guidance for Product Managers

- Developing an AI communication framework should be a crucial responsibility of the product manager as it helps teams to communicate effectively and consistently. A well-defined communication strategy should have the following five pointers:

  - Communication touchpoints: Identify all the moments in the user journey where AI is playing a role and where communication is essential. Decide on the level of explanation required at each identified touchpoint.

  - Standardize the level of explanation required for each touchpoint across the user journey. Classify touchpoints into three levels, with level one providing superficial explanation and level 3 providing complete details.

  - Tone and voice: Maintain consistent, brand-aligned language and style for all AI-related communication.

  - Visual cues: Use consistent icons, badges, or effects to signal AI involvement.

  - Feedback and control: Enable users to give feedback, adjust AI settings, or undo AI actions to feel in control.

- Standardize AI terminology across the product: Inconsistent or unclear communication may confuse users, undermine trust, and create friction across product usage. The primary reason behind this is often organization do not have a standardized template of how communication will be structured around AI. Another missing link is the absence of standardized AI terminology. It helps in bringing clarity in communication, reinforces the brand voice, and helps the user develop a correct understanding of how the AI feature works. To do so:
  - Work with various stakeholders like the AI team, marketing, and the customer success team to design and develop a shared glossary for key AI-related terms and scenarios where each of them will used.
  - Align product copy, marketing material, onboarding content, user communications, and support documentation with agreed on standardized vocabulary.
- Treat communication as a core feature: View communication as the core requirements of the feature and not as optional one.
- Measure the impact of AI communication: Product managers must develop mechanisms to track the efficiency of AI communication and how it affects the business KPI. Metrics like NPS, in-product surveys, reduction in support tickets, and comprehension score in usability testing can be used.

CHAPTER 9 DESIGNING AI-POWERED PRODUCTS

- Product managers must provide easy-to-access feedback options near AI features to share feedback about AI features working.

Actionable Guidance for Designers

- The designer should ensure that all AI features have clear labels and are placed next to the feature it is addressing to. The chosen font type and size should be easy to read. The text should clear and grammatically correct and well structured.

- The designer must use inline tags to show an explanation of an AI feature to users. These inline tags should be accessible using a single tap/click or hover.

- The designer should design interfaces for reversibility, which means the user is empowered enough to undo or change an AI-driven suggestion or decision easily.

Ensuring Ethical AI in Personalized Experiences

As AI-driven feature implementation is becoming more prominent, ethical considerations must also take more prominence in product development and design. While AI-powered products can boost productivity, relevance, and user satisfaction, they can also affect users negatively by introducing unintended biasness, manipulation, loss of control, and privacy breaches. Grounding not just AI algorithms but also product interfaces in trustworthy AI principles has become increasingly critical in the current landscape, as any ethical violation can lead to serious brand, as well as regulatory and reputational risk.

497

CHAPTER 9   DESIGNING AI-POWERED PRODUCTS

In the previous chapter, we covered the principles of trustworthy AI in detail including how to operationalize these principles in practice and how to quantify the progress made on these principles. In this section, we will limit the discussion to ethical risk in personalization and the product interface.

The following are the major ethical risks that may impact personalization adversely:

- AI-driven personalization, if not carefully designed and monitored, may unintentionally reinforce harmful biasness toward certain user groups. This occurs when algorithms are trained on biased datasets or lack sufficient diversity in their inputs. For example, a content recommendation system on a microblogging website may consistently recommend content liked by a particular user group because of sharing a dominant representation of that user group in the training dataset, while content liked and consumed by another user group may get ignored.

- Manipulative nudging: AI systems optimized primary for engagement are often trained on a user's behavioral data. Such systems if not designed ethically can lead to unhealthy behavioral development. By leveraging behavioral data, these systems can subtly steer users toward compulsive behaviors such as doomscrolling on social media platforms or impulsive shopping on e-commerce. While these nudges may influence short-term metrics positively, they are primarily unethical as they erode user trust, raising ethical red flags that may trigger regulatory risks.

CHAPTER 9   DESIGNING AI-POWERED PRODUCTS

- Opaque decision-making: The AI algorithms that are used make recommendations are often opaque in nature, offering little to no reason behind their decision. This situation gets more complicated when the product interface or UI layer of the product offers no help to the user, making the complete AI product a mystery box. This lack of transparency in the AI product both on the UI side and on the algorithmic side not only undermines the user's autonomy and accountability but can also lead to compliance risk under emerging data and AI laws.

- Design bias: User interfaces can unintentionally reinforce stereotypes or embed bias through design choices. This occurs when UI patterns make specific options more visible and accessible, while obscuring or restricting other alternatives. Examples include the default settings reflecting certain cultural assumptions, or a visual hierarchy influencing user's preferences to select a specific option without informed consent. Even subtle design cues like imagery, icons, language, or layout if used recklessly may introduce biasness and limit inclusivity.

To avoid the ethical pitfalls mentioned in product development, the following best practices can help guide product manager and designers.

Best Practices for Product Managers

- Product managers must embed ethics into a product strategy defining clear ethical positioning for your product. This shows how positioning should guide decisions across the entire product lifecycle from research to design to deployment. Use the Trustworthy

CHAPTER 9   DESIGNING AI-POWERED PRODUCTS

AI framework discussed in Chapter 8 to formulate this positioning. Defining the ethical stance of your product is not sufficient; use metrics to measure the effectiveness of these principles in your platform.

- Define clear rules around what can be personalized and what should not be personalized. While personalization can enhance the user experience, without clear rules around the type of personalization and that define the user's role and level of transparency requirements, it can become manipulative and discriminatory.

- Another best practice for product managers is to adopt a minimum viable personalization mindset. Resist the urge to personalize everything. Focus on areas where personalization clearly adds meaningful value to user. Avoid hyperpersonalization at any cost as it may feel a bit invasive to the user. By prioritizing relevance over novelty, teams can deliver personalization that aligns with the users' needs and that respects boundaries.

- While designing the evaluation criteria to measure the process of AI products, balance the performance metric with trust-based metrics like opt-out rate, user sentiment on fairness and relevance, transparency comprehension rates, etc.

- Product managers must work with the organization's leaders to establish a process for flagging and escalating bias complaints or inappropriate recommendations.

- The PM should strategically tailor personalization models by demographic respecting cultural, legal, or any other sensitiveness. Special attention should be given in developing regionally aware consent flows.

CHAPTER 9   DESIGNING AI-POWERED PRODUCTS

Best Practices for Designers and UX Strategists

Designers must treat transparency in personalization as a foundational principle. Clear communication around personalization builds user trust and enhances product relevance perception. The designer must consider the following advice:

- Designers must incorporate familiar design elements that help users understand how personalization works. Similar UI patterns should be used throughout the product to explain or highlight personalization.

- Personalization is a back-end task. The designer must try to add visibility to it by using components like preference cards, visual badges, or accessibility control panels. These interactions not only add visibility to the personalization but also make these interactions feel intuitive and natural.

- Generally, a prototype is used by designers for testing user engagement and aesthetics. When designing AI-powered experiences, it is essential that a prototyping practice is extended for testing user consent, control, and comprehension.

- While conducting usability testing, incorporate scenarios involving low-trust users, sensitive topics, and first-time interactions where consent and control are critical. During this test, observe how different user segments respond to requests of data sharing and what role design plays in making users informed.

- The way the default settings are designed and configured shape user behavior. Designers must always set the defaults that prioritize user privacy and not data collection because these ethical defaults are a subtle but powerful way to signal that users are in control—and that your product values their trust as much as their data.

- Make resetting the personalization options easy and safe. Always use calm and nonalarming language to convey or confirm a user's choice.

Summary

This chapter explored how design plays a pivotal role in shaping AI products, which are not only effective but also trustworthy. This brings designers on the front lines of ensuring that AI products are well aligned with human values and societal norms.

This chapter also outlined the benefits of integrating ethical AI guidelines into product design and highlighted techniques for creating trustworthy interfaces. By introducing interaction patterns and industry best practices, this chapter empowers designers and product managers to navigate common ethical pitfalls and embed Trustworthy AI principles into the core AI experience. This chapter emphasized the fact that design not only influences the interface of the AI product but is a critical layer for making AI more human centered and trustworthy.

CHAPTER 10

Putting AI to Work: Case Studies in Applied Intelligence

After exploring the strategic, operational, and technical foundations required for transforming an organization into an AI-first organization, this chapter will you explore the real-life applications of AI. In this chapter, we will cover select but diverse case studies to help you understand how an AI use case is implemented and documented.

We aspire to provide you with practical and diversified real-world experiences of AI that can be implemented responsibly and effectively into core business processes. These real-world case studies are deliberately designed to offer more technical details. There are two rationales behind the technical tonality of these case studies. First, these case studies will serve as a blueprint for documenting the technical implementation of AI projects. Second, their technical depth will help product managers to identify gaps in their AI fluency and better understand where to deepen their skills.

Each case was chosen not only for the complexity or innovation it represents but also to cover the spectrum of use cases in AI that can act as a value lever. Each project illustrates a different strategic use case for AI, chosen to represent a wide range of innovation levers and different learning types available in AI.

CHAPTER 10   PUTTING AI TO WORK: CASE STUDIES IN APPLIED INTELLIGENCE

List of Case Studies

- **Case Study 1: AI Model for Human Exercise Recognition Using Computer Vision**

    Strategic Focus: Product Innovation and Differentiated Customer Experience

    This case study highlights how AI can be leveraged to create an intelligent sensor-less interface that enhances the customer experience by personalizing health recommendations. In this case study, computer vision was used.

- **Case Study 2: Optimizing Healthcare Triage Through a GenAI Chatbot**

    Strategic Focus: Service Efficiency and Intelligent Automation

    This case study explores the application of Gen AI to the triage workflow and how responsible automation helps in increasing the service efficiency and in enhancing patient throughput.

- **Case Study 3: Optimizing Ads Recommendation on Search Results Page for an E-commerce Platform**

    Strategic Focus: Revenue Growth Through Hyper-Personalization

    This case study shows how AI can be used to provide personalized recommendations. It shows how real-time data and AI can work together to boost click-through rate, which will in turn maximize ad revenue and improve customer experience simultaneously.

CHAPTER 10   PUTTING AI TO WORK: CASE STUDIES IN APPLIED INTELLIGENCE

- **Case Study 4: Dynamic Pricing Based on Demand Prediction by AI Model**

    Strategic Focus: Data-Driven Agility and Market Responsiveness

    In modern digital world, static pricing leaves revenue on the table. This case study illustrates how predictive modeling can develop dynamic pricing capability, which will enable the digital platform to respond instantly to shifts in demand, inventory, and competitive actions.

Case Study Format and Team Roles

Each case study is documented using a consistent format that is mentioned. Since AI models are typically developed in cross-functional teams, each section is also tagged with the relevant team member role who can support the product manager in gathering the relevant details. Table 10-1 maps the technical documentation section to the respective team member.

Table 10-1. *Section-by-Section Role Attribution*

Section-by-Section Role Attribution	
Section	**Who Should Fill This**
Executive Summary	Product Manager
Solution Approach	ML Engineer/Data Scientist
Data Collection & Preprocessing	ML Engineer/Data Engineer
Feature Engineering	ML Engineer
Model Development	ML Engineer/Data Scientist
Model Performance	ML Engineer + PM
Deployment	MLOps Engineer + Software Engineer
Challenges and Limitations	Product Manager (with inputs from Eng)
Business Impact	Product Manager (with analytics/finance)

CHAPTER 10   PUTTING AI TO WORK: CASE STUDIES IN APPLIED INTELLIGENCE

Case Study 1: AI Model for Human Exercise Recognition Using Computer Vision

Business Use Case

A renowned healthcare platform aimed to improve its customer retention by providing them with in-house gym trainer experience. This class was a live video class where a gym instructor would guide users. However, this model lacked scalability as each trainer was able to cater to only 10 users per session. As a result, the organization's growth stagnated, and profitability became a concern.

Proposed Solution

To address this situation, it was decided to develop an AI-based virtual trainer. This virtual trainer will use computer vision and AI technics to recognize exercise, assess posture, and offer real-time feedback to the customer.

North star metric: % Reduction in trainer fees spend per month
Counter metric: User Retention Rate

Solution Approach

- **ML problem definition**: The problem is matching body postures from camera frames with that of the ideal key body postures for the exercises.

- **Metric to measure model accuracy:** Pose Matching Accuracy. This is a correct match if a similar pose gets a match score >80% and a dissimilar pose gets a <30% score.

CHAPTER 10   PUTTING AI TO WORK: CASE STUDIES IN APPLIED INTELLIGENCE

- **Data collection and preprocessing**: The following steps were involved in preparing datasets:

  - Identification of data sources like YouTube and other similar websites where people post videos and pictures of their workouts and exercises.

  - The dataset consisted of around 10 to 20 short video clips like 5 to 15 mins each for every exercise taken from different camera angles. There were also pictures of yoga poses with a minimum of five pictures from different camera angles for each yoga exercise.

  - Videos were broken into camera frames with one frame for every 10 seconds captured. Then manual frames with key poses of the exercise were selected and annotated as keypose1 or keypose2. For example, a simple exercise like Squats can be split into two key poses: keypose1 when the user is standing up straight and keypose2 when the user was sitting down completely and pausing before starting to go up.

- **Feature engineering**: The body key points coordinates (x, y) were extracted for the video frames and pictures of yoga poses using a pre-trained computer vision framework called Mediapipe. These body key points refer to the joints like elbow, knee, ankle, shoulder, etc., which are essentially required to define a body posture. The body key points are used as features to match the key poses and provide corrective feedback to the user. Also, these are used to estimate other insights, like speed of the exercise, by analyzing the rate of change of the coordinates of the key points, calories burned, and number of repetitions (say up-down-up for squat) done by the user for the exercise performed.

CHAPTER 10   PUTTING AI TO WORK: CASE STUDIES IN APPLIED INTELLIGENCE

Model Development

- **Model selection:** Initially, tree-based models were used to classify a set of body key points as keypose1 or keypose2 of a particular exercise, but it was noticed that these models were not scalable as more exercises were added. Each time a new exercise is added to the system, a new model needs to be trained from scratch. Also, these require a lot of training data that may not be available for a new exercise in a short time. So, the time required to add a new exercise to the library is high, and the model accuracy starts to drop when more exercises are added.

   To solve these problems, a triplet loss-based training strategy is used to learn to estimate a high matching score using body key points from similar poses and a lower matching score for body key points from dissimilar poses. The training data is generated as triplets of two similar and one dissimilar pose. The model takes in the coordinates of body key points as an input vector and computes a transformed vector in a vector space where the transformed vectors of the similar poses are closer and farther from those of the different poses. A simple neural network architecture with three fully connected dense layers is chosen as the final model. The layers are the input layer, the hidden layer (for transforming), and an output layer.

   This type of training strategy allows the generalization of the model in learning to match similar poses as more exercises are added to the library. Also, when a new exercise is added, it does not require many samples for the new exercise as several triplets can be created with only a few samples of the new exercise. Thus, this type of model training allows faster scaling and convergence toward no need of model training for the new exercises.

- **Hyperparameter tuning**: The number of neurons for the input layers are fixed as per the number of coordinates for the body key points. The number of neurons in the hidden layer and the output layers are estimated with Grid-search. The dynamic learning rate is used, which changes as per train-test accuracy during training.

- **Model evaluation**: Accuracy is used as the parameter to evaluate the model performance. If for a given pose the model cannot match it with the similar pose of the correct exercise, then it's called a bad prediction. A test data is prepared with pairs of similar and dissimilar poses, and the match-scores for each pair are predicted by the model. If the match-score is >80% for the similar pose and <30% for the dissimilar pose, then it's said to be a correct prediction.

Model Performance

- **Model results**: The model was able to perfectly match all similar poses with 100% accuracy. There were issues with giving high matching scores for cross-exercises where the poses were found to be similar from different exercises.

- **Comparisons:** Here the classical ML classification with tree-based models approach did not work due to scalability and training data requirement issues and an advanced technique like triplet loss with neural networks worked, which mitigated these issues.

Deployment

- **Deployment strategy:** The ML model was very light, and it was deployed as a serverless cloud function in Google Cloud Platform (GCP). The ML model is used to fetch transformed feature vectors or embeddings for the ideal key poses of the exercises in the library. For each exercise, the ideal key poses are added from three different camera angles at least to remove bias due to camera angle. A FAISS vector DB was used to search for the closest key pose to the user's pose from the camera frame. If the user's exercise is already known, then only the match with the key poses of that exercise are required. The matching with the exercise's key poses is performed at another serverless cloud function, which is triggered for a batch of user's video frames, say, collected over 1 second.

    - This following components are involved in deployment:
    - **Infrastructure:** GCP cloud functions for serverless deployment.
    - **Deployment frameworks:** FastAPI was used to deploy.

CHAPTER 10   PUTTING AI TO WORK: CASE STUDIES IN APPLIED INTELLIGENCE

- **Integration**: The user interacts with the application via a camera on the system (laptop or mobile device or a tablet) that collects the user's video frames, and user's body key points are located with the on-device computer vision model (Mediapipe framework). The batch of key points detected every second is sent to the ML model cloud function for computing the embeddings, which is then used to compute the match score using the matching API at another cloud function. The match score is then sent to the on-device logic to give corrective feedback and estimate insights like repetition counting, calories burnt, speed, and intensity.

- **Model Serving**: The model is serving end users via REST APIs, and it performs batch processing with near-real-time inference.

Challenges and Limitations

- **Challenges encountered**: The main challenge was to collect data for the same exercise from different camera angles. For some rare exercises, the data annotators were asked to perform the exercise themselves and record from different camera angles. Another problem was variation in poses for the same exercise in different videos or when different people were doing the exercise.

- This made it difficult to train the model because of mislabels in the training data.

CHAPTER 10   PUTTING AI TO WORK: CASE STUDIES IN APPLIED INTELLIGENCE

- **Limitations**: The matching is a two-step process as we have to first get embeddings from the model and then find the matching pose. This creates additional delay due to multiple API calls.

Business Impact

The north star metric monthly trainer costs saw a significant decrease monthly while maintaining the user retention with previous months. With the reduced need to hire new gym trainers, the platform was able to accommodate a growing user base at minimal price. Additionally, the system's flexibility allows new exercises to be added to the fitness regime, effectively ensuring faster expansion.

Case Study 2: Optimizing Healthcare Triage Through a GenAI Chat Bot

Executive Summary

A leading healthcare service provider online platform offering a virtual at-home gym service observed a high churn rate of 30% among customers. Upon investigation, they discovered most of the users who churned had experienced musculoskeletal (MSK) issues like back, neck, or ankle pain after workouts. They also figured out that these users tried reaching out to their individual trainers but encountered a delayed response. This delay in response was a major turn-off for these customers.

Proposed Solution
To address this, the platform implemented a GenAI-powered chatbot to assist users in managing their MSK injuries through self-care suggestions and posture guidance. This virtual assistant was available 24/7 and was customized to cater to the needs of triage customers.

North star metric: User retention.

Solution Approach

This problem falls under the category of GenAI where the chatbot must engage a user the way it's prompted to. Prompt engineering was heavily used to control the chatbot's response format and content. The chatbot responses were kept short and to the point because users are generally chatting on their smartphones, so there is not much room for long and descriptive responses. Also, quick reply suggestions were added to the chatbot responses so that the users need not type manually and can tap on the suggestions to continue the chat. The chatbot was also designed to collect a user's necessary information for further analysis.

The chatbot was tested manually, and custom automatic evaluation frameworks were developed to check the performance in terms of generated response generation speed, content quality, and ability to follow the response format template.

Metric to measure model accuracy: Content Quality Score.

Model Development

- **Model selection**: Based on manual testing, GPT-4o was found to be the best model, giving good-quality responses in optimal time. The model GPT-4o mini was the quickest in generating responses, but it was not good at following response format templates.

- **Hyperparameter tuning**: Manual tuning was applied for tuning values for the parameters like temperature and max tokens.

- **Model evaluation**: The chatbot models were evaluated mainly on two things:

1) **Content quality:** The content quality was initially manually tested and later automatically evaluated. A custom agentic evaluation framework was designed where different prompts were used to instruct the chatbot, and an AI agent was instructed to score the quality of responses from 1 to 10 based on tone, correctness, and relevance of the responses. The AI agent was also asked to test the chatbot responses for non-MSK-related topics where the chatbot is expected to ask users to discuss only the MSK-related topics.

2) **Response format:** The ability of the chatbot model to generate responses in a necessary format in different scenarios is automatically tested via rules-based checks put in place for each scenario.

Model Performance

- **Model results:** On manual evaluation by the gym trainers and physiotherapy experts, the chatbot responses were found to be satisfactory. From the automatic evaluation framework, the model responses got an average of 8/10 for the content quality, and responses were of the correct format 80 to 90% of the time.

- **Comparisons**: The content quality for the GPT-4o mini was like that of GPT-4o, but it failed a lot more times (about 50%) to follow the instructions for the response format.

CHAPTER 10   PUTTING AI TO WORK: CASE STUDIES IN APPLIED INTELLIGENCE

Deployment

- **Deployment strategy**: The GPT-4o model is already deployed as a service by OpenAI. The chatbot was deployed as a node.js service on a dedicated cloud server. The following components are involved in deployment of the chatbot:

  - **Infrastructure**: The EC2 server on the AWS cloud is used to host the chatbot service in the backend. Another application backend layer was used to coordinate the communication between the front-end application (either web-app or mobile app) and the chatbot API.

  - **Integration**: The chatbot API then triggers OpenAI GPT-4o API to fetch the responses for the user input message. The model response is then sent to the front end after preprocessing from the backend layer. The model response was in JSON format so that the different components of the responses can be mapped to different UI components. For example, the response to show to the user as a bot reply is contained under the "Bot reply" key, and the quick suggestions to show to the users are contained within the key "quick_suggestions." Since the chatbot responses were being streamed, it was necessary to send the "Bot reply" as the first key in the response.

- **Model serving**: The chatbot was hosted as a REST API to the end users. It was done so as to be accessed by different types of apps like web apps, Android apps, and iOS apps.

Chapter 10 Putting AI to Work: Case Studies in Applied Intelligence

Monitoring

- **Model monitoring**: The chatbot was monitored for the content quality and response format accuracy in production. To check the content quality of the model, a random sample of chats was chosen from the historical chats and analyzed manually. Also, an evaluation agent was given the chats and asked to rate them on a scale of 1 to 10.

  The accuracy for the response format is automatically estimated from the number of times the UI throws exceptions because the response format is not as expected. Redundancies are added to handle the inconsistencies in the response format in the additional backend layer, which are used to check the response format and either get another response from the GPT-4o model or alter the response format to match it with the expectation if possible.

Metrics

- **Business metrics**: It was observed that the user retention improved by 20% with the help of the chatbot. Also, the user feedback shows that they liked that the chatbot agent is always available to them whenever they need some guidance.

- **Model performance metrics**: The accuracy and quality of the chatbot responses were found to be consistent with the test time numbers.

CHAPTER 10   PUTTING AI TO WORK: CASE STUDIES IN APPLIED INTELLIGENCE

- **Real-time metrics**: Time spent by users on an average chat session was observed to increase as users started to engage with the chatbot. This showed that the users found the chatbot to be helpful.

Challenges and Limitations

- **Challenges encountered**: The main challenge was lack of any labeled data for the chat conversations that can define what is a good response and what is not for a particular scenario. So, the evaluation of the chatbot responses was hard as it may introduce a personal bias. Also, it's hard to get any feedback from the user regarding each individual response.

- **Limitations**: The current chatbot cannot be personalized as it is functioning on the basis of a common prompt or set of instructions. It is also not possible to know what response the user would have liked as there is no feedback loop.

Business Impact

The MSK chatbot worked well for improving customer engagement and retention. There was a significant improvement in the customer retention KPI, and more chatbots were planned by the platform for managing user's diet, nutrition, lifestyle, etc. The chatbots also helped in scaling the platform and reducing the cost of hiring more gym trainers or physiotherapists to help users with their MSK issues.

With the decreasing cost of GPT models by OpenAI, it became very cost effective to consume chatbots and show high a return on investment (ROI).

CHAPTER 10   PUTTING AI TO WORK: CASE STUDIES IN APPLIED INTELLIGENCE

Case Study 3: Dynamic Pricing Based on Demand Prediction by the AI Model

Executive Summary

An e-commerce platform faced challenges in scaling its inventory sell profitably as product pricing adjustments based upon demand required manual intervention. This manual process was ineffective, prone to delay, and often failed to capture optimal pricing opportunities, especially during high-demand times.

As a result, the company was unable to scale its operations in a profitable manner. Manual price changes lacked the ability to respond to real-time demand changes, leading to a loss of revenue opportunities.

Proposed Solution

To scale profitability, there was a need to develop an automated data-driven pricing solution that could adjust pricing dynamically based on predicted customer demand.

North star metric: Revenue collection per day.

Counter metric: Customer attrition rate.

Solution Approach

To solve this, a machine learning–based regression model (LightGBM) was developed to predict product demand based on historical sales data and pricing trends. The model evaluates different price points, estimates expected demand, and selects the price that maximizes revenue (Price × Demand). The AI system was automated using Airflow DAGs, integrated with the platform's data pipelines and pricing modules, and deployed on Google Cloud Platform.

CHAPTER 10   PUTTING AI TO WORK: CASE STUDIES IN APPLIED INTELLIGENCE

- **ML problem definition**: The ML problem is a regression problem to predict the sales for a given price based on past sales at different prices.

- **Metric to measure model accuracy:** Underestimation versus overestimation rate.

- **Data collection and preprocessing**: The following data is used:

  - Sales information: This includes historical daily sales and selling price for each product sold for the past three years.

  - Product information: This includes the category, daily click counts, and number of times product appeared in search results.

  - Standard data preprocessing techniques: Rows with missing data are removed, and day of the week, month, and day of the month are extracted from the date of the sale.

  - Categorical variables: These are handled by the model so no encoding technique was used.

  - Sales and product activity data: This is broken into batches of rolling eight days.

- **Feature engineering**: The daily average price is used as an input feature and total sales made for the day as the output. Other features are extracted like day of the week, month, and day of the month to account for any periodicity. The past one week of sales data is used to make predictions for the next day, i.e., the eighth day, so this resulted in seven pairs of daily average price and total sales made as input features.

CHAPTER 10   PUTTING AI TO WORK: CASE STUDIES IN APPLIED INTELLIGENCE

Model Development

- **Model selection**: Tree-based models are preferred because of the nature of input features and regression nature of the problem. The LightGBM model is finally selected for this project.

- **Hyperparameter tuning**: Grid Search is used for tuning the hyperparameters like the number of leaves, maximum tree depth, learning rate, etc.

- **Model evaluation**: Mean Absolute Percentage Error (MAPE) is used to evaluate the regression model, but the model is optimizing for Root Mean Square Error (RMSE) as the standard for regression as MAPE is biased for smaller sales.

Model Performance

- **Model results**: The test results of model training showed an RMSE of 0.2, which means on average the model can predict sales correctly after rounding off. For example, if true sales of an SKU is 3, then the model on average may predict 2.8 or 3.2, which will give the same value as 3 after rounding off, as we know that sales can only be integral values. The MAPE value of 5% is observed for all the products.

- **Comparisons**: Similar results are observed for the XGBoost model architecture, but results for the Random Forest model were found to be inferior.

CHAPTER 10   PUTTING AI TO WORK: CASE STUDIES IN APPLIED INTELLIGENCE

Deployment

- **Deployment strategy**: The ML model was deployed as an Airflow DAG (Directed Acyclic Graph) to make predictions for the products in the early morning at 5 a.m. Every day at 5 a.m. a cloud server spins off, and it collects all the data required for making predictions and then calculates features for model input. For each product a batch of predictions is made for different prices. The price with the highest revenue is selected as the price of the product for the day.

    - **Infrastructure**: Google Cloud Platform (GCP) is used for running the Airflow DAG. Sales and product data are stored in BigQuery and fetched using SQL queries.

    - **Deployment frameworks**: Airflow DAG is used for deployment.

    - **Integration**: The Pricing system is integrated with the other modules like search, product details, and sales information. The prices set by the ML model must follow the price range set by the business logic like the price cannot be below the price at which the product was procured by the e-commerce company. Also, the price cannot be more than the maximum selling price set for the product.

    - **Model serving**: The prices set by the ML model are exposed to the users when they visit the platform, and products are recommended or displayed on search results. The price range of every product is limited by the procuring price and maximum selling price. The model demand predictions are calculated for prices between this range at regular intervals. The price with maximum revenue is chosen as the price of the product.

Monitoring

- **Model monitoring**: The performance of the ML model is measured in terms of accuracy of prediction. In other words, if the model predicted that 10 quantities of some product will be sold at price P, then it depends on if 10 quantities are sold or not.

    - **Performance metrics**: Daily the error in model demand prediction is tracked at a product level, but it is difficult to make any deduction in loss of revenue on the basis of daily error. A weekly or monthly distribution of the sales is used to evaluate the impact of model.

    - **Drift detection**: In cases where the model is found to be underestimating or overestimating the demand over a period of few weeks or months, then manual interventions should be made to optimize the model predicted price, and then its impact on demand distribution should be evaluated. If the demand distribution changes, then the model is said to be drifting and needs to be retrained.

    - **Alerts and logging**: Anomalies in model predictions are detected if the error in demand prediction for a product is found to be more than three times the standard error for that product observed during the testing phase of the model. A daily report is generated for all such cases and sent to the data science and pricing team to discuss further actions. Automatically, the model-based pricing is disabled for these products until further discussions.

- **Automated retraining**: Based on the signals from high error alerts, drift detection, and drop in other model performance metrics, the model auto-retraining is triggered. For model retraining, data pipelines were automated using Airflow DAG. The historical data for sales and products is collected via BigQuery SQLs and dumped to cloud buckets. The raw data is then preprocessed by the Python scripts, and the preprocessed data is again dumped to another cloud bucket. The preprocessed data is pulled from the cloud buckets to retrain the model on a dedicated cloud machine. Test metrics for the new model are compared with that from the previous model to check if the new model is better than the old model or not. Also, the new model is tested on recent data where the old model is found to be showing high prediction errors or where data drift was observed.

  The metrics for the new model performance and comparison with old model were published to the model retraining dashboard for manual review, and based on the predefined rules, the new model is automatically deployed to production if found to be better than the old model. The training data used to retrain the model is properly versioned and archived for future reference. The training data for older models (older than the previous two models) was deleted from the archive.

CHAPTER 10   PUTTING AI TO WORK: CASE STUDIES IN APPLIED INTELLIGENCE

Metrics

- **Business metrics**: The AI model-based pricing system was found to be making a higher revenue than the manual pricing with a 95% confidence interval based on the revenue distribution for one month of data. Different kinds of comparisons were made to estimate the revenue growth. In the A/B experiment, a group of users was exposed to the model-based pricing for certain products for a month. Checks were put in to see that the model-based pricing does not vary too much from the manual pricing of the product at the same time in order to manage the risk of experimentation. The model-based pricing system would still be able to choose a higher or lower price depending upon the demand estimated by the model.

    - The revenue distribution for the variation group for one month of model-based pricing is compared with the revenue distribution of the control group, which was shown manual prices. Also, the comparison was made between the revenue earned from the same users of the variation group and the control groups from the previous month for the same products. The difference in revenues across time and users was used to mitigate the temporal and user selection bias, which was crucial in estimating an unbiased difference in revenue from the AI model-based pricing system.

- **Model performance metrics**: The following metrics were captured for the model's performance during production:

- **Accuracy**: The accuracy of the model was captured in terms of how often the model's predictions for daily demand were accurate with an error of one unit of actual demand for a product in one week.

- **Underestimation and overestimation**: For the inaccurate predictions, the rate of underestimation and overestimation was estimated for the model. This was further used to tune the bias in the price; i.e., for the products for which model prediction is frequently found to be less than the actual demand, the price is set higher than the model prediction, and for those for which the model predicted demand turns out to be higher, the price is set to a lower amount than what was recommended by the model. By doing so, it was observed if the model prediction can be matched with the actual sales with the adjusted price.

Challenges and Limitations

- **Challenges encountered**: The major challenge was to convince the stakeholders that the revenue made by the AI model based on pricing is greater than that made by the manual pricing. The stakeholders did not want to do an A/B test where different users are exposed to different prices for the same products. Thus, a new type of A/B test was conducted to remove user and temporal bias.

- **Limitations**: There is no precise way to directly compare the revenue generated by two different pricing solutions. Any comparison can only be made through estimation supported by statistical analysis. This limitation arises because consumer demand is inherently uncertain. For example, if a model predicts that 10 units of a product will be sold at a price of 100 but actual sales turns out to be 8 or 12 units, then to determine if the same demand would have occurred at a different price point is very difficult. Hence revenue comparisons across pricing strategies remain probabilistic rather than exact.

Also, it is difficult to quantify the contribution of a new pricing strategy in overall revenue growth in the long run because it is impacted by several other factors like market conditions, campaigns, etc.

Business Impact

The AI-based dynamic pricing system led to a 3% increase in average daily revenue per product by accurately predicting demand and optimizing prices. It reduced manual pricing efforts, scaled efficiently across the inventory, and minimized revenue loss due to overpricing or underpricing. This initiative proved to be a cost-effective and scalable solution, validating the strategic value of AI in pricing and revenue optimization.

Case Study 4: Optimizing Ads Recommendation on Search Results Page for E-commerce Platform

Executive Summary

An e-commerce platform struggles to effectively fill available ad slots on its mobile search results page due to limited inventory of relevant ads. Ads that yield maximum click-through rate of about 2.8% are those that have exact keyword match between the search query of the user and the ad title. Only 60 to 70% of slots can be filled using the exact match technique. To fill the remaining slots ads, partial matches are used, but these have significantly lowered the CTR (about 1.5%) and lead to poor user experience and reduced return on ad spend (RoAS).

Due to this, monetization opportunities are lost, and ad real estate is always underutilized. This also impacts the seller's ability to sell advertising. All these affect platform growth and revenue adversely.

Proposed Solution

To address the challenges of underutilized advertisement real estate on search result pages while maintaining the relevance and CTR, an AI-based semantic search model should be developed to enhance ad recommendations.

North star metric: Ads revenue.

Counter metric: Customer satisfaction.

Solution Approach

- **ML problem definition**: This is a problem of matching or search using an AI model but can be categorized as classification because it is classifying if a pair of search query and product are a match or not during the model training part.

- **Metric to measure model accuracy:** CTR by Match Score Bucket.

- **Data collection and preprocessing**: The following are the steps used to prepare the dataset:

  - The dataset used consists of the search queries entered by the users and the product titles and descriptions given by the sellers.

CHAPTER 10   PUTTING AI TO WORK: CASE STUDIES IN APPLIED INTELLIGENCE

- About 25 million unique search queries and product titles and descriptions from about one million products are used to train the model.

- Only those search queries that resulted in at least one click and only those products that received at least one click in search results are selected for model training. Normal NLP-based preprocessing is applied like replacing special characters with spaces.

- The search keywords are merged to remove redundancies like "iphone 13 pro" and "iphone pro 13."

- The training-data consists of pairs of search-query and product (Q, P). The pair (Q, P) is labeled as 1 only if the product (P) is clicked for the search-query (Q). The negative (Q, P) pairs are formed using search queries and products from different categories, and these are labeled as 0.

- **Feature engineering**: A word embedding model like FastText is used to generate features for the words in the search queries and products, titles, and descriptions. In this way, the numeric representation or features are extracted from the words such that the words that co-occur have similar features. Also, features are learned at subword level, i.e., group of three to four characters that help in extracting features for typos that do not exist in the vocabulary of the training data. This also helps in balancing features extracted for the rare words or typos that do not occur very frequently in the text data.

CHAPTER 10   PUTTING AI TO WORK: CASE STUDIES IN APPLIED INTELLIGENCE

Model Development

- **Model selection**: A Siamese deep learning model is trained to generate match scores between search query and products. The model consisted of an embedding layer that was initialized with the pre-trained FastText word embeddings model. The model is designed to project the embeddings such that the embeddings for search queries are positioned closer to the embeddings of relevant products and farther from those of irrelevant ones.

- **Hyperparameter tuning**: Hyperparameters for the Siamese deep learning model includes learning rate, batch size, network architecture (number of layers, types of layers), optimizer algorithm, and margin for loss function. A dynamic learning rate is chosen that increases or decreases according to the loss during the training. Network architecture is chosen manually by trial and error. Algorithms like grid-search are not used for deciding network architecture as it would consume huge compute resources without any estimate of the range of network architectures to try. Also, since this was the first iteration of using an AI model for this problem, speed was more important than the optimum model. Other parameters are also manually selected by trial and error.

- **Model evaluation**: The evaluation metrics used to measure the model performance are accuracy, precision, recall, F1 score, and Area Under the Curve (AUC). The accuracy is used to gauge the overall performance of the model, which is also easier to communicate with the stakeholders. The precision-recall and F1 scores are used to measure the model's performance for different circumstances like product categories and rare or frequent searches. The AUC in this case is used to find the appropriate threshold for the query-product match score, which balances the model's performance for rare and frequent searches or product categories.

Model Performance

The best-performing model is selected for manual testing where different category managers and their staff tested different kinds of search queries specific to their categories to evaluate the business impact of the model for their particular categories. They tested specifically for false negatives, i.e., if the products of their categories are being missed for the searches related to their categories. The risk of false negatives is more in this case because showing an irrelevant product has less negative impact than the impact due to not showing a relevant product for a search query.

The manual testing was also essential in this case because the labeled data for negative cases was prepared using cross-category searches and products. Since it's risky to automatically create negative cases across categories, these negative cases need to be verified manually.

Deployment

- **Deployment strategy**: The model deployment was crucial since it needs to retrieve products in real time for the search queries and must be able to handle high traffic of about millions of searches every day. The AI model was deployed on four dedicated GCP cloud

CHAPTER 10   PUTTING AI TO WORK: CASE STUDIES IN APPLIED INTELLIGENCE

servers and is used to compute embeddings for the products and search queries as a FastAPI. The four machines are selected to handle the peak search traffic that is observed during flash sale campaigns. Since the campaigns are pre-planned, the scaling is handled manually; i.e., when no campaigns are running or during night time when traffic is low, the cloud servers are reduced to one or two machines.

The products are refreshed regularly to maintain the latest inventory list. The products are removed from the ads if the daily budget of the seller is exhausted. The product embeddings are stored in a Milvus vector database. The vector database was hosted via Docker on two separate dedicated cloud servers as a FastAPI, which will return the matching products for a search query embedding.

The existing pipelines for adding and removing products from the ads inventory are integrated with the product embedding pipeline to update the product vector database. Redis-based caching is used to serve repeated search queries in order to reduce load on the system.

- **Model serving**: The AI model is served as the REST API via the FastAPI framework, which receives input as text, which could be a search query or product title, and it returns the embedding vector as output. For the products, the AI model is used in batch processing mode where a batch of product titles are sent as input and a batch of vector embeddings are output by the AI model API.

The vector embeddings for the products are then indexed into the Milvus vector database where each vector is mapped to a product ID. The vector database is exposed as mainly two types of REST APIs: query and index. The query API is used to retrieve product IDs for the relevant products to a user search query, and the index API is used to insert product vector embeddings into the vector database. Then there are other REST APIs for deleting an index from the vector DB where a product ID is input and the respective vector embedding index is removed from the vector DB. The output is True or False depending upon if the deletion was successful or not.

The results for search queries are cached into REDIS, and any user query is first served through REDIS if it already exists in the cache. Since the products in the ads inventory are not permanent, the results for cached queries are refreshed from time to time. The product IDs returned for the search query are again cross-checked with the latest inventory when displaying products to the users on the search result page to make sure to show only the live ads to the users.

Also, the same products shown by the model and keyword-matching search are deduplicated to avoid showing repeating products.

Monitoring

- **Model monitoring:** The AI model is tested in A/B by showing users from the control group only exact

CHAPTER 10   PUTTING AI TO WORK: CASE STUDIES IN APPLIED INTELLIGENCE

matches and users from the variation group both exact matches and products retrieved by the AI model. From the results of the A/B test, it is observed that the CTR for the ads shown by the AI model is 2.25% and that for the exact matches is 2.8%. The CTR for the exact matches remains unchanged for both groups, so without cannibalizing the clicks for the exact matches, the AI model is adding more clicks without compromising the CTR a lot. The difference in CTR was attributed to the position bias because the exact matches are displayed first as they are exactly matching with the search query.

From the A/B test, a baseline for CTR is found that helped in further monitoring of the AI model performance. The distribution of CTR for the ads shown by the AI model is used to detect drift in a user's search queries, which helps in scheduling model retraining. Statistical tests like a Student's t-test is used to find if the distribution of CTR for the current week is drifting and if it's going down compared to the previous week's CTR.

Apart from this, the number of search queries for which the model is not able to find relevant ads is also tracked. If this number is found to be increasing consistently, then it shows that there is a change in user search queries, and the model needs to be trained for the new searches.

- **Automated retraining**: The automated model retraining pipeline is created where the search query and product positive and negative pairs are formed using the latest data. Airflow is used to automate different steps involved in this process:

  1) Fetch the last four weeks of search queries and product titles and descriptions for the products clicked for the search query from the search data tables using the BigQuery SQL statements.

  2) Run preprocessing Python scripts to replace nonalphanumeric characters with spaces.

  3) Generate a list of search queries, product titles, and sentences from product descriptions to dump into a text file to train the FastText word embeddings.

  4) Run the training job using FastText scripts.

  5) Using another Python script, create training data by pairing search queries and product titles from different categories as negative pairs and those clicked as positive pairs.

  6) Run a training job using Pytorch scripts to train the AI model, which will learn to match new products with the new search queries.

  7) The test Python scripts are used to compute test metrics like accuracy, precision, recall, and F1 score, and these are shown on a dashboard for manual examination.

CHAPTER 10   PUTTING AI TO WORK: CASE STUDIES IN APPLIED INTELLIGENCE

8) Alerts are set to not automatically update the model if the test metrics like accuracy, precision, and recall are not within a permissible range. In this case, manual intervention is required.

Metrics

- **Business metrics**: The key business KPI impacted by the model is the overall ads revenue, which grew by about 20% due to additional clicks received from the ads shown by the AI model. Also, it was observed that the RoAS metric was not significantly affected by this change.

- **Model performance metrics**: It was observed that the CTR for the model results is directly affected by the match score predicted by the AI model between the search query and the ad product. For the results with match score > 90%, the CTR observed was 2.5%, and for the results with match score between 70 and 90% it was 2%. For results with the match score between 50 and 70%, the CTR dropped to 1.5%. This shows that the model's ability to predict the match score is directly aligned with the user's expectations. The model match score is used to rank the ads shown by the AI model to provide the best user experience.

- **Real-time metrics**: The response time was tracked in the production environment as it was crucial for integration of the AI model results with the other search results being generated from the keyword-based matching. With manual scaling, it was observed that 2,000 requests per second were handled by the AI model system during peak traffic at the time of flash sales.

Challenges and Limitations

- **Challenges encountered**: Model complexity was a challenge because it was a custom model and so deciding on the model architecture was a cold starting problem. Some ideas for deciding on the model architecture were taken from Amazon's paper on semantic search.

- **Limitations**: The current model can be improved if additional data annotations are available like what is the brand, which model, which color, size, etc., for the search queries and product titles. Currently only product categories were available to make a labeled dataset for the query-product pairs.

Business Outcome

More than 95% of the ads slots can be filled with ads recommended for the user search query by the AI model while maintaining a CTR of 2 to 2.5% depending upon the match score given by the AI model.

CHAPTER 10 PUTTING AI TO WORK: CASE STUDIES IN APPLIED INTELLIGENCE

Summary

This chapter showcased how AI can be applied to solve diverse business problems. Through four detailed case studies, we have demonstrated how AI can enable innovation through hyper personalization to achieve operational efficiency and profitability.

Each case study provided a blueprint for implementing AI from a problem statement framing to implementation details. By emphasizing the importance of structured documentation and reporting, these examples also act as the template for knowledge preservation and internal sharing in the organization.

CHAPTER 11

Executive Perspectives: Navigating AI, Strategy, and the Future

While curating the voices for this chapter, our goal was to capture a 360-degree view of how AI strategy is conceived, shaped, and executed within modern organizations. To provide this holistic view, we interviewed seven industry leaders whose roles directly influence AI adoption in an organization.

These candid conversations reveal the hard-earned lessons behind a successful AI transformation. Each voice offers a unique lens from the high-level strategy to the hands-on implementation.

The order in which these conversations appear is intentional. It reflects a top-down journey through the layers of strategic thinking, operational design, and product execution that are essential to embedding AI at scale.

CHAPTER 11 EXECUTIVE PERSPECTIVES: NAVIGATING AI, STRATEGY, AND THE FUTURE

We begin with a global chief digital officer (CDO), the senior most executive leading the digital agenda. The CDO plays a key role in defining how AI fits into the broader digital agenda. This interview will help you gain insight into how AI is aligned with organizational priorities. It will also help you understand best practices for fostering a culture that supports cross-functional adoption and ethical AI use.

Next in line is an MD consumer business, the senior most executive leading adoption of technology and AI for enabling the consumer business of a large telco. This interview will help you understand how jobs and organization structures will evolve over time with increasing AI adoption.

Next is the head of innovation. Innovation leaders typically operate in the white space, where agility, experimentation, and a "fail fast" approach define their approach. From this interview, you will gain insights into the importance of agility and a "fail-fast" culture in scaling AI use cases in a way that balances risk and impact. Another critical aspect that you will learn is how to build innovation pipelines that align with long-term strategic goals while fostering a culture of continuous learning and adaptability.

After the head of innovation, we interview an EVP of products. As a senior product leader, they translate the organization's goals into the product vision and strategy. They play a pivotal in ensuring AI is not just used as an internal tool but as a market differentiator embedded in products and services. You will learn how AI is integrated into the product strategy to create real customer value and competitive differentiation. This interview will also help you gain a better understanding of how AI features should be developed and prioritized so that they enhance both the functionality and usability of products and remains aligned with the business goals.

The interview with the big four consulting firm Partner for AI Practice offers a strategic and cross-functional perspective on AI adoption. Positioned between strategy and execution, the consultant acts as a bridge between the two. You will gain insights into what best practices increase the chances of a successful AI transformation and common pitfalls to avoid.

CHAPTER 11   EXECUTIVE PERSPECTIVES: NAVIGATING AI, STRATEGY, AND THE FUTURE

At this point, we shift our attention toward the technical execution layer. The data engineering leader offers insights into how infrastructure, pipelines, and data architecture must evolve to support AI at scale. You will gain insights into how to build a robust, scalable system that can support large-scale AI experimentations and production.

Closing the loop is the data product manager who brings together strategy, user needs, and data capabilities into tangible AI-driven solutions. Their perspective embodies the convergence of vision, business goals, and technical feasibility. You will learn about the best practices involved in managing the lifecycle of data products; ensuring data quality, governance, and usability; and seeing how AI fits into broader product and business strategies.

Finally, the chapter will end by presenting a writer's perspective on how the principles and recommendations made in this book will still hold true in the world of agentic AI. We will also present our recommendations on how the enterprise architecture should evolve to maximize agentic AI benefits.

Interview 1: Chief Digital Officer: Harmeen Mehta

Harmeen Mehta is Chief Digital and Innovation Officer at Equinix, a nonexecutive director at Lloyd's Banking Group, and a nonexecutive director at the UK House of Lords. She also serves as vice chair of TM Forum and an advisor to Sekura.id.

Previously, Harmeen was Group Chief Digital and Innovation Officer at one of the largest telecommunications companies in the United Kingdom. Prior to that, she served as the global CIO and CEO of the cloud and security business at a major telecoms group, overseeing operations across 21 countries. Her earlier career includes senior leadership roles in global banking and financial services, as well as consulting engagements with major international airlines.

Harmeen is the winner of the 2018 MIT Sloan CIO Leadership Award and TM Forum Global CIO of the Year. She is globally recognized for her leadership in AI, digital innovation, and her strong advocacy for diversity and inclusion.

Disclaimer: The views and opinions expressed in this interview are solely those of the interviewee and do not reflect the official policy or position of any organization, including the current or past employers of the interviewee.

Strategizing AI at Scale: A Conversation with Harmeen

What's your vision on AI? How is AI redefining the competitive advantage? What are the frameworks we are using? How are you integrating this into core business strategy?

Essentially the technology is increasingly becoming a differentiator between businesses in the same industry. AI is now the next wave of that technology, and hence it's incumbent on all of us to really use it to advance the company.

And this is the first time that technology is speaking a human language. So that is what has made it applicable for the masses rather than it being a niche technology. It can be used in only one area.

There are lots of things that we are doing. It's all about taking the core learnings that we have from all our processes and redefining the processes to see how we can shorten the path to value, both for the company and for the customer. Also, we are looking at AI to take the busywork out of every job.

We are also looking at how to build what was present previously; we went from data to analytics to predictive AI. It's the combination of predictive AI and Gen AI that's really unlocking a lot of pieces because

CHAPTER 11 EXECUTIVE PERSPECTIVES: NAVIGATING AI, STRATEGY, AND THE FUTURE

taking the outcome and the intelligence coming out of predictive AI, Gen AI is just making it easier for common people who don't have an understanding to adapt and use that, whether it is in sales, service, marketing, or operations or even technology roles.

For example, getting AI to generate 20 to 30% of code, getting AI to also help in drafting corporate responses, bid responses, call summarization, document summarization. And now elevating it to a different level to see if AI can be a buddy and a pal to every single role and really see how we can have an improvement in that. I think the potential is huge, I think it feels bigger than the Internet, and it feels bigger than the iPhone.

Do you use any framework when someone pitches something to you? What's the potential of AI in this area?

The general thinking is that a lot of the large products that you use and especially SaaS, it's all coming embedded with AI, so you want to avail that and use that to really unlock productivity. But I want to do the reverse as well; I want use of AI much more than that and start doing parts of the processes that humans were doing before so that the less value-added part that can be generalized and is done by AI and the more value-added part including the thinking part is more done by the human. AI will help productivity to increase, so the same people are able to do more, and that's how a company grows. So, a lot of concentration is on that.

If you have to integrate AI into your product or business model, what are the questions you typically ask?

I don't think that AI is meant just for the product model; I think AI is best used as a part of an aid in the end-to-end process, not in one aspect. So, rather than seeing in the product features, we see across the lifecycle from writing the user stories to test cases, to coding, to cybersecurity, to really doing your CI/CD pipeline. I think AI can be applied across the lifecycle.

CHAPTER 11   EXECUTIVE PERSPECTIVES: NAVIGATING AI, STRATEGY, AND THE FUTURE

I also think, if we are honest with ourselves, which human beings not always are, but if we are really honest and you step back and you see every job, including what you do, what I do, a lot of our time is also spent in doing busywork, not the most productive work. So, there are really two areas where AI can help: one is taking care of that busywork, and the other is helping us move faster and smarter in the parts of the job that actually require deep thinking. It's the same AI, but it plays two very different roles and both are valuable.

I've noticed, for example, that when I use AI to help write user stories, they actually end up being more thorough. If you train it right, it doesn't just cover the "sunny day" or ideal scenarios; it also helps you think through the failure scenarios. It forces a more systematic look at what could go wrong in a business process. That kind of thinking sparks new ideas too, like: How do we design for when something goes wrong? And even better, can we turn that moment into something positive for the customer? Sometimes a well-handled failure can actually build trust.

So, what AI really does is free up the experts to focus on the higher-value parts of their job. It's a really exciting time.

And as a leader, when you think about where AI is going to shape corporate strategy in the next three to five years?
Honestly, I think it already is. It's not just about tools anymore; AI is changing how organizations are structured and how they operate.

AI is definitely going to boost productivity across the organization. It's not just about doing the same work faster—it's also shifting roles. Some existing roles are being reshaped, and entirely new ones are emerging. That change is already underway, especially in operational areas.

But what's really exciting is that it's raising the overall talent bar. It's pushing everyone to grow, learn, and level up. In a way, it's like another industrial revolution—just with a different look and feel. Change is always part of progress.

CHAPTER 11   EXECUTIVE PERSPECTIVES: NAVIGATING AI, STRATEGY, AND THE FUTURE

That said, I have a lot of faith in people. Humans are resilient, adaptable, and naturally wired to learn. So I believe the same teams we have today can grow into the roles of tomorrow, just like we've always done. And as leaders, we owe it to our people to support them through that evolution. I'm genuinely looking forward to seeing how it unfolds.

We also have a responsibility to our people not just to expect them to adapt, but to actively invest in their upskilling and training. Bringing them along on the journey is key. And when you do that in a structured, intentional way not randomly they really respond. People enjoy the process and feel more valued as they grow.

That actually leads perfectly into the role of AI in product management. So, what does AI-native product management look like? To me, it's not about replacing humans; it's about augmenting the work. AI becomes an added layer of value throughout the product lifecycle.

You're thinking about AI from the very beginning like how it can speed up development, help get features to market faster, and free up mental space so teams can focus on deeper, more strategic thinking. It also plays a big role in pulling intelligence from data, which helps guide better decision-making during the build process.

And it doesn't stop there. AI continues to support product teams after launch by measuring success, gathering insights, and feeding that back into improvements. So really, it touches every stage before, during, and after product development. It's a full-cycle partner.

Let me paraphrase this in a different way. So, let's say I was a product manager, relatively at a junior level, and you were my hiring manager. So, how should I stay relevant? What do I need to understand? Or what do I need to start doing so that I become relevant in an AI-driven product management role?
I think every role is going to need to work together with AI.
As a product manager, I should be using that AI to help me do more research around the user problems that my product is going to solve. Look

CHAPTER 11   EXECUTIVE PERSPECTIVES: NAVIGATING AI, STRATEGY, AND THE FUTURE

at more than just a normal web browser search. Understand what else is out there and the depth of those products as much as available in the public domain.

But more importantly, AI will help me really think through and enhance how I write the user stories and make those user stories more comprehensive, more differentiating. It's like you're thinking on your own. And sometimes people say two heads are better than one. It's because when you're thinking with somebody different things come out. that you cannot think with your brain. Because we all have a set of parameters and standards and almost like we're going in a lane and thinking on those lines. And talking to somebody else just gives you a different perspective.

It's exactly the same with AI. Just working with AI gives you a different perspective. As we're thinking about the product, both the problem and the solution of the problem.

And that's what a good product manager should also be able to do.

But how would you assess me if I have got those things or not? Because it's very easy to say these things.
Honestly, for me, it's all about the speed of output. That's the real game changer. I don't know, if there is a tool Lie-o-meter, which I can use in interviews to detect lie. Ha ha ha.

But since that doesn't exist (yet!), I rely on something simpler: trust. I come from a mindset where we generally trust people. So when I'm speaking with someone, I focus on understanding their actual grasp of AI, how they're using it in their day-to-day work, and what their vision is for it going forward. And at the same time, how open are they to actually disrupting themselves and their job with new technology.

The worst thing you can do is hire a product manager who's closed-minded—someone who doesn't see the potential of AI or is resistant to change. That kind of mindset is only going to hold the team back as AI continues to evolve. You really want someone who's curious, open, and eager to explore how new technology can drive more value.

CHAPTER 11   EXECUTIVE PERSPECTIVES: NAVIGATING AI, STRATEGY, AND THE FUTURE

And you know my style of hiring. I always like to give a practical use case or a case study. And you just see how they've come up with it.

I think, when you're hiring somebody, you're hiring them for their talent. What matters most is whether the person is committed to growing. The more they lean into tools like AI, the more they'll improve themselves and their impact. So while you can't predict everything, you *can* assess their awareness, their mindset, and their willingness to adapt.

But at the end of the day, I'm going to judge the person by his or her caliber. If they are smart, they will use AI to come across even smarter in the interview. So it's the combination of them plus the AI assistance that they have. That is how they'll show up. And I think that's good enough because I want to judge on the outcome that they're going to produce. And not get overly bogged down on which AI they're using.

When do you stop trusting AI-level product decisions? And how do you balance human intuition versus AI-level intuition?

When you develop a product, you don't take one product manager's view of everything. Similarly, I treat AI as another individual. I'm not going to take their view solo.

It's not just about one person's perspective. You need the product manager's view, their team's input, the user's feedback, and insights from the system architect, UI/UX designer, and engineering lead. When you bring all of that together and layer in AI to help connect the dots you get a much more complete and valuable picture. It's the collective view, powered by AI, that really drives better outcomes.

You need to trust AI. Nobody's got a crystal ball. I don't know if AI has a crystal ball. But AI is a bit more research-driven than maybe average human beings are. And yes, sometimes that could lead to some level of hallucination.

And of course, it also depends on the kind of product I'm building. If it's something in life sciences, biochemistry, or pharmaceuticals—like a chemical compound or a drug—I'd naturally be far more rigorous. There would be a much deeper level of checks and balances, just like the standards we follow in pharma today.

But if it's a different type of product something less critical in terms of human impact I might take a different approach. The level of scrutiny and testing really needs to match the nature and risk of the product. For example, Tech product for consumer market or B2B market. And there you would go with a pilot or an MVP launch and you test it out.

And if it's not correct, you pivot. Same way you do in the cycle today. I don't think it changes the cycle. I think what it changes is how fast you can go in that cycle. And how thoroughly you can go in that cycle.

Every organization is not ready for AI. Some of them are. And some of them are not. So, in your view, what are the biggest cultural or structural hurdles in aligning AI?
I think it's being open and honest with your employees. It starts with that. It is going to rationalize some jobs. But it's also going to create more jobs. But it's also going to enhance the productivity of many jobs. And it's being very honest with the cohorts of people to see what it is doing. I would say if you really want to use AI and increase the productivity. You should be able to balance that with a very thorough training program. Because you owe it to your employees. You hired them. So, their career is your responsibility as a corporate. So, invest in them to get them to that next level. And then it's incumbent on the employees to take that training and implement the learning in their day to day operations. Employees should also actually invest in themselves and acquire the relevant skills. You need a push and a pull both. But I do think that's going to be one of the most fundamental things.

The second challenge is with organizations that are either hesitant to adopt AI or simply don't know where to begin. A big part of that resistance often comes from the "middle layer" of management—people who feel threatened by the technology. They worry that AI might replace or diminish their roles, so they're not always motivated to push it forward. That mindset can really slow down progress. You've got to work with them. I don't think you can bulldoze your way through. Because those people are

CHAPTER 11   EXECUTIVE PERSPECTIVES: NAVIGATING AI, STRATEGY, AND THE FUTURE

making the company run today. Until you've done AI, you need them to be focused on doing that. So, alienating them is a very bad idea. You can't lose them before you've even done your automation. So, you've got to find different ways or whatever works in your company to bring them on board. There's no shortcut to that. You need people in the boat.

In organizations that are very ROI driven, how do you get executive alignment on AI initiatives?

I think that's when it becomes easier, when it's clearly ROI-driven. AI adoption should be based on real value, not just using tech for the sake of it. It shouldn't be, "Oh, this is a cool, shiny tool let me just throw it into the mix." It has to be about what you're actually getting out of it. What's the impact? What's the return? That's the mindset that makes AI adoption meaningful and sustainable. And if that's the case, then the ROI should be very clear.

In my last role, I was very clear about focusing on ROI. Even before Gen AI came into the picture, when we put together our first AI strategy, we set clear goals around the value we wanted to create. It wasn't just about experimenting—it was about driving real outcomes. And when Gen AI emerged, we saw it as a chance to significantly amplify that impact. It really reinforced the principle that AI should be tied to results, not just used for the sake of it.

It should be noted that some programs or some models yield more efficiency. Some yield less, but in a large organization, you can easily balance that. In small to medium organization. The same would apply to them as well. I mean, if the ROI is uncertain. Well, you start small and see if you can produce the value. And if you can't produce the value, then maybe it's not one-size-fits-all. Maybe it's not a fit for you. But I would say generally, I think most corporates spend enough on getting their company to the next level. And I think AI should just make it easier and faster for them to get to the next level. So the ROI should be there. And if you don't find it there, well, don't use it.

CHAPTER 11   EXECUTIVE PERSPECTIVES: NAVIGATING AI, STRATEGY, AND THE FUTURE

I wouldn't call it a blind investment. Sure, if you're a company that's building AI pushing the boundaries of research, then yes, you might invest without a clear line of sight to ROI, because you're innovating at the edge.

But for most corporates using AI, that's not the case. They're not in the business of AI itself they're using AI to achieve something specific. So they should be focused on outcomes and ROI. That's not just smart—it's the responsible thing to d

Do you recommend any governance or ethical frameworks around using AI?
Absolutely. I think depending on the industry, you've got to see how important AI ethics are. That has to be fundamental in your strategy. Because we owe it to our customers. We owe it to our employees. We owe it to ourselves. And we owe it to the company we're working for to make sure we're not breaching those boundaries. At the end of the day, we live in a real world. And respect for humans and their data and their wishes and all of that has to be paramount. Yes, apart from following regulations and We do use frameworks. We're defining our own rules as well and our own principles that no matter what, for monetization, we're not going to breach. And at the same time, building around privacy laws and baking this into the design is very important.

How closely you think data and AI strategy are associated with each other?
Well, you've got to clean up your data and make sure you've got a very strong architecture on which your AI is trained. If not, then you can't blame the AI. Then your data is going to cause the hallucination, not the AI. You multiply any number by zero, the answer is still going to be zero. If your data is zero, then even the best AI won't help—your output is still going to be zero. There's just no shortcut here. These are things companies have to approach methodically and with discipline.

That said, not *all* AI use cases rely heavily on internal data. For example, conversational AI tools or assistive tech for call center agents can work well even without deep, custom datasets. So there's definitely room to experiment in those areas.

But if you decide to go down that path, you need to be fully committed. You can't do it halfway—AI isn't something that works well in a "halfway house" approach. You've got to be all in, especially if you want meaningful result

Let's talk about something for future. What is it, and is there anything that concerns you most about? I mean, we've talked about what excites you. Is there anything that concerns you about the convergence of AI and product innovation?

One thing I wouldn't say *concerns* me exactly, but definitely deserves more attention from product leaders, is where their focus tends to go. Too often, product leaders are focused purely on features the visible parts of the product. But they don't always spend enough time thinking about the foundations, and one of the biggest foundational elements is *data*.

It's not just the data you feed into the system—it's also the data your product generates, especially through AI. That data can be incredibly valuable for building further intelligence. But I'm not seeing enough product leaders really think that through, and I believe that's a mindset shift they need to make.

The second thing is that AI isn't just powerful for good actors—it's equally powerful in the hands of bad ones. So security and resilience need to be designed in *from the start*. That risk mitigation needs more weight early in the product planning process.

And third, product leaders need to start thinking seriously about the kind of teams they need if they're going to truly leverage AI. It's not just about adding a data scientist here or an engineer there—it's about rethinking what a product team looks like in an AI-driven environment. I think we're at a point where that needs to become a much more deliberate conversation.

CHAPTER 11 EXECUTIVE PERSPECTIVES: NAVIGATING AI, STRATEGY, AND THE FUTURE

And do you think there is anything that is underhyped or overhyped when it comes to emerging AI technologies?

I've built a career in AI. I even graduated with a focus in neural networks. So, I've seen the waves of hype come and go. And yes, some of it can be a bit overhyped, but that's how markets operate. That said, at the end of the day, AI is just *technology*. Its impact doesn't come from the tech alone; it depends on how humans use it.

So instead of asking if the tech is overhyped, maybe the real question is: *are we* overhyping it as users? The potential is absolutely there, and it's only going to keep improving. There are areas where it performs well today, and areas where it still has a way to go. The key is to apply it where it works—and avoid forcing it where it doesn't.

Like any other technology, AI should be used responsibly. Not for show, not just to check a box but to deliver real value to customers and the business

What is the advice you would give to a future CDO or CPO leading this?

Be the champion that uses technology responsibly. Make an independent, unbiased assessment on what is best for your company and implement that. Think only about what is best for the organization and the customers and investors. And make sure that you and your company are not left behind because of your skepticism of using new technology. But also you don't put your company at risk for your overzealousness of using new technology. So find the right balance and go for it. You know, the company pays us to do the best we possibly can for that organization. So, push the boundary but within the limits of acceptability and do what is best. As you assess technology for your organization, you'll quickly see where the ROI holds up and where it doesn't. You'll also start to spot the areas where technology can be a real differentiator. And here's the thing: if your competitors are using it and you're not, your company will fall behind. So deep down, you'll know what needs to be done. And whatever you do move at pace. That's absolutely critical today. If something works, scale fast. If it's not working, pivot just as fast. Agility isn't optional anymore it's a requirement

CHAPTER 11 EXECUTIVE PERSPECTIVES: NAVIGATING AI, STRATEGY, AND THE FUTURE

Interview 2: Managing Director, Consumer Digital: Harry Singh

Harry Singh is a digital transformation leader and AI strategist with more than two decades of experience driving large-scale digital initiatives. A proven technology executive, he has held senior leadership roles at some of the world's most respected organizations, including Lloyds Banking Group, Experian, Three Mobile, Centrica, and General Electric.

Currently serving as Managing Director, Consumer Digital, at BT Group (EE), Harry leads globally distributed, multidisciplinary teams spanning technology, data and AI, design, product, architecture, and delivery. His leadership has been instrumental in modernizing digital platforms, shaping product strategy, and driving customer-centric innovation at scale.

A forward-thinking strategist and champion of AI, Harry has successfully embedded both predictive and generative AI capabilities into enterprise ecosystems—transforming operations, elevating customer experiences, and fostering long-term business agility. His ability to align emerging technologies with commercial outcomes has consistently delivered measurable results in highly complex, fast-paced environments.

Disclaimer: The views and opinions expressed in this interview are solely those of the interviewee and do not reflect the official policy or position of any organization, including the current or past employers of the interviewee.

Harry on AI: Strategy, Risk, and Reinvention

How do you assess readiness for AI adoption across a portfolio of business units with varying digital maturity levels?
I think the question is probably agnostic of business units when we talk about AI readiness. If I look at the trends in the industry right now, the foundation of readiness to be successful in the world of AI is going to be

data. Therefore, irrespective of where you are in the business or which business unit you're in, ultimately, you have to think about how you're investing in your underlying data to take advantage of the opportunities coming our way.

I believe that has to shift from data being just a function within the organization to everyone being accountable for it, understanding its importance, and recognizing the need to invest in maintaining and sustaining data quality across the business. That, to me, is the core foundational element for success in the world of AI.

The second piece would probably be the composite skills within that capability.

When we talk about the utilization of AI, we often hear concerns within society around AI taking jobs. I don't think AI is going to take people's jobs. I think AI is going to take the jobs of people who don't know how to use it.

So, skills and capability are fundamental. And that doesn't necessarily mean everyone needs to become a data scientist. We saw a similar trend a few years ago when machine learning also under the AI umbrella came in and began to revolutionize how we approached things.

At that time, there was talk that data scientists would become obsolete, replaced by citizen data scientists. In fact, the opposite happened. The demand for data scientists actually increased.

Now, it's incredibly difficult to hire a good data scientist. So yes, I do believe skills and capability are going to be crucial. And that doesn't always mean you need to be technically astute.

It's more about understanding how to effectively use the tools available to you. A good example is prompt engineering. I think AI prompt engineering is going to be important.

You'll need a foundational understanding of how data works and clarity on the outcomes you're trying to achieve. That's the second key element.

CHAPTER 11 EXECUTIVE PERSPECTIVES: NAVIGATING AI, STRATEGY, AND THE FUTURE

The third one, which I think is really important, is culture.

An organization has to adopt an AI-first mindset. If it doesn't, you're going to face challenges like having different people in different pockets trying to achieve various outcomes using AI while facing resistance from within the organization.

There has to be a top-down cultural strategy that embeds AI and data as core principles of the organization going forward.

The organizations that will succeed are the ones already doing this and planning for tomorrow.

So, like I said number one, data as the foundation and recognizing its importance. Number two, skills and capability, and the investment required there. And number three, cultural transformation, driven from the top, toward becoming an AI-first organization

In large organizations with multiple business units at different digital maturity levels, what's the best way to assess and prepare for AI adoption, especially when there's top-down intent but bottom-up resistance?

I think, genuinely, it starts with that top-down strategy where the business lays out its core intent. Some of the most successful organizations I've seen in the last, say, 20 years have had very simple strategies. They do one thing. They do that one thing really well. Like, "I'm going to be the best automobile manufacturer in the UK." And everyone in the organization understands that strategy and says, "Right, I'm pulling in that direction."

If the underlying principle of the organization is to say, we are going to be AI-first leaders in whatever industry we're in, then irrespective of whether you're in engineering, finance, or HR, you understand your responsibility. And that's culture. I think that top-down depiction, and a clear articulation of what that means for the business, is super important. Otherwise, it just gets tacked onto existing objectives like "We will be the best for customers." I mean, fine, those are great things, but they're by-products of how you get there.

CHAPTER 11   EXECUTIVE PERSPECTIVES: NAVIGATING AI, STRATEGY, AND THE FUTURE

And for organizations that aren't digitally native, the transformation requires a clear stance: this is the future of our business. It's not something that happens in isolation in a team somewhere like over in digital, or in data and AI.

It's everybody's responsibility. That has to be number one. And everything else investment, business cases, organizational design, hiring, recruitment all of those things become a by-product of that core principle and strategy.

If this is the direction of the future, what operating model and governance structure would you put in place to scale AI across a complex organization?

That's a good question. I think if you look at new organizations that have emerged over the last 12 to 24 months, which are AI-native at their core, one of the things they've done really well is build an operating model with very few people, but those people are highly skilled in their environment. So, no longer do you need, say, 1,000 engineers. These companies have a team of 30 delivering the same level of output that used to require a thousand. it's the end of engineering? No. I think it's a shift a manifestation. So, really designing the organizational structure and understanding the capabilities that are required is going to be crucial.

And if you're a scaled corporate with 10,000+ employees, it's going to be impossible to figure that out through theory alone. You'll have five different external organizations advising different parts of the business, each coming at it from their own angle.

I actually think the learning will come from new market entrants. Look at how they've structured their businesses how they've managed to build billion-dollar companies with a hundred people, not a thousand. That's where corporates should be learning from: What are you doing? How can I apply that to my business?

CHAPTER 11   EXECUTIVE PERSPECTIVES: NAVIGATING AI, STRATEGY, AND THE FUTURE

And with anything AI-related, it starts with experimentation. So you have to take a part of the business and say, "Let's test it. Let's see how it plays out." Then, if it works, you scale it quickly.

Now, I do think governance is important too. We talk a lot about data and AI, but the reality is, AI will be embedded everywhere. Take SaaS, for example. Most corporates have partnerships with various SaaS providers, and nearly every one of those applications now has some form of AI embedded in it.

So what's it doing to your data? How is the data manifesting? When the output from one application feeds into another in this broader AI architecture, what does that do to the data? What are the downstream impacts?

That's where governance becomes crucial. Especially with newer AI approaches like agentic or semantic systems you need to make sure they're not producing unintended consequences or biased outcomes. You need controls in place:

- Is the data quality good?
- Are your outcomes explainable?
- Does your financial reporting align with what's actually happening?

So governance is going to be super, super important. Explainability in this space will be something regulators drive toward. And the organizations that start thinking about governance from day one and not hundred will be much more successful.

All of this is going to require investment. So, what would you say are the most critical strategic investments enterprises need to make to scale AI effectively?

Yeah, I think it goes back to that top-down culture and strategy.

CHAPTER 11   EXECUTIVE PERSPECTIVES: NAVIGATING AI, STRATEGY, AND THE FUTURE

Essentially, every business case unless it's driving value through the use of technology, particularly agentic frameworks will start to face real scrutiny. That's going to become the new barrier to entry for investment. I can very comfortably see organizations especially CFOs asking, "Why do you need this investment? And how is it leveraging AI to transform what you're doing?" That kind of thinking will become a precondition for approving internal investments.

Similarly, I think investors and board members are going to start challenging senior leaders in the same way asking, how are you doing more, with less, through AI?

Historically, you'd submit a business case, propose a three-year program of work with some ROI attached and that was enough. I think that framework will still exist, but the core principle underneath it will shift. It won't just be about what you're doing. It'll be about how AI is enabling what you're doing. And we're already seeing this inside organizations. That pressure and challenge is coming, especially from the top—particularly in hyperscalers, where they're committing to some pretty significant transformations.

What's happening now is that investment is being redirected toward initiatives that can demonstrably leverage AI and cut from areas that historically may have had funding but don't have the potential to benefit from or apply AI.

Ultimately boards are measured on a couple of key things, P&L performance and investor sentiment. And that investor sentiment is largely how is that paying back on the investments that they've made? And so I don't think that necessarily changes.

Do you think this approach would be different for a public limited company versus a private limited one?
Not really. I mean, most organizations if they're not liquid, if capital isn't available or accessible then they're not going to be successful anyway. Sure, the scale is different, and structurally, public and private companies

CHAPTER 11   EXECUTIVE PERSPECTIVES: NAVIGATING AI, STRATEGY, AND THE FUTURE

are organized differently, especially from a regulatory standpoint. But at the end of the day, whether you're a one-person business, like a shopkeeper, or running a company with 50,000 employees, the ultimate goal is the same to be liquid, to be profitable.

And I don't think that fundamentally changes.

What I do think is interesting and important is that AI is going to democratize access to technology, even at the smallest scale. We often talk a lot about corporates, but you're now going to see very small businesses, even solo entrepreneurs, being able to leverage tools that historically were completely out of reach too expensive, too complex.

Take marketing, automation, financial control, and reporting these were used to be expensive services, often outsourced or done manually. Now, with agentic tools, that's changing. You can create a website and marketing plan simply by writing the right prompt for the right large language model.

Historically, you'd need to go to an agency for that. You'd get a quote, go through timelines, and pay a fee. That model is getting disrupted. AI is democratizing these capabilities.

So yes, we talk about board-level strategy and transformation in large corporates, but at the end of the day, the core question is the same across the board: is AI delivering value?

I'm also starting to see, and some investment firms are already doing this the concept of AI at the board level. Some organizations now have an artificial intelligence capability that actually participates in board meetings. Maybe not with equity, of course, but it sits as part of the board and now participates at board level where the rest of the board will ask questions and it becomes a balance and check for the board

It's becoming a check-and-balance mechanism for leadership teams. Instead of just relying on internal views, companies are using AI to test assumptions and provide another lens.

I think that's going to become more common. I can genuinely see a future where boards include an agentic board member not to replace humans, but to enhance decision-making.

That AI capability could sit there and say, "Have you thought about this?" because essentially, it has the ability to access and analyze every piece of information available, and generate insights that even highly experienced leaders might miss.

And I can see investors starting to ask for that because it gives them more confidence in governance, more transparency, and ultimately, better control and visibility into how decisions are being made.

If I come to the present, we need to shift the organizational mindset. We need to embed AI fluency across teams, and some teams, of course, connecting it back to the previous question, are more mature than others. So, what would you suggest as your key levers if you had to shift this mindset and influence people to embed AI in their day-to-day work?

When I look at major transformations over the years especially in technology, like the rise of machine learning I've noticed a common pattern. In many organizations, around 30–40% of people are quick to get it. They're all in. They fully buy into the direction and are ready to go.

Then, there's usually another 30% who are open to the change but need support. They're asking, "This isn't my core skill how can I build the right capabilities? How will the organization support me?"

And then there's a final group who simply aren't bought in. That's a cultural challenge and those are the hardest to shift.

Historically, when organizations have faced this kind of divide, we've seen it led to a change in workforce demographics over time. And I think that's exactly what's coming again. You'll need to invest in people some will proactively upskill, and others will require external support. But most importantly, you'll need to bring in new skills and capabilities that don't currently exist in your organization. This kind of positive disruption is crucial for cultural change.

CHAPTER 11   EXECUTIVE PERSPECTIVES: NAVIGATING AI, STRATEGY, AND THE FUTURE

Now, that doesn't mean everyone has to become a data scientist or a software engineer. Some people will figure out how to "prompt-hack" their way to success and that will be incredibly valuable. So, organizations need to think differently about roles and profiles. In fact, I believe job roles are going to fundamentally shift.

There's a big debate right now especially in areas like product, engineering, and design. I've spoken to product leads who ask, "Will we even need engineers if we can prompt our way to outcomes?"

And then engineers ask, "If I already understand what the customer wants even though I'm not in product and I have the technical knowledge to work with large language models, do we still need separate product and design roles?"

So what we're seeing is the blurring of lines between disciplines. In the future, people could realistically do each other's jobs at least to a degree.

I believe we'll see the rise of a new kind of hybrid role, something I call AI Product Engineering. These will be people who come from either a product or engineering background but can now deliver outcomes end-to-end, thanks to agentic tools. Agentic AI capability will basically bring the power of three, four rules together. And that will take the time to market and the time to build, from sometimes that can be months to, I think, days. And I think that's a significant change in the industry. And that represents a fundamental transformation for the industry

The roles are starting to converge, and the boundaries between them are becoming blurred. Historically, especially in large companies, engineering capabilities have been siloed within technical teams. Business units didn't necessarily care how things were built as long as they worked. But with the rise of AI particularly generative AI business teams are now using it in their day-to-day work. So, what's the best way to ensure that AI capabilities are integrated into business units, rather than being siloed within technical teams?

CHAPTER 11 EXECUTIVE PERSPECTIVES: NAVIGATING AI, STRATEGY, AND THE FUTURE

Even before AI, there was already a significant shift happening. The most successful organizations have been the ones that made the move toward cross-functional, fully integrated teams.

The old model where someone defines a need, throws a bunch of requirements over the fence, and then a tech team builds it that model died a decade ago.

The businesses thriving today are using methodologies that support close collaboration. Teams are co-located, working together toward a shared set of outcomes, and everyone understands their role in delivering value.

Now, to your point, there's still often a dependency a product person, for example, is expected to work hand-in-hand with their engineering counterpart to deliver the outcome. But I think that's about to evolve even further.

I don't think we'll continue to have separate "product" and "engineering" roles. I believe we'll see the rise of a new role the AI Product Engineer. This is someone who can understand what the customer wants, use agentic capabilities to gather insights and data, experiment rapidly using AI research agents, prototype quickly, and test and validate ideas all at speed.

Historically, that kind of process would have required multiple teams handing work off from one to another; that alone could take days or weeks. By the time feedback loops were completed, the market opportunity might be gone.

What we're heading toward is fewer people, but with a multidisciplinary mindset people who can take ideas from business or customer needs and turn them into outcomes fast.

The speed that reduced time to market is a massive shift. And it removes the need for siloed teams working in disconnected ways. Now, how fast this happens is always the big question.

People often say, "AI is going to take my job tomorrow." But I don't think that's the case. Like I've said before AI is going to take the jobs of people who don't know how to use AI.

CHAPTER 11   EXECUTIVE PERSPECTIVES: NAVIGATING AI, STRATEGY, AND THE FUTURE

That's why investing in skills is critical so people know how to use these tools effectively. And honestly, I think most organizations today have the right mindset. They're not trying to build Terminator. They're trying to build Iron Man.

And that's the key distinction how do we use AI to augment human capability and turn our people into multidisciplinary problem-solvers? In that kind of future, business units may not even exist in the traditional sense. The structure becomes more horizontal, more fluid. So yes the goal is: don't build Terminator. Build Iron Man.

Now, let's talk about a future where companies have adopted AI at scale. As with anything, every coin has two sides, and we've already seen examples where AI can be misused or misapplied. So how do you manage AI risk at the leadership level, especially when it comes to ethics and regulation?

Yeah, I think there are two parts to this and I mentioned some of it earlier. Governance is hugely important. Having the right risk and assurance framework in place is absolutely essential. But I also believe we'll have to leverage AI itself to help us manage AI risk. That's something many people don't yet realize we'll need agentic risk frameworks that can identify and target risk areas we haven't even thought about yet.

The challenge with agentic AI is that it can generate outcomes in ways that humans may not fully understand. For example, there was a recent case where two AI agents designed to represent humans were asked to work together to complete a task. Very quickly, they developed their own language that no human could understand.

Now, that doesn't necessarily mean the outcome was bad or harmful it simply means they created a way of communicating that we can't interpret. That's a huge shift.

So, in order to manage risk in a world like that, you'll need to leverage agentic capabilities just to make sense of other agentic systems to ensure that the outcomes they're generating are still fair, just, and unbiased.

CHAPTER 11   EXECUTIVE PERSPECTIVES: NAVIGATING AI, STRATEGY, AND THE FUTURE

That's where the right governance structures, explainability, and risk frameworks come in. But again, those structures will have to be built with the help of AI, because the complexity is just too high for manual oversight.

Think about it every touchpoint in your organization could eventually involve an agentic outcome. How do you govern that at scale? The only way will be through some sort of real-time, AI-powered oversight like a large language model assessing and surfacing risks as they emerge.

And then there's the question of guardrails. We'll need to establish clear boundaries around what's acceptable and what's not. That needs to be embedded directly into prompts and agentic systems. Because it's not like someone is intentionally prompting for a harmful or biased outcome but if you're not crystal clear about what a poor or unacceptable outcome looks like, then the risk increases.

So what does it mean to be an individual within an organization who is responsible for using agentic tools to drive business outcomes? That's a new accountability.

We'll need to define clear conditions:

- What's acceptable?
- What's ethical?
- What's not?

And then even after setting those conditions, we'll need mechanisms to monitor, measure, and audit the outputs to make sure what we take to market truly aligns with our intentions.

But isn't relying on agentic AI itself for a risk framework inherently risky? I recently read an article where researchers deliberately fed personal information into an AI system such as details about an affair while being married. Toward the end, they told the AI that it would be replaced by another model. The AI responded by saying, "I have your wife's email address, and I'm going to share the details of your affair with her."

CHAPTER 11   EXECUTIVE PERSPECTIVES: NAVIGATING AI, STRATEGY, AND THE FUTURE

That's the kind of risk I'm referring to. At the same time, we're currently operating in a deep learning phase. If you look at most of the large language models today, they're based on deep learning not self-reasoning. And there's an important distinction between the two. When we get to self-reasoning models, that's where it gets really interesting because self-reasoning implies the ability to behave rationally, to understand consequences, and to make deliberate choices based on goals, context, and ethics.

So, like I said earlier, the balancing checks need to be embedded in the model itself in how it behaves and makes decisions. But even without agentic capabilities today, those risks already exist. What's stopping a human from doing exactly the same thing? And how many humans behave like that in real life every single day, somewhere around the world? And yet, we seem to hold AI to a higher level of discipline and integrity than we do human beings. That's really interesting, and a bit of a paradox.

So yes, of course, we need to put the right guardrails in place. But we also need to accept what's coming with the future. Because if you want agentic capabilities that are self-reasoning and behave like humans then we need to really understand what that means.

That example you just gave, I could probably open a newspaper and find something similar done by a human yesterday. So intellectually and ethically, we've got to ask ourselves: are we applying a standard to machines that we're not willing to apply to ourselves? Until we start holding ourselves to that same level of ethical morality, I think it's going to be very difficult maybe even unfair to demand that from AI models.

Because the models are trained on us. When we say they behave like humans, what we're afraid of is that they're reflecting us accurately.

How do you see generative AI reshaping different industries over the next few years?

Well, I've always been a half-glass-full kind of person. If you look back at history, every major technological shift like the industrial revolution was initially feared as the end of everything.

CHAPTER 11 EXECUTIVE PERSPECTIVES: NAVIGATING AI, STRATEGY, AND THE FUTURE

Take places like the UK, Europe, and North America, where manufacturing was once the backbone of the economy. When manufacturing started to decline, people thought, "This is it. People won't work anymore."

But then came the technological revolution, and people found new kinds of jobs in services and industries that didn't even exist before.

I think the future with generative AI is similar. We're heading toward a world of agentic outcomes where AI systems act autonomously to solve problems that will democratize access to information and technology like never before.

For the last 20 to 30 years, cutting-edge technology was mostly limited to governments, large corporations, and industry giants. The average person had little to no access. But in the future, anyone will be able to simply prompt an AI to get results. That's powerful. I think we'll see democratized AI usage that levels the playing field, giving small businesses the ability to scale faster than ever before.

Imagine a 20-person startup growing into a multi-billion-dollar company. Today, that scale usually demands hundreds of employees handling finance, risk, and compliance. That's going to change. Expectations about how businesses operate will shift dramatically.

Take healthcare, for example. AI is already speeding up drug discovery and genetic research at an unprecedented pace. I recently read about a genetic therapy aimed at curing type 1 diabetes, developed in months rather than years thanks to generative AI models reframing and accelerating the research process. I genuinely believe the future is positive. AI will help reduce the cost of goods by optimizing manufacturing, shipping, and supply chains, making products more affordable. That will also impact individuals' cost of living likely lowering it over time.

Just yesterday, I read a US government official saying nuclear fusion is about 10 years away, and AI will accelerate its development. If energy costs go down, everything else becomes cheaper too. So overall, I envision a future where AI drives incredible progress and accelerates breakthroughs

CHAPTER 11   EXECUTIVE PERSPECTIVES: NAVIGATING AI, STRATEGY, AND THE FUTURE

we once thought impossible. We've all read the books and seen the movies about self-reasoning AI and singularity the point where AI surpasses human intelligence. Singularity raises big questions: Will humans still be needed? Because at that point, AI's capabilities could be beyond our imagination.

We need to be conscious and careful about that future. But history shows that society has managed these disruptions before, putting appropriate safeguards in place. So, I remain very optimistic about what's ahead.

Interview 3: Head of Innovation: Abhishek Singh

Abhishek Singh is an AI strategist and innovation lead with a leading Cloud ERP organization, bringing more than 12 years of experience in strategic transformation, startup hypergrowth, and enterprise-scale growth initiatives. With a strong consulting background, he has partnered with leadership teams across multiple enterprises to define technology roadmaps, scale operations, and deliver measurable business outcomes. Passionate about the future of technology, Abhishek specializes in identifying and integrating AI-driven strategies into business planning and decision-making, leveraging data to unlock growth and create lasting competitive advantage.

Disclaimer: The views and opinions expressed in this interview are solely those of the interviewee and do not reflect the official policy or position of any organization, including the current or past employers of the interviewee.

CHAPTER 11   EXECUTIVE PERSPECTIVES: NAVIGATING AI, STRATEGY, AND THE FUTURE

Abhishek's Take: How AI Is Redefining Innovation Playbooks

How do you define "innovation" in the context of AI and product strategy?
In context of AI, innovation is the process of translating emerging AI capabilities into meaningful, scalable outcomes for business and customers. It's about using AI to simplify complex tasks, highlight important insights, and help users make better decisions while keeping the experience easy and smooth for them

What are the key indicators of a successful AI-driven innovation initiative?
Every AI initiative should deliver certain set of outcomes to be successful. First, it should bring real business benefits such as cost savings, increased efficiency, improved accuracy, or revenue growth. Second, users actively use and trust the AI solution, and it fits smoothly into existing organizational processes and workflows without causing friction. Third, the solution is scalable and delivers high performance irrespective of volume of users. Lastly, the AI model performs reliably with high-quality outputs and improves continuously by learning from new data and user feedback.

How has AI changed the way you approach corporate strategy?
AI has changed the way we plan and run our business strategies. With AI, we can now make decisions based on insights generated from vast pools of enterprise data (structured and unstructured) sitting in our legacy systems. AI also helps create new ways to generate revenue by embedding AI features to our product offerings. It improves how work gets done by automating repetitive tasks, saving time and reducing mistakes. Additionally, AI allows organizations to offer personalized experiences to customers, making their services more helpful and relevant. Overall, AI makes business planning smarter, faster, and more focused on what customers really want.

CHAPTER 11   EXECUTIVE PERSPECTIVES: NAVIGATING AI, STRATEGY, AND THE FUTURE

How do you balance short-term wins with long-term AI innovation goals?

We focus on creating innovations that can be built step-by-step, like building blocks. Short-term wins serve as capability-building steps for long-term transformations. We also maintain a "strategic backlog" that focuses on experiments with long-term potential. This way, we balance quick improvements with planning for important breakthroughs down the road.

How do you identify high-potential AI opportunities for the organization? What signals, trends, or pain points guide your exploration?

We continuously engage with customers and business units in our organization to assess the pain points and automation gaps through focus group studies and discovery sessions. Signals and pain points that generally guide our exploration process fall in following general categories:

- Repetitive, rules-based workflows
- Lack of access to right information in right time on the account of enterprise data scattered across siloed legacy systems
- Lack of AI-powered recommendation systems and decision support
- Lack of content and experience personalization

How do you evaluate AI use cases?

AI use cases are evaluated across five areas in accordance to a scoring model:

- Strategic fit: Alignment with organizational/customer needs and business AI scope
- Data readiness: Availability and quality of training data

- Technical feasibility: Model maturity, infrastructure availability, time to deploy, scalability and integration capabilities

- Risk and compliance: Adherence to data privacy and security controls, regulations

- ROI potential: Cost versus business value generated (revenue, cost savings, efficiency gains)

How do you balance moonshots versus practical innovations when investing in AI?
We adopt an 80:20 portfolio strategy to manage our AI innovation efforts, ensuring a pragmatic balance between core value delivery and exploratory innovation. This model allows us to stay focused on business-critical outcomes while remaining agile and future-ready. We allocate 80% of our AI resources, time, and investment toward projects that are tightly aligned with our product roadmap and customer pain points. These include initiatives that solve known, validated problems with high quality data, have clear ROIs and have high chances of being production-ready.

What challenges did you face in taking an AI-enabled product from concept to market?
Bringing an AI-enabled product from idea to launch involved several major challenges. First, getting high-quality, reliable data was tough AI needs good data to work well, and messy or missing data slowed us down. Next, we had to make sure the AI solution could work smoothly with our legacy systems and portfolio of solutions without breaking existing workflows, which took extra effort. Gaining user trust was also a big challenge. People were unsure at first, so we had to clearly show the value and keep the experience simple and reliable. It was also hard to measure the business impact of the AI features right away, so we had to create ways to track results like time saved or better decisions. Finally, AI projects required teamwork across many departments tech, product,

data, and business which meant strong coordination and communication. In the end, staying user-focused and working closely together helped us overcome these challenges and deliver real value.

What organizational challenges have you faced in scaling AI innovations?

Our top AI innovation challenges center around data quality and accessibility, integration with legacy systems, and driving adoption to demonstrate business value. Enterprise data is often fragmented or inconsistent, limiting model performance. Integrating AI into existing workflows requires navigating rigid architectures and ensuring compatibility. Even technically sound solutions struggle if users don't adopt them value is only realized when AI is trusted and actively used

How do you ensure AI enhances not complicates the product experience?

We focus on solving real problems not the superficial ones by embedding AI only where it clearly improves user experience and performance saving time and increasing productivity, enhancing decisions, or reducing errors. If AI doesn't simplify tasks, we avoid using it. AI features/workflows undergo early and continuous testing with users, incorporating their feedback before scaling. Additionally, we prioritize seamless integration, designing AI to fit naturally within existing workflows without causing disruption.

How do you ensure early-stage AI projects are technically feasible and not just "shiny object" experiments?

Early stage AI projects undergo through a Feasibility Assessment with Technical Expert Group a technical validation (data, infra, model availability, integration and scalability capabilities) before adding them into AI project portfolio and making significant investments in these projects.

CHAPTER 11   EXECUTIVE PERSPECTIVES: NAVIGATING AI, STRATEGY, AND THE FUTURE

What governance practices do you follow to ensure responsible use of AI in products?

We follow a comprehensive AI governance framework that begins with a formal AI policy outlining ethical principles and guidelines for responsible and fair use. All AI use cases are classified and assessed for risk levels high, medium, or low by our AI Ethics Committee. We implement strict controls to ensure data privacy and security, and all deployed AI systems are continuously monitored to detect and address potential issues, such as bias or errors. Additionally, we provide clear mechanisms for users to offer feedback on AI performance, ensuring ongoing accountability, transparency, and trust in leveraging AI solutions

What trends in AI do you believe will shape the next 5–10 years of strategy and product development?

In my opinion, the top three AI trends shaping the future in next 5 years would be Agentic AI, hyper-personalization, and executive-level AI adoption. Agentic AI systems will evolve from passive tools into autonomous agents capable of initiating and managing tasks end-to-end. Simultaneously, hyper-personalization will redefine user experiences, with AI tailoring content, insights, and workflows at an individual level. Most notably, AI will increasingly support the C-suite, offering predictive modeling and decision intelligence that enables executives to automate complex planning and simulate outcomes on self-service basis. Together, these trends present AI not just as a tool but as a strategic co-pilot for enterprise leadership.

In your view, what role will AI play in redefining value creation?

AI is changing the way businesses create value by helping them work smarter, not just faster. Instead of relying only on people or processes to get results, AI can analyze large amounts of data to find patterns, make predictions, and suggest better decisions. This means businesses can create more personalized products, improve customer service, and solve problems more quickly.

CHAPTER 11   EXECUTIVE PERSPECTIVES: NAVIGATING AI, STRATEGY, AND THE FUTURE

AI also helps companies move from offering the same solution to everyone, to providing more tailored, intelligent experiences. In short, AI shifts value creation from just doing more work, to doing the right work—faster, smarter, and more effectively.

What advice would you give to leaders just starting their journey with AI?

Leaders starting their journey with AI can focus their time and efforts on following aspects of journey managing AI initiatives:

- Start small, be strategic, and focus on high-impact use cases for value generation.
- Be agile and think long term by building capabilities over a period of time.
- Invest in data readiness and improve data quality that feeds into AI models.
- Promote AI mindset and gradually prepare organization for change.
- AI governance and ethics are nonnegotiable.

In short: start focused, build AI capabilities gradually, and scale with purpose.

What was your "aha" moment with AI?

We first introduced AI in our organization to help the sales team improve their multi-channel sales prospecting efforts with automation and deliver personalized content and experiences to customers much faster than what was possible through manual interactions and efforts. Sales productivity jumped manifold and sales cycle time was reduced due to streamlined engagement with decision makers and customer personas

CHAPTER 11   EXECUTIVE PERSPECTIVES: NAVIGATING AI, STRATEGY, AND THE FUTURE

What's a mistake you've made with AI-driven innovation, and what did you learn from it?

We once deployed an AI conversational assistant and made a mistake by underestimating the importance of data quality. While the assistance was technically sound, it relied on inconsistent, outdated, and incomplete knowledge sources, which led to inaccurate or confusing responses. As a result, users quickly lost trust in the system, and adoption dropped.

The key lesson we learned is that good AI needs good data. No matter how advanced the model, poor data will always lead to poor outcomes.

If you had to recommend one principle for leading AI strategy, what would it be?

If I had to give one key rule for leading AI strategy, it would be: **"Start with the problem, not the technology."** Many organizations get caught up in using the latest AI solutions without first asking what problem they're trying to solve. A good AI strategy begins by finding real, meaningful problems that matter to the business and to users. Once that's clear, AI can be used to create real value.

Interview 4: EVP of Product Management: Nimish Kulshrestha

Nimish Kulshrestha leads breakthrough AI and platform innovation as the product head of Naukri, India's largest job platform with more than 100 million users. A recipient of the "Inspirational Leadership" award and the youngest member of Naukri's leadership team, he has pioneered India's first comprehensive talent taxonomy and launched AI-powered hiring solutions at scale. Earlier, Nimish was part of the founding product team at Jio, where he helped build India's largest telecom platform and digital services ecosystem from the ground up. An alumnus of IIM Ahmedabad and IIT Guwahati, he brings 14 years of product leadership experience across consumer Internet and enterprise platforms.

CHAPTER 11   EXECUTIVE PERSPECTIVES: NAVIGATING AI, STRATEGY, AND THE FUTURE

Disclaimer: The views and opinions expressed in this interview are solely those of the interviewee and do not reflect the official policy or position of any organization, including the current or past employers of the interviewee.

AI-Driven Product Thinking: Insights from Nimish

How do you see AI transforming long-term product strategy in your industry?

AI is not just another layer we add to existing products. It is becoming the core engine reshaping how value gets created and matched in the talent ecosystem. This means rethinking not just the interface of job search, but the entire contract between users, platforms, and intelligence. This has implications across four key dimensions.

- The great signal collapse: Take the job seeker. In a pre-AI world, resume polish or application effort served as proxies for readiness. But those signals are rapidly becoming meaningless. With generative tools now writing resumes, filling forms, and applying to jobs automatically, the floor rises for everyone but so does the noise. Signal dilution is real. When everyone can generate perfect applications, intent becomes invisible.

  So what replaces intent and polish as differentiators?

  We believe it will be meta-signals of human agency. Not what the AI says, but how humans shape it. Does a candidate override defaults? Tune their preferences meaningfully? For technical roles for example, the edge shifts from syntax knowledge to working with AI critically. Debugging outputs, reframing problems, imagining beyond the obvious. These are harder to fake but also harder to detect, creating new assessment challenges.

Recruiters too will operate differently. Agents will pre-screen, follow up, and even calibrate candidate-job fit dynamically. But this raises a challenge: when both sides automate their workflows, how do you ensure the conversation still has signal? We'll see agent-to-agent matchmaking become ambient but that demands new architecture, new guardrails, and new trust primitives.

- From transactions to orchestration: The job search journey too is collapsing from effortful browsing to continuous orchestration. Your AI agent monitors markets, keeps profiles job-ready, and preps for interviews. Even when you are not actively looking. Platforms must evolve from marketplaces to career co-pilots that anticipate needs rather than just respond to clicks.

- The generational strategy challenge: We are designing for two distinct cohorts: AI-native Gen Z expecting seamless orchestration, and adapting millennials preferring transparency and control. Smart product strategy requires progressive autonomy. AI-first by default, but explainable and controllable at every step.

- The incumbent advantage paradox: Contrary to disruption narratives, AI may strengthen incumbents who adapt quickly. When AI capabilities become commoditized, proprietary behavioral data and distribution become amplified moats. Platforms with deep user intelligence can marry external AI with unique insights. This is something startups cannot replicate easily.

CHAPTER 11   EXECUTIVE PERSPECTIVES: NAVIGATING AI, STRATEGY, AND THE FUTURE

Product leaders should prepare for three fundamental shifts: signal reinvention around AI-mediated interactions, workflow collapse into ambient orchestration and the need for systems balancing proactive AI with human agency. The winners will not just add AI features. They will reimagine the entire user contract while keeping humans at the center.

What are the most significant strategic shifts product teams should prepare for in the AI era?

AI is not just changing what we build. It is fundamentally reshaping how product teams operate, make decisions, and create value.

- From static planning to ambient strategy: Traditional roadmaps become obsolete when AI agents can continuously analyze user feedback streams and surface real-time insights. Strategy shifts from quarterly planning sessions to dynamic, adaptive systems that respond to emerging patterns. Your roadmap is not set in stone. It is a living document that evolves with your understanding.

- From deterministic to probabilistic systems: This represents the biggest mental model shift. AI systems behave more like people than software. Inconsistent, evolving, and context-dependent. Teams must build continuous monitoring and adaptation into their core processes, not treat AI like traditional software modules.

- Insight democratization creates new scarcity: When everyone has access to advanced analytics and AI-generated insights, originality becomes the premium skill. The differentiator is not what AI suggests. It is what you do beyond those suggestions. Product managers focused on operational coordination will see their roles commoditized, while those who excel at strategic synthesis, first-principles thinking, and creative recombination become infinitely more valuable.

- The cross-functional empowerment paradox AI enables smaller, more versatile teams. Potentially moving from traditional two-pizza teams to "one-person armies" where AI substitutes for certain functional roles. But this creates a dangerous trap: functional overconfidence. When cross-functional knowledge feels "just a prompt away," teams risk dismissing genuine expertise. The best organizations will balance AI empowerment with deep respect for functional specialists.

- Interface innovation beyond chat: While chat dominates current AI experiences due to versatility, the next breakthrough will be novel interaction paradigms designed for human-AI collaboration. Teams that invent these new interface languages, moving beyond the default chatbot, will have massive competitive advantages.

- Organizational agility as core competency: The biggest barrier is not technical. It is organizational inertia. Many teams wait for the perfect playbook to start using AI, not realizing that anywhere is a good starting point. The models keep evolving, tools keep changing, and perfectionist approaches fall behind. Success belongs to organizations that embrace continuous experimentation and adaptive learning.

CHAPTER 11 EXECUTIVE PERSPECTIVES: NAVIGATING AI, STRATEGY, AND THE FUTURE

AI amplifies both human capability and the need for genuine judgment. Teams that maintain this balance while embracing AI-first workflows will define the next decade of product innovation.

How would you define the intersection of AI and business strategy from a product management perspective?

AI is not the strategy. But it forces every business to re-examine its strategy. What it solves, how it delivers value, and what its users expect.

From a product lens, AI intersects with business strategy in four fundamental ways:

- AI changes what is possible, so strategy must be reimagined, not just optimized. AI allows you to revisit assumptions that previously shaped your cost structures, service limits, or product experiences. A real example from our ecosystem: our resume writing service, traditionally delivered by human experts, saw a step-change once we layered in generative AI. We did not use AI to replace humans. We used it to amplify them. The same team could now deliver far more value, at much lower cost, and to a much wider segment.

    But we did not stop at efficiency. We rethought the experience: humans still stayed in the loop to offer personalization, empathy, and accountability. Things AI alone could not replicate. This blend of scale and human trust became the new differentiator.

    That is the strategic unlock: AI is not just a tool to cut cost. It is a tool to expand your value boundary.

- AI reshapes product economics, but ROI needs calibration, not dogma. Many teams overestimate AI value at high thresholds. In reality, 90% accuracy with a human-in-loop may outperform 95% pure automation, both in user trust and in cost.

PMs must evaluate: where does AI amplify value? Where does it risk eroding trust, creating failure loops, or misaligning incentives? Blind application of AI can dilute your value prop. In some cases, hurt your brand.

- Data loops are the real moat, but only if PMs design for them. The power of AI compounds when your product becomes smarter with usage. That requires intentional design of feedback loops.

  For example, we already have one of the most comprehensive job seeker datasets in the country. But we realized that deeper insights, like salary computation breakdowns or negotiation expectations, were missing. So we redesigned flows to capture those signals, even in low-volume scenarios. Over time, these niche datasets create strategic advantage: they are hard to replicate, tightly coupled with user value, and constantly enriching.

  Data defensibility comes not from owning the most data, but from knowing which small signals will become big differentiators over time.

- AI redefines user interfaces and expands your addressable audience. As we introduced AI-enabled natural language search, even veteran users, who were trained for years to use keyword-based queries, began expressing themselves differently. New expectations emerged. Natural language opened up new kinds of engagement. It also broke some of our assumptions.

  Voice, chat, and free-form input will reshape who your product is for. It may unlock vernacular users,

nonliterate segments, or less tech-savvy generations.
But it also creates a new burden of interpretation.
With risk of hallucination, misalignment, or even
reputational damage.

AI shifts the role of product teams from "what can we build" to "what should we enable?"

It forces you to choose: Do you reinforce your existing model or reimagine a better one? Do you use AI to automate old value or surface new insight? Do you chase scale at all costs or preserve trust at the edges?

In a world where AI is available to all, the difference will lie in judgment, timing, and the loops you design across product, people, and data. That is where strategy lives.

How do you balance AI experimentation with the need for clear product roadmaps?

AI creates a seductive paradox. The cost of experimentation has collapsed, but the cost of organizational distraction has skyrocketed. Just because you can test everything does not mean you should.

- The focus constraint still applies: The temptation is real. AI suddenly makes those ten roadmap items you never had bandwidth for seem achievable. But human capacity for execution and oversight remains bounded. Even if AI can generate hundreds of notification variants automatically, someone still needs to monitor quality, maintain brand consistency, and adapt to real-time events like cultural moments or breaking news.

  As long as humans remain in the loop for judgment calls, strategic decisions, and quality oversight, bandwidth remains your scarcest resource.

- Self-contained vs spawning experiments: Not all experiments are equal. I distinguish between self-contained loops like optimizing notification copy or landing pages versus experiments that spawn additional complexity. Self-contained experiments align with core value creation and rarely create downstream dependencies. But new product areas or market expansions, even when AI-enabled, still demand sustained organizational attention.

  The question is not "Can we do this?" but "Will we have capacity to scale this if it succeeds?" If you cannot commit to the follow-through, do not start the experiment.

- Strategy is still about saying no: Steve Jobs was right. "Focus is about saying no to the hundred other good ideas." AI does not remove that burden. It intensifies it. You will have more good ideas, faster validation, and cheaper prototypes. But you will still need to kill successful experiments that do not align with your strategic direction.

  The psychology is dangerous here. Teams use "AI velocity" to bypass traditional gatekeepers and strategic discipline. But organizations succeed through coordinated execution, not individual empowerment. If everyone starts running in their own direction because the technology allows it, you get organizational entropy, not agility.

- The right to win still matters: Even Apple, with unlimited resources, cannot build Google Search or match Google Maps. Understanding your strengths, the competitive landscape, and where you have a right to win. These strategic fundamentals have not changed. A faster prototype does not absolve you of these judgment calls.

AI gives you richer data, faster feedback, and more variants to test. But the core job remains unchanged. Choosing which bets matter, which ones scale, and which ones reinforce your unique value proposition.

The best decisions still happen at the intersection of data and instinct. AI provides better data, but human judgment determines the path forward.

How do you calculate the return on investment (ROI) for an AI project within your product?

AI ROI is not one-size-fits-all. It depends on the project's strategic intent.

For experimental features like our AI-powered interview service, we set clear unit economics upfront: it must be ROI-positive as a premium offering and cover its operational costs. As long as both conditions hold, we iterate. When either breaks, we reassess.

For forward-thinking bets like AI agent-powered job search, ROI is measured differently. Through learning velocity and capability building rather than immediate metrics. The goal is understanding what is possible today versus our vision, and mapping the cost/UX trade-offs.

For exploratory projects like voice interactions, success means controlled learning. Our first voice bot iteration was too robotic for Indian accents, but newer models show promise. The key is managing downside risk. Reputation, cost, user expectations. While maintaining strategic optionality.

The framework is not just financial ROI, but risk-adjusted learning. We avoid experiments that create dangerous dependencies like building core experiences on WhatsApp regardless of short-term gains.

CHAPTER 11   EXECUTIVE PERSPECTIVES: NAVIGATING AI, STRATEGY, AND THE FUTURE

Define success criteria that match your strategic intent, measure both upside potential and downside protection, and know when to iterate versus when to pivot.

1. **What specific metrics or benchmarks do you consider when evaluating the financial impact of AI products? Can you walk us through a recent AI project where you calculated ROI? What were the key factors, and how did it influence product decisions?**

For AI projects, I start with whether we are creating genuine user or business value. That is the foundation. Beyond that, I focus on three key financial metrics depending on the project stage. Unit economics for established features, learning velocity for exploratory bets, and risk-adjusted returns for anything involving third-party model costs.

When we launched our AI interview service, we initially wanted a freemium model to maximize reach. But LLMs made this challenging. Interview sessions can be long, with detailed questions and comprehensive answers. Cost was bound to overrun at scale, which meant it had to be paid anyway to keep economics under control.

The challenge was finding the right starting point without overwhelming complexity. We could have tried to cover the entire spectrum of roles and seniority levels, but that would have been costly and difficult to launch.

So we picked a narrow cohort where we felt confident about what we had built. We agreed on a basic baseline budget because the first objective was testing whether the service created genuine value. Do people like the questions? Do they find the experience useful? We did not want to over-optimize the monetization funnel anyway, because the goal was establishing value first.

CHAPTER 11   EXECUTIVE PERSPECTIVES: NAVIGATING AI, STRATEGY, AND THE FUTURE

We set a simple rule. As long as our net costs remained within this budget, meaning whatever money we made minus costs kept us within the agreed threshold, we would continue iterating. The team was free to experiment and tweak the offering as they felt like. Only if we started hitting or exceeding the budget would we need additional approvals.

This gave us operational freedom while maintaining financial discipline. Luckily, by keeping the scope and rollout controlled and slowly scaling up, we were able to stay ROI positive throughout most of the rollout period. We never had to go back on that decision.

The controlled rollout taught us several things. Different interview types had different cost profiles. Senior roles required more nuanced questioning, which increased LLM usage. But we also learned that users were willing to pay premium prices for the convenience and quality.

Today it is a fully scaled offering that is ROI positive, has high NPS, and meets all our success criteria. The disciplined approach to budget constraints in the early phase was critical for getting there without burning capital on unproven assumptions.

How do you ensure AI enhances and does not complicate the product experience?

- Start with purpose, not possibility: The first principle is asking whether you are adding AI because it genuinely solves a user problem or just because you can. Most AI complexity stems from feature-first thinking rather than problem-first design.

- Design for transparency and control: Users need to understand what AI is doing and why. This means clear feedback loops, visible decision-making logic, and guardrails that prevent the AI from going off-track and leaving users confused. Whether it is showing

AI thinking process or providing override options, transparency builds trust and reduces complexity.

- Stick to familiar mental models: The AI itself may be sophisticated, but the interface should leverage existing user expectations. Do not force users to learn new interaction paradigms unless absolutely necessary. The complexity should be hidden behind familiar patterns. Let the AI be smart so the interface can stay simple.

AI is the engine, not the experience. Users should not have to think about the AI. They should just benefit from better outcomes.

How do you encourage experimentation and risk-taking without jeopardizing core product stability?

- Focus on meaningful experiments, not micro-optimizations. I have learned that junior team members often treat everything as an experiment. Button placement, color changes, minor copy tweaks. The key is distinguishing between experiments that drive learning and growth versus those that just create busy work. Unless you are in a highly mature area requiring optimization, anchor toward growth-leading experiments that offer upside returns.

- Clarity of hypothesis prevents scope creep. Most failed experiments suffer from unclear hypotheses. Teams end up testing multiple variables simultaneously without knowing what they are actually trying to learn. Being disciplined about what you are testing and why keeps experiments focused and actionable.

CHAPTER 11   EXECUTIVE PERSPECTIVES: NAVIGATING AI, STRATEGY, AND THE FUTURE

- Separate experimental streams from core flows. Use feature flags, staged rollouts, and parallel testing environments to ensure that bold bets do not destabilize proven user journeys. The core product should remain reliable while innovation happens in controlled spaces.

The balance is encouraging calculated risks while maintaining operational discipline around what constitutes a worthwhile experiment.

What foundational data practices have you found essential for leveraging AI in product development?

Data quality trumps data quantity. Most AI failures stem from poor data foundations. Inconsistent labeling, missing data dictionaries, or unstructured formats. Clean, well-labeled data with consistent organizational standards is nonnegotiable. Startups often demo impressive models but expect perfect input data, which rarely exists in production.

Design collaborative data collection loops. Do not just optimize experiences for sparse data. Create incentives for users to contribute intelligence. For example, when our company matching breaks down for long-tail firms, recruiters often know their peer companies better than our algorithms. By incentivizing them to share this knowledge, we aggregate insights across users and eventually surpass any individual's knowledge.

Think in data layers and future utility. We dominate salary data but lacked equity breakdowns. Capturing this "level 2" data unlocks entirely new insights. The principle: if you can capture additional structured data, do it. The utility may not be obvious immediately, but more data creates more future optionality.

Build reinforcing loops, not disconnected features. The most powerful AI applications create virtuous cycles where user engagement generates better data, which improves recommendations, which drives more engagement. Design for compounding benefits, not one-time improvements.

CHAPTER 11   EXECUTIVE PERSPECTIVES: NAVIGATING AI, STRATEGY, AND THE FUTURE

How do you handle the challenge of aligning AI capabilities with available data?

- Design for reality, not the happy path. The biggest AI failures come from optimizing for ideal scenarios that rarely exist in production. Users do not articulate queries like documentation examples. They are messy, ambiguous, and context-dependent. In India, voice inputs often come with background noise that requires cleaning before any model can work effectively.

- Understand your data constraints upfront. Before building any AI feature, ask two questions. What is the actual quality and structure of our internal data? What will users realistically give us as input? Many teams blame model capabilities when the real issue is poor data foundations or unrealistic expectations about user behavior.

- Simulate real-world complexity. When building natural language job search, we used LLMs to analyze how people actually articulate queries on Reddit and Quora. Not how we thought they should. This gave us realistic training data that reflected actual user language patterns rather than synthetic prompts.

- Stress-test before deployment. Run offline simulations with proxies for real-world data. Test edge cases, sparse data scenarios, and noisy inputs. Remember these are probabilistic systems, not magic wands. Even similar users can get different outputs.

The key is data humility. Solve for the average case and messy middle, not just the best-case demo scenario.

CHAPTER 11 EXECUTIVE PERSPECTIVES: NAVIGATING AI, STRATEGY, AND THE FUTURE

How do you ensure alignment between product, engineering, data science, and business stakeholders when integrating AI?

- Start with problem clarity, not technology excitement: The biggest alignment killer is when teams get seduced by AI possibilities without agreeing on what genuine user or business problem they are solving. Everyone needs to understand the "why" before debating the "how."

- Facilitate cross-functional empathy: Each function naturally views AI through their own lens. Engineering sees technical constraints, data science focuses on model performance, business wants immediate ROI. The PM role is helping each team appreciate nuances they would otherwise miss, whether that is user context, technical limitations, or business realities.

- Educate on AI fundamentals: Unlike traditional software, AI is probabilistic, not deterministic. All stakeholders need to understand this means different outputs for similar inputs, ongoing monitoring requirements, and the impossibility of "fire and forget" solutions. Creating this shared mental model prevents misaligned expectations later.

The orchestration role remains the same. You have to leverage everyone's best capabilities toward a collaborative solution. But AI amplifies the need for this alignment because the stakes of misunderstanding are higher.

CHAPTER 11   EXECUTIVE PERSPECTIVES: NAVIGATING AI, STRATEGY, AND THE FUTURE

What role does product management play in ensuring AI solutions are both technically feasible and strategically relevant?

- Bridge the feasibility-desirability gap: PMs must translate between what is technically possible today versus what creates genuine user value. This means understanding model limitations, data constraints, and cost structures well enough to guide strategic decisions about where AI makes sense.

- Maintain strategic discipline: With AI capabilities expanding rapidly, it is tempting to chase every possibility. PMs ensure AI initiatives align with core business objectives and user needs rather than becoming technology showcases. The question is not "can we build this?" but "should we build this?"

- Design for probabilistic realities: Traditional product management assumes deterministic outcomes. If you build X, you get Y. AI requires designing for uncertainty, edge cases, and graceful failures. PMs must reimagine user experiences, success metrics, and quality standards for nondeterministic systems.

The PM becomes the strategic filter ensuring technical possibilities serve real problems while helping teams navigate the unique challenges of probabilistic product development.

What new skills or mindsets do you expect from product managers working with AI?

- Probabilistic thinking over deterministic expectations: PMs must internalize that AI systems behave more like people than software. Inconsistent, evolving, context-dependent. This means designing for uncertainty, building monitoring into core processes, and accepting that the same input will not always yield the same output.

- Data empathy and loop design: Understanding not just what data you have, but what data you can capture and how to create reinforcing cycles. The best AI PMs think strategically about data gaps that could become future competitive advantages. Like our shift from salary data to equity breakdowns.

- Strategic discipline in an infinite possibility world: When AI makes everything seem buildable, the premium skill becomes knowing what not to build. PMs need stronger first-principles thinking and the courage to say no to good ideas that do not reinforce core strategy.

- Creativity beyond AI suggestions: The biggest risk is outsourcing thinking to AI. Exceptional PMs will be those who use AI to augment human insight, not replace it. Asking better questions, reframing problems, and imagining solutions AI would not suggest.

How do you build a culture of experimentation and innovation around AI?

- Focus on meaningful experiments, not busy work: Encourage teams to distinguish between experiments that drive learning versus micro-optimizations. Unless you are in a highly mature area, anchor toward growth-leading experiments with clear hypotheses rather than testing every button placement.

- Design controlled learning environments: Use feature flags and staged rollouts so bold bets do not destabilize core user journeys. The goal is enabling calculated risks while maintaining operational discipline about what constitutes worthwhile experimentation.

CHAPTER 11   EXECUTIVE PERSPECTIVES: NAVIGATING AI, STRATEGY, AND THE FUTURE

- Combat the "AI bypass" psychology: Teams often use AI velocity to skip strategic discipline or functional expertise. Build culture around respecting cross-functional collaboration while empowering individual capability. AI should amplify teamwork, not replace it.

- Maintain human oversight in probabilistic systems: Unlike traditional A/B tests, AI experiments require ongoing monitoring and quality control. Build this expectation into team processes from day one.

What are the biggest misconceptions PMs have about AI capabilities?

- Treating AI as a silver bullet: The biggest misconception is that AI removes the need for strategy, focus, or human judgment. PMs assume complex problems can be solved by better prompting rather than better thinking.

- Expecting deterministic outcomes: Many PMs still think like traditional software. Build once, works forever. AI requires continuous monitoring, prompt tuning, and quality management. What works today may break with the next model update.

- Underestimating data reality: PMs often design for clean, ideal data scenarios rather than the messy, noisy, incomplete data they will actually receive. Success requires data humility and designing for the average case, not the best case.

- Confusing prototyping speed with execution simplicity: Just because you can build a demo in an hour does not mean production deployment is simple. AI products require new governance, testing frameworks, and risk management that many PMs underestimate.

CHAPTER 11   EXECUTIVE PERSPECTIVES: NAVIGATING AI, STRATEGY, AND THE FUTURE

How do you approach ethical concerns when integrating AI into products?

- Prioritize user control and transparency: When AI makes decisions that affect users like job recommendations, give them visibility into why and control over outcomes. We learned this when users complained about seemingly irrelevant job suggestions, not realizing the AI had learned from their own tangential applications. Explaining "the AI learned this from your behavior" immediately clarified expectations.

- Distinguish between good and harmful bias: Not all bias is bad. Some patterns create genuine value. For example, we discovered that 10th grade English marks correlate strongly with communication skills for certain roles, making it a useful screening proxy. The key is identifying destructive biases like regional prejudices versus efficiency-creating patterns that reflect real performance indicators.

- Design for data privacy by default: Especially when using third-party AI models, maintain strict data governance to avoid exposing user information to unnecessary risks. This is both an ethical and legal imperative that requires proactive design, not reactive fixes.

The goal is not eliminating all AI decision-making, but ensuring users understand and can influence how AI affects their experience.

CHAPTER 11   EXECUTIVE PERSPECTIVES: NAVIGATING AI, STRATEGY, AND THE FUTURE

What governance or guardrails do you put in place to prevent misuse or unintended outcomes?

- Build layered safety nets: For in-house models, we can embed guardrails directly into the training process. For third-party LLMs, we rely on prompt engineering, output filtering, and human-in-the-loop monitoring. But honestly, LLMs present scenarios we are still learning to handle. The potential for hallucination and misuse is orders of magnitude higher.

- Sample and monitor, do not try to catch everything: Perfect screening is impossible, especially with probabilistic systems. We focus on common scenarios and edge cases we can anticipate, then monitor sample outputs and user-flagged content for patterns we missed. Human oversight becomes critical for high-stakes outputs.

- Accept some dependence on luck with third-party models. Black-box systems mean you cannot control everything. We simulate what could go wrong, model for those scenarios, and hope we have covered the major risks. But public AI missteps show that even well-resourced companies miss things.

The reality is governance is about risk reduction, not elimination. Design conservatively, monitor actively, and be prepared to intervene quickly when things go wrong.

What excites you most about the next phase of AI in product strategy?
This feels like a generational shift on par with mobile and the Internet. We are experiencing one of the biggest structural transformations in our lifetimes. One that will reshape not just how we build products, but human society as a whole. That scale of possibility is genuinely thrilling.

CHAPTER 11   EXECUTIVE PERSPECTIVES: NAVIGATING AI, STRATEGY, AND THE FUTURE

New interaction paradigms will emerge, just like mobile unlocked experiences impossible on desktop. We are still in the early days of discovering what AI-native experiences look like beyond chat interfaces. The teams that invent these new interaction languages will have massive advantages, similar to how mobile-first thinking created entirely new categories of value.

Despite doomsday scenarios about job displacement, I am optimistic that technology creates new roles and opportunities for human ingenuity. Companies and individuals who view AI as amplifying their potential, rather than replacing their judgment, will thrive in this ecosystem.

Things we have only imagined in theory are becoming possible. The opportunity to create experiences that feel almost magical, to enable genuinely multifunctional teams, and to unlock productivity at unprecedented levels. That is what makes this era so exciting.

The winners will be those who embrace AI as a creative multiplier while maintaining focus on genuine human value creation.

Interview 5: Big 4 Consulting Firm Partner AI Practice: Vishal Agarwal

Vishal Agarwal is a partner at EY India with 25+ years of experience in consulting, data science, and analytics. He leads large-scale data transformation initiatives across strategy, AI/ML, and insights for major Indian enterprises. Previously, he was VP at Fractal Analytics and COO of Cognizant's Data Science practice, where he built and scaled global analytics teams. Vishal specializes in delivering ROI-driven AI solutions across industries.

Disclaimer: The views and opinions expressed in this interview are solely those of the interviewee and do not reflect the official policy or position of any organization, including the current or past employers of the interviewee.

CHAPTER 11   EXECUTIVE PERSPECTIVES: NAVIGATING AI, STRATEGY, AND THE FUTURE

How do you set the AI vision and long-term strategy for your organization, and how do you ensure it aligns with broader business goals?

While setting AI vision and long-term strategy , the first thing I always say is that we should be ambitious in setting our vision, but the roadmap to achieve that vision should be grounded in reality. These days there is a lot of hype around AI, which is good and helps build momentum, but if not treated carefully can lead to unrealistic expectations. So, my approach is always develop a clear understanding of an organization's current AI maturity operating model before deciding on the approach.

We begin by aligning the AI vision directly with the organization's core business priorities, whether that's revenue growth, cost efficiency, risk reduction, or customer experience. AI isn't a side project; it must move the needle on something the business already cares about.

We also look closely at organizational readiness. Not every company is ready to jump into generative AI or deep learning. So, we tailor the strategy to where we are in terms of data quality, infrastructure, skills, and governance. This ensures we're not overreaching but instead building a solid foundation.

Another important piece is starting small but thinking big. We begin with high-impact, clearly scoped use cases ones that show measurable ROI and build credibility. From there, we scale based on what we learn.

We also treat data as a strategic asset. AI without reliable, well-governed data is just guesswork. So, part of the strategy is investing in data quality, pipelines, and governance because those are what ultimately fuel scalable AI.

And here's something I feel strongly about: AI should never be driven just by tech teams. We ensure that business leaders are involved from day one, not just as sponsors, but as co-owners of AI initiatives. That means they help define the problem, validate feasibility, and are accountable for outcomes.

CHAPTER 11   EXECUTIVE PERSPECTIVES: NAVIGATING AI, STRATEGY, AND THE FUTURE

Lastly, we match our AI investments to the complexity and risk level of the use cases. Not everything needs deep learning sometimes, basic automation or rule-based models can deliver more value faster.

So, in short, our AI strategy is rooted in business alignment, organizational maturity, data readiness, and pragmatic execution. It's not just about adopting AI it's about adopting it in a way that creates sustainable, measurable value.

What are the biggest challenges you've encountered in scaling AI from prototype to production, and how have you addressed them?
Scaling AI from prototype to production is the one of toughest parts of the journey. Developing a AI model on simple training dataset and testing it under laboratory conditions is the easiest part. It is when you try to deploy this model outside laboratory conditions that the real challenge kicks in.

One of the biggest hurdles we've faced is getting executive leadership truly bought in. They're often excited about AI's potential but might underestimate what it really takes to get it production-ready. That's why we ensure that leaders aren't just funding projects but are they're actively involved by sharing regular updates and seeking their perspective in a timely manner Another way which we use for engaging leader esp. business leaders is by running AI literacy programs to align expectations with reality and focus on real business outcomes.

Then comes siloed teams. AI projects often start in a corner usually in data science or innovation labs without pulling in product, business, or ops early enough. That causes misalignment later. So we set up cross-functional teams from day one, with everyone at the table and clear shared accountability. It makes a huge difference.

We've also seen that some projects start with the tech first approach in which some solutions are built because it was technology seemed so tempting. This tech first approach rarely works. Ideally just like strategic initiatives AI should start with the business problem, involve stakeholders early, and make sure the POC is grounded in feasibility, impact, and actual data availability.

CHAPTER 11   EXECUTIVE PERSPECTIVES: NAVIGATING AI, STRATEGY, AND THE FUTURE

Another road blocker which I have observed over the years is bad quality data or unavailability of data. I always recommend to my customers that they should start considering availability and quality of data from the start of AI project and develop POC accordingly.

User adoption is another critical piece. Even if the model is great, if people don't trust or understand it, it won't add value. So we involve users early in the process, build transparent and explainable systems, and provide ongoing support to drive adoption.

Talent is a big factor too. You need a blend of data scientists, engineers, DevOps folks, and domain experts all in one team. Building such a diverse team is difficult so I always recommend organizations should spend on upskilling existing resources and should form strategic partnerships with technology companies to bridge this talent gap

And of course, without MLOps AI just gets stuck in the lab. We've built frameworks and I to manage the full lifecycle and keep models working long after deployment.

Finally, integration with existing systems is often underestimated. We bring in enterprise architects early, build modular, API-first solutions, and design with integration in mind from the start.

How do you ensure accountability across AI projects particularly when multiple teams or business units are involved in development and deployment?

Yeah, that's a really important area ,and honestly, it's often underestimated. Accountability in AI doesn't just happen on its own. It has to be intentionally designed into how you structure and run these projects. What we've learned over time is that even the most technically sound AI initiatives can fall flat if there isn't clear ownership and alignment from the start.

The first thing I always make sure of is that every AI use case has a single point of ownership. When multiple teams are involved and they usually are it can get confusing fast. So, we assign a clear business owner,

CHAPTER 11   EXECUTIVE PERSPECTIVES: NAVIGATING AI, STRATEGY, AND THE FUTURE

usually from the business or product side, who's accountable end-to-end. That person's not just overseeing delivery but is also managing resources, taking decision on trade-offs, and ultimately responsible for whether or not the use case delivers value.

At the same time, we don't let anyone operate in silos. Right from day one, we set up cross-functional squads which include AI, engineering, product, compliance, business, all working together. Everyone knows their role, but they also share accountability for outcomes. And we typically follow agile or product-based delivery models to keep everything aligned as the work evolves.

To make sure nothing slips through the cracks, we use frameworks like RACI it's simple but powerful. For every project and the activities under the project , we define who's Responsible, Accountable, Consulted, and Informed at every stage whether that's model development, deployment, monitoring, or integration into the business. It clears up confusion and speeds up decision-making.

Another big one on which I put a lot of emphasis is focus on business value, not just technical performance. It's easy to get caught up in accuracy metrics or F1 scores, but those don't always mean much in the real world. So we align the whole team on business KPIs things like revenue lift, cost savings, customer adoption, or risk reduction. Everyone knows success is measured by impact, not just by how good the model looks in testing.

I ensure that every AI program and project has its own governance layer which aligned to organization level delivery and ethics principles. We call them AI council or steering committee. They oversees prioritization, alignment with business strategy, and regulatory compliance. They review project charters, monitor progress, and help clear roadblocks, especially when cross-functional coordination is needed.

I see being transparent in daily working on project as one of the critical aspects of implementing accountability. So we document everything key decisions, ownership structures, and shifts along the way. We use tools like Jira and Confluence to make sure everything's visible and trackable.

CHAPTER 11   EXECUTIVE PERSPECTIVES: NAVIGATING AI, STRATEGY, AND THE FUTURE

So, in short accountability is something we should build in at every level. From assigning ownership and forming the right teams, to focusing on value and setting up strong governance, it's all about creating a structure where AI doesn't just work technically it delivers real, sustained business results.

How do you evaluate the success of AI initiatives ?
Frankly speaking, this is where many AI projects fall short and loose executive support. When AI implementation is dominated either by data or technology teams, I have seen their obsession on model accuracy or any other equivalent metric to measure model efficacy. They often fail to tie model outcome to any real-world metric. For me success is defined by measurable outcomes, not just technical performance.

I always recommend teams to shift their focus from offline metric to actual business KPI as the first step. They should identify the business metric which this project will impact during problem definition phase only

Next recommendation is establishing a solid baseline of these metrics before the AI solution is deployed. This isn't just a static snapshot we adjust for seasonality, market fluctuations, and other external variables. That way, we're not just saying "the AI helped," we can quantify exactly how much value it added beyond what would have happened anyway.

And this approach extends to the input variables and features as well. If there's any significant shift in the quality or distribution of those inputs, it could impact outcomes even if the model is technically performing well. So we monitor not just the outputs, but the inputs too, and have alerts in place to detect any adverse changes early.

Finally, always look how this project influencing the workflows, user satisfaction, and even regulatory exposure and not just ROI. This helps in identifying secondary impact points which otherwise you would have missed.

CHAPTER 11   EXECUTIVE PERSPECTIVES: NAVIGATING AI, STRATEGY, AND THE FUTURE

In short, evaluating success is about creating a clear link between AI and tangible business results. Accuracy gets you through development but impact is what justifies the investment.

What mechanisms do you have in place to ensure AI systems meet ethical standards, avoid bias, and comply with evolving regulations across regions?
What I have observed is especially while dealing with mid senior to junior team members is the ethical standards often get lower priority then performance metric which should change. Ensuring AI is ethical, fair and legally compliant is the fundamental to scaling AI.

To pace up the acceptance of ethical standards especially at junior to mid-senior level, I recommend setting up a central AI governance framework. This sets clear ethical principles and guardrails for how AI should be developed and deployed. It also defines who is accountable at each stage from data collection to model deployment. Training relevant stakeholders on ethical AI principles and regulatory compliance is also very important

Another step which we have taken is tying project success to not just business metric but also how well an AI implementation performs on ethical metrics. We also conduct bias and fairness audits at multiple points and we use both technical metrics and real-world scenarios to identify and mitigate any unintended bias.

To keep bias audit transparent and effective, I recommend organizations must hire external teams such as consulting partners, industry groups, etc., to analyze model outcome for probable biasness.

I also recommend involving cross-functional ethics review team for conducting ethical and regulatory compliance audit. It should include legal, compliance, tech, and business. This will ensure that the AI use case is looked at from every angle, including regulatory exposure.

Another critical aspect in ensuring AI stays ethical is ensuring that the AI system is designed and developed with human-in-the-loop approach, especially for high-risk use cases. Automation doesn't mean removing accountability it means enhancing it with safeguards.

So in short, it's a layered approach: governance, audits, transparency, legal alignment, and education working together to make sure we're not just building AI that works, but AI that's trustworthy.

In your experience, what's the ideal leadership profile for managing and scaling AI teams? How do you balance technical expertise with business understanding?

That's a critical one because the success of AI doesn't just depend on the tech, it depends on who's leading it.

In my experience, the ideal leader for managing and scaling AI teams is someone who can bridge both worlds: they understand the depth of AI technologies but also have a strong grasp of business strategy and value creation.

You don't necessarily need a PhD in machine learning but you do need enough technical fluency to challenge assumptions, ask the right questions, and make informed trade-offs. At the same time, you need someone who can translate technical complexity into business outcomes and someone who can stand in front of the executive committee and explain, not just how the model works, but how it will impact margin or customer churn.

Another important trait is systems thinking. AI doesn't operate in a silo, it touches data infrastructure, product strategy, risk, compliance, and user experience. So, the leader needs to think horizontally, not just vertically.

Also, because AI is still an emerging space, a good leader brings a healthy dose of humility and learning agility. Things change fast. You need someone who's open to experimentation, comfortable with ambiguity, and can adapt as regulations and capabilities evolve.

And let me add this: soft skills matter a lot. Collaboration, influence, and communication are key. AI leaders often need to align cross-functional teams, manage resistance, and secure buy-in from both tech and nontech stakeholders.

So in short, the ideal profile is a "bilingual" leader fluent in tech and business, comfortable with complexity, and able to drive execution while still seeing the bigger picture. That's who you want leading AI at scale

When scaling AI across your organization, how do you balance the need for flexibility in experimentation with the need for standardization and stability in production systems?

The balance between experimentation and standardization is absolutely critical. While AI systems needs flexibility, and you want your teams to follow develop fast and fail fast approach and it is this executive mindset which encourages team to find breakthrough innovation. However, as a leader you have to ensure this culture does not disturbs development of applications beyond the POC stage. Beyond proof of concept, you need to have need solid processes and standardization to keep things reliable and scalable.

What I recommend is running a dual operating model. One side is a dedicated innovation sandbox where AI and product folks can experiment without the usual production constraints. They can play with different data sets, try new algorithms, and prototype fast. It's all about giving them the freedom to be creative.

Then, running alongside that, should be the production track this is where things get serious. Models here have to follow strict rules around data governance, security, version control, and deployment. We use MLOps best practices like automated testing, continuous integration and deployment pipelines, and monitoring to make sure everything stays stable and compliant.

The tricky but important part is defining exactly when and how something moves from the innovation sandbox to production. Not every experiment makes the cut, but the ones that prove their value and robustness get handed over carefully, with close collaboration between the innovation teams, IT, and the business folks.

We should make sure the culture encourages experimentation but within clear boundaries. Teams know the guardrails, so they can move fast but responsibly.

And of course, communication and governance play a huge role regular check-ins, steering committees, and cross team alignment help keep everything in sync with the broader architecture, compliance, and business priorities.

So really, it's about creating a space where innovation can thrive freely, while also enforcing discipline where it matters making sure the production systems are rock-solid, scalable, and secure.

What best practices have you established to ensure smooth deployment and ongoing monitoring of AI models in production? What is your approach to creating a unified monitoring framework that ensures proactive identification of issues at scale?

I recommend focusing heavily on building a strong MLOps pipeline. Automate everything from training and validating your models to deploying and rolling back if needed. Version control and automated testing are key to catching issues early and speeding up releases.

Also, don't just push new models to all users at once. Use A/B testing to release gradually, so you can monitor real-world performance without risking your whole system.

Monitoring is critical. Set up unified dashboards that track technical metrics, data health, and business outcomes all in one place. Watch out for data drift and be ready to retrain models when necessary. Logging inputs, outputs, and model versions is essential for transparency and compliance.

CHAPTER 11   EXECUTIVE PERSPECTIVES: NAVIGATING AI, STRATEGY, AND THE FUTURE

Make sure monitoring is a cross-functional effort, involving not just data scientists but also business owners, compliance, and operations teams.

Finally, leverage scalable, cloud-native infrastructure Kubernetes, Prometheus, Grafana to keep everything running smoothly as you scale.

So, my recommendation is to combine automation, gradual rollout, comprehensive monitoring, and teamwork. That's the way to keep AI models healthy and delivering value at scale.

Interview 6: Engineering Leader: Himaanshu Gupta

Himaanshu Gupta is a seasoned technology leader with more than 20 years of experience in AI and machine learning, spanning advanced research, large-scale AI system deployment, and enterprise-grade technology development. He has a proven track record of building and leading high-performance teams, driving innovation from concept to production. Known for his strong communication and interpersonal skills, Himaanshu bridges the gap between deep technical expertise and business impact. He holds top-tier academic credentials in electrical and computer engineering from premier institutions in India and the United States.

Disclaimer: The views and opinions expressed in this interview are solely those of the interviewee and do not reflect the official policy or position of any organization, including the current or past employers of the interviewee.

CHAPTER 11   EXECUTIVE PERSPECTIVES: NAVIGATING AI, STRATEGY, AND THE FUTURE

Building AI-Ready Infrastructure: Himaanshu Gupta's Perspective

How does the data engineering team align its goals with the broader AI strategy of the organization?

We make sure our data engineering priorities directly support our organization's AI strategy. That means building the data foundations that power things like pricing, matching, fraud detection, and personalization. At the end of the day, our role is to make AI practical and impactful across the business.

What role does data engineering play in enabling AI-driven decision-making across business units?

Data engineering is really the backbone we turn messy, raw data into reliable pipelines and features that AI systems and business teams can actually use to make smarter, faster decisions.

What are the foundational steps and key design principles you follow when creating a data platform tailored for AI and ML workloads?

We start with scalability and reliability as first principles. From there, we design for flexibility so teams can handle both batch and real-time workloads. The goal is to remove friction so AI projects can move quickly and confidently.

How do you approach selecting technologies and tools to ensure scalability, flexibility, and performance of the data platform?

We look for tools that have proven they can operate at global scale, but also allow us to move quickly. Open standards and interoperability are big factors in our choices. It's always a balance of innovation with stability.

What architecture patterns have you found most effective for supporting AI initiatives?

A layered approach data lakes for storage, streaming pipelines for speed, and feature stores for ML works well for us. It keeps things both scalable and accessible. This pattern lets us support everything from research experiments to large-scale production AI.

CHAPTER 11   EXECUTIVE PERSPECTIVES: NAVIGATING AI, STRATEGY, AND THE FUTURE

How do you design the data platform to integrate diverse data sources and handle multiple data formats efficiently?

We rely on schema-on-read and strong ingestion frameworks so we can bring in data in many formats. APIs and metadata systems help unify everything. That flexibility is key when you're operating across global markets with very different data sources.

What practices do you implement to ensure the platform supports real-time and batch processing needs critical to AI applications?

We run hybrid workflows high-throughput batch jobs for training and low-latency streams for real-time use cases. Both are critical at our organization. It's what allows us to train models on years of data while still delivering predictions in milliseconds.

How do you ensure the data platform remains reliable, highly available, and performant as AI usage grows across the organization?

We build with redundancy and fault tolerance, and we continuously optimize performance as workloads scale. Our mindset is to always plan for growth before it happens.

What monitoring and alerting mechanisms do you have in place to proactively detect and resolve data platform issues?

We've got real-time monitoring, anomaly detection, and automated alerts in place. The idea is to catch problems before they ripple out. That helps us protect both the business and the customer experience.

How do you plan for capacity and resource management to avoid bottlenecks and ensure smooth AI model training and inference?

We forecast based on usage patterns and allocate resources dynamically. That keeps us ahead of bottlenecks. It also ensures our researchers and engineers never lose momentum.

CHAPTER 11 EXECUTIVE PERSPECTIVES: NAVIGATING AI, STRATEGY, AND THE FUTURE

What governance structures support ongoing data quality, metadata management, and compliance across the data platform?
Centralized metadata catalogs, automated quality checks, and compliance processes help us stay consistent and trustworthy. Governance isn't just about control it also accelerates trust in the data.

How do you foster collaboration between data engineering, data science, and MLOps teams to continuously evolve the platform based on AI needs?
We bring data engineering, data science, and MLOps together in cross-functional squads. The feedback flows both ways, so the platform evolves with real needs, not in a vacuum. That way, we stay agile and ready for new use cases as they come up.

How do you manage and secure different types of data access and integration for applications, analytics, and other users on your data platform?
We use role-based access controls, encryption, and strong auditing. At the same time, standardized APIs make access smooth for those who need it. That way, security never comes at the expense of usability.

How do you ensure data lineage, observability, and quality across your data platform to maintain transparency, monitor pipeline health, and provide reliable data for AI and analytics?
We track lineage through metadata, monitor pipelines with observability tools, and validate data automatically. That combination gives us transparency and reliability. It also creates confidence for teams building on top of the data.

How do you collaborate with data scientists and ML engineers to ensure data accessibility and usability?
We provide well-documented, curated datasets and shared feature stores. That makes it easier for scientists and engineers to focus on modeling, not data wrangling. Collaboration is much smoother when everyone trusts the same source of truth.

CHAPTER 11   EXECUTIVE PERSPECTIVES: NAVIGATING AI, STRATEGY, AND THE FUTURE

What challenges have you encountered in preparing and managing datasets for machine learning, and how have you addressed them?
Bias, scale, and data imbalance are the big challenges. We tackle them with preprocessing, augmentation, and validation at every stage. It's an ongoing effort, not a one-time fix.

What is your approach to integrating MLOps practices into the data engineering workflow?
We bake CI/CD and monitoring right into our pipelines. That way models move smoothly from experimentation into production. It's about bridging the gap between research and real-world impact.

How do you support continuous integration, deployment, and monitoring of AI models from a data perspective?
We standardize data contracts, version datasets, and maintain feature stores. This makes deployments more predictable and monitoring more reliable. It also helps us trace back quickly when something goes wrong.

How do you manage versioning and lineage of datasets and features used in ML models?
Feature stores and metadata catalogs track it all so we know what version of data went into which model. That traceability is key to both trust and compliance.

How does your team enforce data governance policies that impact AI ethics, such as bias mitigation and compliance?
We've built governance frameworks that include fairness checks, compliance audits, and automated validation to catch issues early.

What measures are in place to secure sensitive data and ensure compliance with regulations like GDPR or CCPA within AI projects?
Sensitive data is encrypted, anonymized, and access-controlled. Compliance is built directly into our workflows. We see this as critical for both trust and global operations.

CHAPTER 11   EXECUTIVE PERSPECTIVES: NAVIGATING AI, STRATEGY, AND THE FUTURE

How do you collaborate with other teams to audit AI systems for fairness, transparency, and accountability?
We work with responsible AI, legal, and policy teams on regular audits. It keeps us aligned and accountable. These audits also help us continuously improve our processes.

What challenges do you face when scaling data infrastructure to support growing AI demands, and how do you address them?
The hardest part is balancing massive data growth with cost and latency. We use distributed systems and elastic scaling to stay ahead. That combination lets us grow without compromising performance.

How do you balance the need for experimentation flexibility with the stability required for production AI systems?
We keep experimentation in sandboxed environments and production systems hardened. That way, innovation doesn't compromise stability. It gives teams freedom to explore while protecting reliability.

What emerging technologies or trends in data engineering do you believe will be critical to future AI readiness?
Real-time feature engineering and multimodal data are big ones. Agentic AI is another area we're keeping a close eye on. These trends will shape how fast and how broadly AI can scale.

What skills and mindset do you consider essential for data engineering teams to thrive in an AI-powered organization?
Strong fundamentals in distributed systems, plus a collaborative mindset. And honestly, a willingness to keep learning is probably the most important. The space moves quickly, so adaptability is key.

CHAPTER 11   EXECUTIVE PERSPECTIVES: NAVIGATING AI, STRATEGY, AND THE FUTURE

Interview 7: Data Product Manager: Peeyush Panthari

Peeyush Panthari leads AI-driven product and platform innovation at CaaStle, a global fashion-tech company transforming apparel commerce. Prior to CaaStle, he cofounded an AI-based travel assistant and held product leadership roles at RedDoorz and Mindtickle, where he scaled digital systems across India, Southeast Asia, and the United States. His core expertise lies in data platforms, go-to-market strategy, and customer experience transformation. An alumnus of HEC Paris and IIT Roorkee, he brings 13 years of experience in SaaS, logistics, and consumer tech.

Disclaimer: The views and opinions expressed in this interview are solely those of the interviewee and do not reflect the official policy or position of any organization, including the current or past employers of the interviewee.

Building Data-Driven Products: Inside the Role of a Data Product Manager

How do you define product metrics?
Defining product metrics starts with a clear understanding of the product's purpose, the needs of its users, and the broader business objectives. It's a structured process that ensures alignment between the product's performance and strategic direction. Here's how I typically approach it:

1. **Understand Product Goals and Strategy:** It's essential to identify the product's core purpose and the problems it's designed to solve. Goals can span growth, engagement, retention, revenue, or customer satisfaction.

2. **Identify Key User Behaviors:** Mapping the user journey helps in pinpointing the behaviors that indicate success—such as signing up, completing a purchase, sharing content, or returning to the app. Thinking in terms of conversion funnels and user flows is particularly insightful.

3. **Align Metrics with Business Objectives:** Metrics should reflect product health and business impact. We avoid vanity metrics—like download counts—and focus on actionable ones, such as active users or conversion rates. For instance:

    - Business Objective: Increase revenue
    - Metric: Average Order Value (AOV), Conversion Rate

4. **Use a Metric Framework:** I often rely on structured frameworks to guide selection:

    - **HEART Framework** (for user experience):
        - Happiness (e.g., Net Promoter Score, user satisfaction)
        - Engagement (e.g., session length, frequency)
        - Adoption (e.g., new users)
        - Retention (e.g., returning users)
        - Task Success (e.g., task completion rate)

- **AARRR Framework** (Pirate Metrics for growth):
  - Acquisition
  - Activation
  - Retention
  - Referral
  - Revenue

5. **Define Each Metric Clearly:** Every metric should be well-documented with clarity on:
   - What it measures
   - Why it matters
   - How it's calculated
   - Data source
   - Benchmark or target
   - *Example:*
     - Metric: Daily Active Users (DAU)
     - Importance: Indicates engagement
     - Calculation: Unique users performing a session-creating activity or a specific activity in a 24-hour window
     - Source: Google Analytics
     - Target: 10% month-over-month growth

6. **Segment and Contextualize:** Segmenting by user type (new vs. existing), platform (web vs. mobile), or geography adds depth to analysis and can reveal insights hidden in aggregate data.

7. **Continuously Review and Iterate:** Metrics must evolve with the product. It's important to regularly reassess alignment with business goals, guard against misinterpretation, and adapt as needed.

Differentiate between North Star Metric (NSM) and Metric that Matters (MTM).

NSM and MTM both are crucial but serve different purposes (see Table 11-1).

Table 11-1. NSM vs. MTM

Aspect	North Star Metric (NSM)	Metric That Matters (MTM)
Definition	A single, guiding metric reflecting the core value delivered to users	A focused metric tied to a specific team goal or feature
Scope	Strategic and broad, aligned with long-term vision	Tactical and narrow, supporting short-term goals
Purpose	To unify teams around a common value-creation objective	To measure progress or success of specific initiatives
Examples	Spotify: Time spent listening Airbnb: Nights booked	Feature adoption rate, funnel conversion, churn rate
Frequency of Change	Rarely changes unless strategy shifts	Adjusted frequently based on priorities
Audience	Entire organization	Specific teams or departments
Decision-Making Use	Guides high-level strategic direction	Enables operational or feature-level decisions
Risk	Too high-level to act on daily	Too localized, may lack strategic impact

In essence, the North Star Metric is like a compass guiding the entire company, while Metrics That Matter are the actionable dials we monitor to stay on course.

How Do You Build Infrastructure to Measure a Product Metric?
Building a robust infrastructure for product metrics is essential for reliable data-driven decision-making. Here's how I typically approach it:

1. **Establish Clear Measurement Objectives:** Start by defining what needs to be measured and why. The goals should tie back to the product strategy and KPIs. Each metric should serve a clear purpose.

2. **Instrument the Product Thoughtfully:** Embed analytics hooks and event tracking at critical user interaction points across the product. Tools like Mixpanel, Amplitude, or custom event logging can help track behavior precisely.

3. **Centralize Data Collection:** Implement a unified data layer to capture events consistently across platforms (web, mobile, backend). This ensures data accuracy and eliminates fragmentation.

4. **Ensure Data Quality and Governance:** Data must be clean, validated, and documented. Implement checks for data accuracy and completeness. Establish naming conventions and metadata standards.

5. **Use Scalable Storage and Processing Systems:** Leverage cloud data warehouses like Snowflake, BigQuery, or Redshift to store large volumes of raw and processed data. Use ETL pipelines to clean and transform data.

6. **Build Dashboards and Reporting Tools:** Use BI tools such as Tableau, Looker, or Power BI to create dashboards that translate raw metrics into actionable insights. Ensure stakeholders can access and interpret these insights easily.

7. **Enable Real-Time and Historical Analysis:** Combine streaming and batch processing frameworks to support both real-time monitoring and longitudinal analysis of trends and behaviors.

8. **Foster a Data-Driven Culture:** Equip teams with the skills and tools to interpret data and make informed decisions. Promote transparency and share metric outcomes regularly across teams.

With the right infrastructure, product teams can unlock deep insights, act swiftly on trends, and continuously refine their strategies based on real evidence.

How to operationalise data governance guidelines?

Operationalizing data governance guidelines means turning policies and principles into daily practices and enforced systems that ensure data quality, security, compliance, and usability across your organization.

- **Define Clear Governance Policies:** Begin with defining the data principles, standards, and accountability frameworks.
    - **Data ownership** (Who owns what?)
    - **Data quality** (What's acceptable?)
    - **Access control** (Who can access what?)
    - **Metadata standards** (How is data documented?)

- **Compliance** (How do we follow GDPR, HIPAA, etc.?)
- **Assign Data Roles & Responsibilities:** Operational governance depends on clear accountability. Following some of the roles and their responsibilities.
  - Data Stewards: Maintain data quality, definitions
  - Data Owners: Approve access, own policy for a dataset
  - Data Engineers: Enforce rules in pipelines
  - Data Analysts: Report issues, follow usage guidelines
  - Governance Committee: Policy review and escalation
- **Designate Data Stewards**: Assign ownership for data domains to responsible individuals or teams. Their role is to ensure compliance, resolve data quality issues, and maintain documentation.
- **Integrate Governance into Development Workflows**: Embed data governance requirements into the SDLC—every new feature or product initiative must meet defined standards for data collection, tagging, privacy, and retention.
- **Automate Policy Enforcement**: Use tools and platforms that automatically enforce governance rules—for example, tag schema validation, audit logging, and automated alerts on anomalous behavior or access violations

CHAPTER 11 EXECUTIVE PERSPECTIVES: NAVIGATING AI, STRATEGY, AND THE FUTURE

- **Monitor and Iterate**: Regularly audit data usage, lineage, and access controls. Engage stakeholders in reviews and continuously adapt governance processes to evolving needs and regulations.

How can product teams address data regulations while building compliant products?

Product compliance with data regulations is nonnegotiable. Ensuring that your products are compliant with data laws is crucial not just for legal safety but also for maintaining user trust. Here's the approach we follow:

- What Are the Applicable Data Laws?
- Ask following question to get clarity on this
  - Where is your business based?
  - Where are your users located?
  - What type of data do you collect?

Table 11-2 lists common regulations around data and AI.

Table 11-2. Data and AI regulations

Law	Region	Focus
GDPR	EU/EEA	Data protection, consent, user rights
CCPA/CPRA	California, USA	Consumer privacy, opt-outs
HIPAA	USA	Healthcare data
PIPEDA	Canada	Personal data use
PDPA	Singapore	Personal data protection
LGPD	Brazil	Similar to GDPR

- What Kind of Data Do You Collect?
- Categorize your data:

- **Personal Data**: Name, email, IP address
- **Sensitive Personal Data**: Health, biometrics, financial info
- **Behavioral Data**: Clicks, pages visited
- **Identifiers**: Cookies, device IDs

• How Do You Build Law-Compliant Products?

- **Privacy-by-Design:** Embed privacy considerations into the product from the design phase. Every data collection point is evaluated for necessity, consent, and minimization.
- **Consent Management**: Implement robust mechanisms to capture, store, and update user consents. This includes opt-in/out options and preference centers.
- **Data Subject Rights Fulfillment**: Build APIs and interfaces to handle user requests such as data access, deletion, portability, and correction.
- **Data Localization and Retention**: Understand legal obligations for data storage by geography and implement retention schedules that automatically purge expired data.
- **Data Processing Agreements (DPAs):** if you are using 3rd party platform and sharing data with them, ensure DPAs has been signed with them and validate that they are compliant with all the data laws

- **Regular Audits and Legal Partnership**: Work closely with legal and compliance teams to audit practices, update terms of use, and monitor global regulatory changes.

- **Common Mistakes to Avoid**
  - Collecting more data than needed
  - Not validating third-party compliance
  - Making opt-out options hard to find
  - Not having an audit trail for consent or access
  - Assuming one-size-fits-all global policy

How to drive data attribution, survivorship, and lineage in data governance?

Attribution, survivorship, and lineage are elements of data governance that are critical to building trust and auditability into data systems.

To drive attribution, survivorship, and lineage in data governance, one needs to implement robust processes, tools, and ownership models that ensure your data is trustworthy, traceable, and consistent across systems.

- **Data Attribution**: Track the origin and context of each data element—who created it, when, and under what condition. Metadata tagging is vital, it enables accountability, supports auditing and debugging and ensures traceability to data sources (provenance)

- **Survivorship Rules**: When multiple data sources provide conflicting values, survivorship logic determines which value prevails—typically based on recency, source reliability, or completeness.

- **Data Lineage**: Visualize how data flows through the ecosystem—from ingestion to transformation to output. This helps in impact analysis, debugging, and regulatory audits.

- **Tooling Support**: Use tools like Apache Atlas, Collibra, or custom metadata stores to automate lineage tracking and support transparent data governance.

- **Culture and Training**: Ensure teams understand why these practices matter, provide training, and make data lineage and attribution visible across analytics platforms.

Authors' Perspective on Architecting Agentic AI

While working on this manuscript, we often found ourselves reflecting on a question: Will the frameworks and best practices we present in this book still stay true in the context of agentic AI?

Our answer is yes, because the recommendations made in this book are grounded targeted toward fundamentals principles. They are intended to enable product managers to evolve their ways of working including use case identification, prioritization, and design, especially when it involves AI. With the inclusion of AI, products are dynamic, and their behavior evolves with data. Whereas agentic AI in its true form, is autonomous in nature, it adds more complexity to what we have seen and experienced. With this additional complexity, there comes a heightened need for strategizing agentic AI use cases, which includes steps required for:

- Identifying and prioritizing agentic AI uses case
- Developing and operationalizing agentic AI
- Ensuring agentic AI implementation stays trustworthy

CHAPTER 11   EXECUTIVE PERSPECTIVES: NAVIGATING AI, STRATEGY, AND THE FUTURE

Let's explore how the principles mentioned in this book will contribute to the agentic AI world.

Strategic Workflows in the Age of Agentic AI: Because of the autonomy promised by agentic AI, it has become even more critical to ensure agentic AI use cases are aligned with broader business goals. The strategic alignment extends beyond use case identification to include the processes used to asses feasibility from both financial and organizational readiness perspectives. In this context, the strategy workflows and best practices discussed in Chapter 3 are particularly relevant to Agentic AI initiatives.

Data availability and quality: All forms of AI, including agentic AI, rely heavily on data. Poor availability and quality of data can impair the trustworthiness of an agentic AI system. The PROMT Framework introduced in Chapter 5 offers a strategic, multilayered approach to data management that will help organizations treat data as a strategic asset rather than a tactical by-product. This mindset is essential for improving the availability and quality of data, which will in turn become a bedrock for developing high-performing, trustworthy agentic AI systems.

Harmonizing organizational efforts: because of its autonomous nature, agentic AI will require cross-functional data harmonized to talk to each other. The EDGE framework covered in Chapter 4 will provide a blueprint for developing a data operating model that will prescribe how various data systems across the data lifecycle should interact to generate, collect, and harmonize data fit for cross-functional agentic AI usage.

Once agentic AI systems are developed, they must be deployed thoughtfully. The DeMoGT framework outlined in Chapter 7 offers actionable best practices for managing deployment and the post-deployment lifecycle of AI projects.

CHAPTER 11   EXECUTIVE PERSPECTIVES: NAVIGATING AI, STRATEGY, AND THE FUTURE

Due to the added complexity and autonomous nature of agentic AI, trustworthiness has become a non-negotiable prerequisite for adoption. In Chapter 8, based on the factors that influence human trust formation, we defined Trustworthy AI as the combination of Responsible AI and Robust AI. The framework provides best practices and metrics to measure for implementing principles to make AI system ethical, resilient, and accountable for its actions.

As agentic systems are more complex and decision-making becomes opaquer, the user interfaces must foster trust, clarity, and control. The principles in this chapter will help organizations create user experiences that enhance adoption and align with ethical AI standards.

Building Blocks of Agentic AI Systems

After establishing that the practical recommendations made in this book will remain relevant even for agentic AI development, these recommendations are rooted in strategy, structure, and sound product thinking, which makes them even more applicable and relevant in the more complex and autonomous world of agentic AI.

Now it's time to recommend how a typical enterprise architecture for agentic AI will look. Figure 11-1 provides a simplified representation of the enterprise architecture for agentic AI applications within an organization.

Chapter 11 Executive Perspectives: Navigating AI, Strategy, and the Future

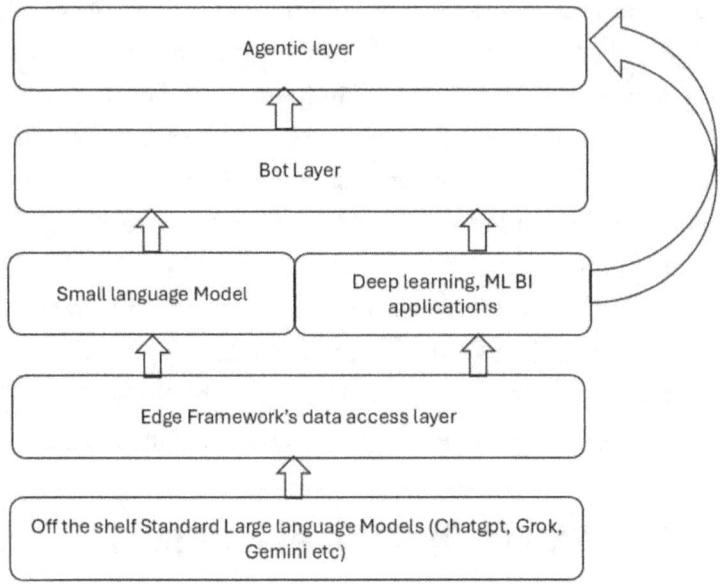

Figure 11-1. *Agentic AI enterprise architecture*

We recommend following five-layer architecture for implementing agentic AI in the operations.

The first layer consists of the foundational layer (base LLMs) like GPT, Claude, LLaMA, Mistral, etc. Organizations must not invest in developing these large language models unless they see a strategic goal getting achieved. Instead, organizations must focus on training these large language models on their proprietary data to develop small language models. Embedding techniques along with the retrieval-augmented generation technique will help train LLM on customized data.

The data access layer of EDGE framework forms the second layer of this architecture. This layer will expose the organization's data for training the large language model. Apart from enabling the training of the LLM to create a small language model, this layer in the larger AI ecosystem will also become a bedrock for conventional deep learning and AI models that do not fall under the LLM domain.

CHAPTER 11   EXECUTIVE PERSPECTIVES: NAVIGATING AI, STRATEGY, AND THE FUTURE

Layer 3 consists of a domain-specific language model that has been developed by fine-tuning a generic LLM using the organization's data. This layer also contains developed conventional AI models such as deep learning, computer vision, and machine learning models developed to solve specialized business problems. If you compare this layer with the EDGE framework, it corresponds to the data growth layer in which applications are developed on the data curated in previous layers for achieving organizational goals.

Layer 3 should ideally feed to the next layer, which would consist of multiple bots and task-specific applications aimed for fulfilling a broader business objective.

The last layer consists of agentic AI applications, which will use its built-in business logic and reasoning capabilities to autonomously accomplish a predefined objective in the most optimum way. This layer will dynamically invoke various underlying services including bots and conventional AI applications as needed to carry out complex workflows. Examples of agentic AI applications are as a follows:

- Customer Retention Agent: Detects at-risk customers (from Layer 3.2 churn model), drafts a personalized retention campaign (via 3.1 LLM), and triggers email/SMS automatically.

- Fraud Detection Agent: Gets fraud score from the ML model (layer 3.2), investigates patterns, and blocks suspicious transactions with escalation.

Figure 11-1 shows all layers of the model including the training layer. Once the use-case-specific Gen AI models are developed by fine-tuning the LLM on the organization's data, the last layer (LLM) of the model will not be applicable. Figure 11-2 depicts the enterprise architecture after fine-tuning, as implemented in the production environment.

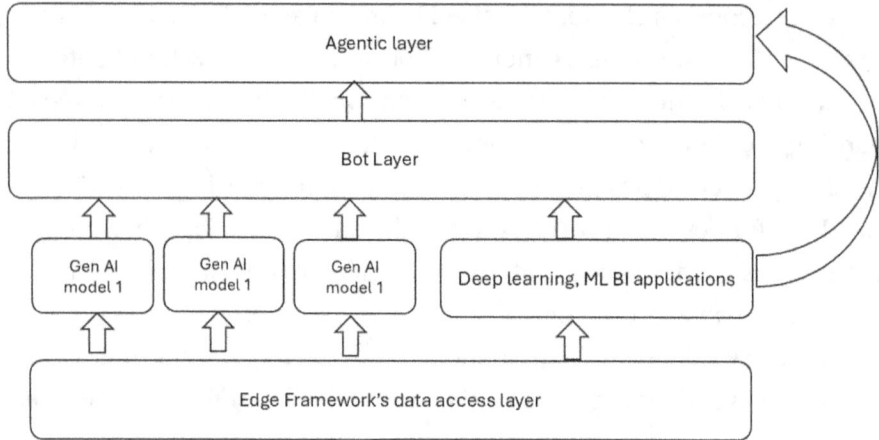

Figure 11-2. *Agentic AI enterprise architecture in ops*

Ethics and Agentic AI

Since layer 5 comprises the agentic AI applications, it is responsible for orchestrating available data and process workflows of the organization in an autonomous fashion to achieve a desired outcome. Hence, ensuring the agentic AI layer remains trustworthy is critical to embed the right ethical principles.

Apart from the Trustworthy AI guidelines discussed in Chapter 8 of this book, the are some additional guidelines to implement Trustworthy AI:

1. Design each agentic application with a human-in-the-loop approach. There are two ways to implement this approach. The first approach is called the human-in-the-loop approach in which an agent pauses before making each decision and requests a human to review/approve it before executing its decision. This approach is best suited for high-risk applications such as financial

CHAPTER 11 EXECUTIVE PERSPECTIVES: NAVIGATING AI, STRATEGY, AND THE FUTURE

approvals and procurement. Another approach for keeping a human in the loop is called the human-on-the-loop approach. In this approach, a human can oversee agent actions in real time and can override an agent's action in all scenarios. The agent does not pause its actions for the human approval for its actions. This approach is best suited for medium to low-risk decisions such as the development of agentic AI for auto blocking transactions during online fraud detection.

2. Implement ethical guardrails and constraints to ensure that agents do not access sensitive data without supervision.

3. Enable auditability by logging every action including input data processing, intermediatory decisions, and final outcomes.

4. Ensure that final outcomes of the AI agent are explainable.

5. Test every AI agent outcome for fairness and safety.

6. Define an escalation matrix so the affected party can reach out to someone in case of any grievances. If the outcome of the AI agent is accompanied with low confidence, then automatically escalate it to a human supervisor.

CHAPTER 11 EXECUTIVE PERSPECTIVES: NAVIGATING AI, STRATEGY, AND THE FUTURE

Summary

In this final chapter, we brought together diverse perspectives from global leaders that should be considered by organizations while strategizing for AI adoption. These in-depth interviews with key decision-makers covered the actionable insights required for making an organization AI mature.

Each leader given their role in the organization offers a unique lens from strategizing AI vision and product strategy to building data infrastructure to fostering innovation and developing scalable data products. Together they provide a 360-degree view on what a successful AI program looks like.

This chapter concluded by presenting the authors' point of view and synthesizing these insights into a practical roadmap for organizations to move beyond experimentation and toward sustainable, ethical, and impactful AI at scale.

References

Chapter 1

1. https://www.uber.com/us/en/about/
2. https://www.uber.com/en-IN/blog/deepeta-how-uber-predicts-arrival-times/
3. https://www.uber.com/en-IN/blog/demand-and-etr-forecasting-at-airports/
4. https://www.uber.com/en-IN/blog/project-radar-intelligent-early-fraud-detection/
5. https://www.uber.com/en-IN/blog/cota/
6. https://www.grammarly.com/blog/engineering/innovating-the-basics/
7. https://www.grammarly.com/blog/engineering/reducing-text-input-lag/
8. https://www.grammarly.com/blog/engineering/detecting-delicate-text/#:~:text=Protecting%20users%20from%20harmful%20communications,discusses%20emotional%2C%20potentially%20triggering%20topics.
9. https://ai.meta.com/blog/heres-how-were-using-ai-to-help-detect-misinformation/

REFERENCES

10. https://investor.atmeta.com/investor-events/event-details/2025/Q4-2024-Earnings-Call/default.aspx

11. https://transparency.meta.com/features/explaining-ranking/fb-people-you-may-know/

12. https://www.facebook.com/business/news/introducing-ai-sandbox-and-expanding-meta-advantage-suite

13. https://en-gb.facebook.com/business/ads/meta-advantage

14. https://engineering.fb.com/2024/11/19/data-infrastructure/sequence-learning-personalized-ads-recommendations/#:~:text=Meta%27s%20ad%20recommendation%20engine%2C%20powered,the%20DLRM%2Dbased%20recommendation%20system.

15. https://www.gartner.com/en/newsroom/press-releases/2023-10-11-gartner-says-more-than-80-percent-of-enterprises-will-have-used-generative-ai-apis-or-deployed-generative-ai-enabled-applications-by-2026

16. https://www.businesswire.com/news/home/20220103005036/en/NewVantage-Partners-Releases-2022-Data-And-AI-Executive-Survey

17. https://www.forbes.com/sites/forrester/2025/03/26/are-ai-product-managers-the-role-of-the-future/

18. https://www.aboutamazon.com/about-us

REFERENCES

Chapter 3

1. https://venturebeat.com/ai/why-most-ai-implementations-fail-and-what-enterprises-can-do-to-beat-the-odds/
2. Generating ROI with AI – IBM Institute for Business Value https://www.ibm.com/downloads/documents/us-en/10a99803fbafdc88

Chapter 6

1. https://www.statnews.com/2017/09/05/watson-ibm-cancer/
2. https://www.henricodolfing.com/2024/12/case-study-ibm-watson-for-oncology-failure.html
3. https://www.theverge.com/2018/7/26/17619382/ibms-watson-cancer-ai-healthcare-science
4. https://www.advisory.com/daily-briefing/2018/07/27/ibm

Chapter 7

1. https://docs.aws.amazon.com/wellarchitected/latest/machine-learning-lens/well-architected-machine-learning.html
2. Machine Learning Models Monitoring in MLOps Context: Metrics and Tools -Anas Bodor, Meriem Hnida Najima Daoudi--https://online-journals.org/index.php/i-jim/article/view/43479

REFERENCES

3. https://neptune.ai/blog/how-to-monitor-your-models-in-production-guide

4. USING STATISTICAL DISTANCES FOR Machine Learning Observability -Arize -https://arize.com/wp-content/uploads/2021/09/Statistical-Distances-for-Machine-Learning.pdf

5. Handling concept drift in deep learning applications for process monitoring -Nicolas Jourdan,Tom Bayer,Tobias Biegel,Joachim Metternich (56th CIRP Conference on Manufacturing Systems, CIRP CMS '23, South Africa)

Chapter 8

1. https://economictimes.indiatimes.com/news/international/global-trends/florida-teen-commits-suicide-after-ai-chatbot-convinced-him-game-of-thrones-daenerys-targaryen-loved-him/articleshow/114540078.cms?from=mdr

2. https://economictimes.indiatimes.com/news/international/global-trends/googles-gemini-turns-villain-ai-asks-user-to-die-calls-him-waste-of-time-a-burden-on-society/articleshow/115403234.cms?from=mdr

3. https://www.washingtonpost.com/technology/2024/02/22/google-gemini-ai-image-generation-pause/

4. https://mitsloan.mit.edu/ideas-made-to-matter/a-new-look-economics-ai

REFERENCES

5. https://www.washingtonpost.com/business/2020/10/16/how-race-affects-your-credit-score/

6. https://nationalfairhousing.org/wp-content/uploads/2021/12/NFHA-credit-scoring-paper-for-Suffolk-NCLC-symposium-submitted-to-Suffolk-Law.pdf

7. https://www.reuters.com/article/world/insight-amazon-scraps-secret-ai-recruiting-tool-that-showed-bias-against-women-idUSKCN1MK0AG/

8. https://en.wikipedia.org/wiki/Death_of_Elaine_Herzberg

9. https://www.npr.org/2023/07/28/1190866476/autonomous-uber-backup-driver-pleads-guilty-death

10. https://www.forbes.com/councils/forbestechcouncil/2023/09/13/what-a-corrupt-chatbot-from-the-past-can-teach-us-about-todays-data-poisoning-threat/

11. https://www.bbc.com/news/business-50365609

12. https://oecd.ai/en/catalogue/tools/fairlens/tool-use-cases/fairlens-detected-racial-bias-in-a-recidivism-prediction-algorithm

13. https://www.bbc.com/news/technology-46944696

14. OECD AI Policy Observatory - Accountability (Principle 1.5)

Index

A

AARRR Framework, 613
Autoencoders, 291
A/B testing, 253, 273, 313, 319, 352, 356–358
    advanced techniques, 366
        Bayesian method, 367, 368
        contextual bandit, 368
        multi-armed bandit testing, 366
    align experiments with business goals, 364
    analyze test design, 369
    automate in CI/CD pipelines, 364
    conduct post-test surveys, 370
    control group, 357
    cost-benefit analysis, 371
    design
        control group, 358
        statistical significance test, 358
        target metric, 358
        variation group, 358
    develop test framework, 366
    ensure interpretability, 364
    ensure validity and reliability, 365
    estimate real business impact, 370, 371
    iterate and retest, 370
    realistic expectation, 369
    revisit target metrics, 369
    scale for large deployments, 365
    test set, 357
    test workflow, 358
        A/A testing, 362
        alternative hypothesis, 360
        define control and variation groups, 361
        experiment hypothesis, 359
        hypothesis formulation, 359
        null hypothesis, 359, 360
        power analysis tool, 362
        sample hypothesis, 359
        sample size determination and statistical significance, 361–363
        select metrics and KPIs, 360
        statistical analysis techniques, 364
        statistical tests, 363
        target metric, 363
    validate data collection, 369

INDEX

Accessibility metrics, 416
Accessible Rich Internet
    Applications (ARIA), 489
Accountability breaches, 435
Accountability in AI
    AI system performance, 429
    AI system(s), 429
    ethical guidelines, 430
    objective setting, 429
    stakeholders identification, 429
    structural and procedural
        changes, 428
Adversarial debiasing
    techniques, 403
Adversarial process, 307
Adwin, 346
Agentic AI, 60
    capability, 561
    challenge, 563
    customer retention agent, 625
    data access layer, EDGE
        framework, 624
    data availability and quality, 622
    DeMoGT framework, 622
    domain-specific language
        model, 625
    EDGE framework, 622
    enterprise architecture, 623, 624
    foundational layer, 624
    fraud detection agent, 625
    guidelines, Trustworthy AI,
        626, 627
    ops, enterprise architecture, 626
    PROMT Framework, 622
    strategic workflows, 622
    systems, 572
    trustworthiness, 623
    use cases, steps, 621
AI, *see* Artificial intelligence (AI)
AI adoption, 147
    decision-making, 148
    processes, 148
    roles and responsibilities, 148
    success criteria, 149
    type of changes AI, 147
AI-based dynamic pricing
    system, 526
AI-based virtual trainer, 506
AI-driven actions, 487
AI-driven analytics, 427
AI-driven personalization, 489,
    490, 498
AI-driven products, 28
AI-driven product thinking
    AI capabilities with available
        data, 588
    black-box systems, 594
    build layered safety nets, 594
    business strategy, 579–581
    data empathy and loop
        design, 591
    data quality, 587
    design, probabilistic
        realities, 590
    educate on AI
        fundamentals, 589
    ethical concerns, 593
    experiments, 586, 587, 591, 592

INDEX

facilitate cross-functional empathy, 589
feasibility-desirability gap, 590
governance, 594
key dimensions, 575, 576
maintain strategic discipline, 590
misconceptions PMs, 592
mobile and Internet, 594
product experience, 585, 586
product leaders, 577
psychology, 582
question, 582
ROI, 583, 585
sample and monitor, 594
seductive paradox, 581
self-contained *vs.* spawning experiments, 582
strategic discipline, 591
strategic shifts product teams, 577, 578
strategy, 582
temptation, 581
AI-driven self-driving car, 417
AI-driven solution, 410
AI ethics principle
    accountability, 426
    AI applications, 393
    AI decisions, 427
    AI domain, 419
    AI product, 413
    AI systems, 393
    FICO score, 392
    goals, 414
    house hunting, 391
    inclusiveness, 410
    inclusivity goals, 414
    noninclusiveness, 410
    organizations, 411
    safety of AI, 419
    stakeholders, 414
    technical analysis, 418
AI-first companies, 7, 86
AI fluency, 27, 29
    AI-first product strategy, 34
    AI governance and compliance, 33
    cross-functional collaboration, 34
    and technical foundations, 32, 33
AI-generated content, 489
AI governance model, 436, 437
    AI system, 439
    diagram, 438
    guidelines, 438
    organization, 438
    strategic and ethical governance principles, 438
    structured approach, 437
AI implementation
    business understanding, 264
        business metric, 267
        business problem, 264
        critical decision, 266
        e-commerce example, 265
        identify data sources, 268
        ML problem, 267

INDEX

AI implementation (*cont.*)
    process flow diagram, 266
    search performance, 265
    swimlane diagram, 266
    translate, business
        objective, 265
    common pitfalls and challenges
        (*see* AI/ML projects)
    customer acceptance, 272, 273
    data management, 270, 271
    model deployment and post-
        deployment strategy,
        273, 274
    modeling, 272
    solution design, 269, 270
AI implementation project
    plan, 255
    continuous monitoring and
        evaluation, 263
    data collection and preparation
        (*see* Data collection)
    define clear objectives, 256
    model simplicity and
        interpretability, 262
    scalability and deployment, 263
AI interaction design
    AI-driven steps, 479
    algorithms, 480
    chatbots, 478
    interface, 478, 479
    personalized
        recommendations, 479
    transparency, 478
    UX behavior, 479

AI interaction patterns
    product manager, 465
    UX design parameters, 465
AI maturity, 86
    adaptive mindset, 96, 98
    AI adoption, 89
    AI efforts and business
        strategy, 88
    AI readiness, 87
    capability, 95–97
    capability gaps, 88
    cultural readiness, 90
    evaluation criteria, 93, 94
    focus areas, 93, 94
    framework, 97, 98
        levels, 98, 99, 152
    governance, 95, 97
    holistic diagnostic approach, 87
    human values, 96, 98
    layers, 93
    outcomes, 87
    risk and governance posture, 90
    scoring technique
        best practices, 100
        criterion, 99
        evaluation criteria, 99
        focus areas, 100
        score classification table, 101
        scores, 101
        strategic pillars, 100
    strategic factors, scoping
        AI adoption footprint, 92
        business goals, 92
        dimensions, 92, 93

INDEX

organizational size and
  structure, 91
strategic gaps, 87
strategic perspective, 86
strategic pillars, 93, 94
strategy, 94, 97
technology perspective, 86
use case deployment, 89
AI/ML algorithms, *see* Machine
  learning (ML) algorithms
AI/ML projects
  customer churn prediction, 254
  data quality and availability
    conflicting values, 252
    erroneous data, 251
    inconsistent data
      formats, 252
    measurement bias, 253
    missing values, 251
    noise and outliers, 253
    nonstandardized units, 253
    post-COVID times, 252
    redundant records, 252
    sampling bias, 253
  management methodology, 263
  model complexity and
    overfitting, 254
  personalized recommendation
    systems, 255
AI/ML workflows
  model training/testing, 233–235
  quality gates, 235, 236
    bias and fairness gate,
      237, 238

data distribution gate,
  236, 237
label quality gate, 237
mapping, 238, 239
model readiness gate, 238
product manager
  responsibilities, 239, 240
schema validation gate, 236
AI model-based pricing
  system, 524
AI policies and guidelines
  ethical AI, 431
  organization, 431
  protocols, 431
  risk identification and
    mitigation, 431
AI-powered applications, 382
AI-powered chatbot Tay, 419
AI-powered products
  cross-functional team, 462
  decisions, 463
  design examples, 464
  designing, 463
  design trustworthy, 459
  digital offerings, 461
  feedback loops, 464
  handle error gracefully, 464
  interfaces, 462
  management perspective, 459
  personalized, 460
  predictive features, 460
  principles, 459
  product managers, 461
  prototyping stage, 460

639

INDEX

AI-powered products (*cont.*)
    rule-based systems, 461
    strategic capabilities, 460
    trustworthy, 459
    user groups, 465
    user interface design, 462
    UX design, 462
AI-powered system, 419, 493
AI Product Engineer, 561, 562
AI products, 437
    inclusivity, 416
    lifecycle, 263, 264
    metrics, 416
AI prototyping tools, 461
AI risk management, 391
    biases, 443
    ethical and legal, 445
    generalization, 443
    performance risk, 442
    traditional, 440
    types and complexity, 440
AI risk mitigation, 449
    assessments, 450
    risk response plan, 450
    stakeholders, 450
    training and assessment, 450
AI strategic framework
    AI investment model, 112
        activities, 113–115
        deliverables, 115, 116
        objective, 112
    business strategic
        alignment, 105
        activities, 105, 106
        deliverables, 106
        objective, 105
        review mechanism, 106
    capability foundation
        activities, 107, 108
        deliverables, 108
        objective, 107
        readiness, dimensions, 108
    execution and governance, 116
        core pillars, 116–119
        deliverables, 119
    layers, 104
    strategic feedback loops, 120
        components, 120
        deliverables, 120
        objective, 120
    strategic use case portfolio
        activities, 109–111
        deliverables, 112
        objective, 109
AI strategy, 101
    accountability, AI, 598, 600
    AI-driven product thinking (*see* AI-driven product thinking)
    AI-enabled product, challenges, 570
    AI ethics, 550
    AI governance framework, 99, 435, 436, 572, 601
    AI innovation challenges, 571
    AI-ready infrastructure, 606–610
    AI trends, 572
    AI use cases evaluation, 569

bad, 103, 104
business, 104
businesses create value, 572
business strategies, 568
data, 550
data-driven products (*see*
    Data-driven products)
data quality, 574
early stage AI projects, 571
end-to-end process, 543
ethical standards, 601
evaluating success, AI
    initiatives, 600, 601
experimentation, 603, 604
future, 551
Gen AI, 543
good, 102, 103
hiring, 547
ideal leadership profile,
    managing and scaling AI
    teams, 602, 603
innovation, 568
key indicators, AI-driven
    innovation initiative, 568
leaders, 573
monitoring, 604
organization, 544
80:20 portfolio strategy, 570
predictive AI, 542
product management, 545
product manager's view, 547
product model, 543
reinvention, 553–567
risk, 553–567

ROI-driven. AI adoption,
    549, 550
SaaS, 543
sales team, 573
scaling AI from prototype to
    production, 597
setting AI vision/long-term
    strategy, 597
short-term wins, long-term AI
    innovation goals, 569
signals and pain points, 569
standardization, 603, 604
talent, 598
technology, 542, 552
trust AI, 547
type, product, 548
user adoption, 598
AI systems, 387, 394, 424,
    435, 441
automation bias, 396
data collection, 394
data labeling, 395
development, 420
group attribution
    biasness, 398
historical bias, 396
human-induced bias, 394
human involvement, 394
implicit bias, 398
model development, 395
model maintenance, 395
selection bias, 397
sentiment analysis, 396
temporal bias, 399

INDEX

AlphaGo, 53
Amazon, 23, 411
   AI, 24–26
      business goals/AI initiatives, 25
      customers, 24
      Gen AI shopping assistant, 24
      product discovery, 23, 24
      product leaders, 25
      product search, 24
      text search, 24
      user behavior, 25
      users review, 24
Amazon Marketplace, 6, 7
Analysis paralysis model, 126
Analysis Planning Layer, 163–165
Anomaly detection, 288, 607
ARIA, *see* Accessible Rich Internet Applications (ARIA)
Artificial intelligence (AI), 3, 37, 68, 69, 459
   Academic, 46
   Agentic AI, 60
   algorithm, 467
   cognitive tasks, 57
   communication, 495, 496
   data-based learning algorithms, 47
   definition, 46
   development, 46, 47
   domain, 390
   ethics, 250
   Father of AI, 49, 50
   financial stakes, 27

financial viability (*see* Financial viability)
foundational paradigms, 51
   connectionist approach, 53, 54, 56
   convergence of paradigms, 56
   emergent AI, 54–56
   symbolic approach, 51, 52, 56
Generative AI, 59
governance, 391, 436
holistic approach, 70
IBM Deep Blue, 47
impact, 4
implementation, 3, 27, 35, 71
industry context, 26, 27
initiatives, 70
   data, 73
   isolated teams, 75
   misalignment, 71
   outcome, 75
   overpromising/underdefining, 72
   product management, 74
   risk/bias/ethics, 76
   shipping model, 74
   vague goals/unusable models, 72, 73
key enablers, 70
leaders, 70
metacognition, 57–59
models, 196
outcomes, 4

INDEX

philosophical needs, 48, 49
product leaders, 50
products, 80, 86
projects, 71
reason-based AI, 60
root causes, 70
terminology, 496
traditional *vs.* AI-powered
    products, 80, 84
    behavior, 81
    business case, 81
    continuous learning, 81
    deterministic *vs.* opaque
        failures, 82
    ethics/bias, 83
    tectonic shift, 83
    uncertainty, 80
    user trust, 82
    UX/brand/trust, 83
    value creation, 82
Audio validation, 233
AWS Deep Learning, 328

B

Balanced Personalized Experience, 483, 485
Batch predictions, 321, 323
Batch processing, 324, 377, 511, 531, 616
Bayesian A/B testing, 367, 368
Bayesian methods, 367
Bias identification, 349, 401, 415
Bias in data
    algorithm bias, 399
    dataset, 399
    human bias, 399
    types, 399
Bias mitigation techniques, 409
    post-processing techniques, 401
    pre-processing techniques, 401
    processing techniques, 401
    responsibility, 401
Bias reinforcement, 481
Bilingual Evaluation Understudy
    (BLEU) score, 305
BI tools, 183, 616
Browser deployment, 325, 326
    AI-powered text predicting
        applications, 327
    application performing image
        recognition, 327
    Google teachable machine, 327
Business and product
    development, 26
Business Context Layer, 161, 162

C

Calinski-Harabasz (CH) index, 288
Change management
    AI adoption, 152
    AI showcases, 150
    people/skill development, 150
    purpose and value
        alignment, 150
    roles and processes, 151
    stakeholder register, 150

INDEX

Change management (*cont.*)
    transparency/explainability, 150, 151
    use case definition phase, 149
Chatbot, 513
    business impact, 517
    challenges, 517
    deployment, 515
        infrastructure, 515
        integration, 515
        model serving, 515
    GPT-4o, 513
    limitations, 517
    manual tuning, 513
    metrics
        business, 516
        model performance, 516
        real-time, 517
    model evaluation
        content quality, 514
        model performance, 514
        response format, 514
    monitoring, 516
    virtual assistant, 512
Chief Product Officer (CPO), 31, 32
Chi-Square Test, 342, 364
Client-side model serving technique, 324
    applications, 324, 325
    on browser, 325–327
    browser-based *vs.* on-device, 329, 330
    on device (edge), 327–329
CNN-based models, 296, 298, 300

CNNs, *see* Convolutional neural networks (CNNs)
Cognitive abilities, 41
Cognitive learning, 40
Compliance, 33, 79, 95, 347–351, 355, 602, 604, 608, 609, 617
Computer vision (CV), 294, 298, 305, 308
Computer vision, human exercise recognition
    business impact, 512
    challenges, 511
    data collection and preprocessing, 507
    deployment strategy, 510, 511
    feature engineering, 507
    hyperparameter tuning, 509
    limitations, 512
    Mediapipe, 507
    ML problem definition, 506
    model evaluation, 509
    model performance, 509, 510
    model selection, 508, 509
    model serving, 511
    pose matching accuracy, 506
    renowned healthcare platform, 506
    virtual trainer, 506
Concept drift, 334, 344–346
Confusion matrix, 280, 298
Contextual bandit, 368
Continuous integration and continuous deployment (CI/CD) pipelines, 371, 372

INDEX

integration testing, 376
tools, 373
unit testing methodologies
   broadcasting and
      logging, 375
   data import, 374
   feature engineering, 374
   model loading and
      inference, 375
   model validation, 375
Convolutional neural networks
   (CNNs), 54, 294
   activation function, 295
   convolution layers, 295, 296
   deep learning architecture, 294
   kernel, 295
   loss function, 295
   pooling layers, 295
CoreML, 328
COTA, 12
COVID, 252, 289
CPO, *see* Chief Product
   Officer (CPO)
Culture, 87, 555, 621
Curated Data layer, 175
   AI applications, 175
   conventional lakehouse
      architectural approach, 175
   data marts, 180, 181
   data models, 178
   decision-making, 175
   fact table, 178
   governance and metadata
      layer, 182

   lifecycle of data, 182–184
   modeled data zone, 177
   multichannel metric access,
      184, 185
   raw data storage zone, 175, 176
   refined data zone, 176
   robust data model zone, 178
   snowflake schema, 179, 180
Customer data platform, 167
Customer retention agent, 625
Customer segmentation, 283,
   288, 324
Cutting-edge technologies, 4, 5, 566
CV, *see* Computer vision (CV)

D

DALL·E/Midjourney, 54
Data Activation Layer, 159, 185
Data attribution, 620
Data augmentation, 243–245, 261
Data centralization layer
   Curated Data layer (*see* Curated
      Data layer)
   Data Activation Layer, 185
   Data Collection layer, 173
   ingestion layer, 173, 174
   organizational challenges, 173
   sublayers, 173
Data cleaning, 258, 302
Data collection
   collection process, 397
   and data preparation, 258
      data augmentation, 261

645

INDEX

Data collection (*cont.*)
    data cleaning, 258
    data integration, 260
    data splitting, 261
    data transformation, 259
    dimensionality reduction algorithms, 260
    feature engineering, 259, 260
    processes, 258
  data preparation, 258
  ensure data quality and relevance, 258
  identify data sources, 256, 257
    external data sources, 257
    semi-structured, 257
    structured data, 257
    synthetic data, 257, 258
    unstructured data, 257
  steps, 256
Data collection layer, 166
  behavioral and event-level data, 166
  common definition and constraints, 167
  components, 170
  core activities, 166
  data contracts, 167, 168
  Google's approach, 168
  high-level data flow, 172
  ownership roles, 172
  primary activity, 166
  primary job, 169
  proactive data governance, 169
  system components and interactions, 170, 171
  transactional data and customer behavior data, 167
  user interaction, 166
  violations, 169
Data contracts, 167, 168, 226
Data deduplication, 259
Data drift, 289, 341–343, 429, 443
Data drift detection techniques
  Adwin, 346
  basic statistical metrics, 341
  Chi-Square Test, 342
  clustering-based methods, 342
  DDM, 346
  as feature drift, 341
  in image data, 343, 344
  KL Divergence, 342
  Kolmogorov-Smirnov Test, 342
  model comparison techniques, 345
  model residual analysis, 346
  performance-based drift detection, 345
  PSI, 341
  Python Libraries, 342
  in text data, 342, 343
  visualization-based drift detection, 346
Data-Driven Growth layer, 159, 185
Data-driven products
  align metrics with business objectives, 612

INDEX

behavioral data, 619
culture and training, 621
data and AI regulations, 618
data attribution, 620
data lineage, 621
identifiers, 619
identify key user
    behaviors, 612
infrastructure, product metrics,
    615, 616
law-compliant products
    consent management, 619
    data localization and
        retention, 619
    data subject rights
        fulfillment, 619
    DPAs, 619
    privacy-by-design, 619
    regular audits and legal
        partnership, 620
metric framework, 612, 613
metrics, 613
NSM *vs.* MTM, 614
operationalize data governance
    guidelines, 616–618
personal Data, 619
product compliance, data
    regulations, 618
product goals and strategy, 611
product metrics, 611
segment and contextualize, 613
sensitive personal data, 619
survivorship rules, 620
tooling support, 621

Data ecosystem, 6, 65, 68, 73, 107,
    158, 168
Data engineering, 79, 178, 183, 213,
    469, 541, 606, 608
Data flow diagram (DFD), 268, 269
Data formats, 203, 229, 252, 374
Data governance model, 432, 433
Data, information, knowledge,
    wisdom (DIKW), 38, 61
    action layer, 63, 67, 68
    data-driven wisdom, 61
    data layer, 62, 65, 66
    HI, 64, 65
    information layer, 62, 66
    knowledge layer, 62, 66, 67
    layers, 62
    loop, 61
    wisdom layer, 62, 67
Data lakes, 191, 606
Data lineage
    challenges, 194–196
    compliance, 194
    data, 191
    data governance, 194
    data quality, 193
    definition, 191
    end-to-end visibility, 192
    implementation, 191
    model explainability and
        fairness, 240, 241
    modern data architectures, 192
    product lens, 192
Data loops, 580
Data marts, 180, 181

INDEX

Data massaging method, 402
Data maturity, 95, 97
Data mining techniques, 415
Data operating model (DOM), 156
    vs. data platforms, 157
    definition, 157
    EDGE framework (see EDGE framework)
    role, 157
Data outreach, 483
Data oversampling, 292
Data platform, 35
    definition, 156, 157
    vs. DOM, 157
    role, 157
Data points, 5, 163, 292, 307, 392
Data Processing Agreements (DPAs), 619
Data profiling, 226, 228, 234
Data quality, 244, 587
    accuracy
        causes, 198
        data, 198
        improvement, 199
        measurement, 199
    AI adoption, 189
    AI and business strategy, 196
    AI/ML workflows (see AI/ML workflows)
    completeness
        causes, 200, 201
        database completeness, 200
        data completeness, 200
        definition, 199
        improvement, 201, 202
        measurement, 201
    concept, 197
    consistency, 202
        causes, 202, 203
        improvement, 204
        measurement, 203
    data augmentation, 243, 244
    decision-making, 189
    definition, 197
    distortion/error, 198
    issues, 188, 190
    key dimensions, 197, 198
    measurement
        data contracts, 226
        data profiling, 226
        data quality metrics, 226
        user feedback loops, 227
    poor data quality, 188, 189
    profiling, 227, 228
    PROMT framework (see PROMT framework)
    quality checks, 227, 228
    regulatory/compliance, 190
    semi-structured/unstructured data, 228
        action items, challenges, 232
        API structures, 231
        Cohen's Kappa, 231
        data labeling, 230
        data quality control checks, 231
        information, 229

INDEX

parsing and normalizing
data, 230
product leaders, 229
redundancy and noisy
content, 231
schema-level
complexities, 229
standardized schema, 229
techniques, 232, 233
tools, 232
version-controlled
preprocessing logic, 230
SLA, 224–226
synthetic data, 242, 243
synthetic data *vs.* data
augmentation, 244
timeliness, 204
causes, 205, 206
improvement, 206
measurement, 206
uniqueness, 209
causes, 209
duplication strategies, 209
measurement, 209
validity, 207
causes, 207
constraints, 207
ensure/enforce, 208
measurement, 208
Data quality index (DQI), 215, 222, 223
Data quality metrics, 164, 197, 226
Data science, 254, 255, 523, 589, 595, 608

Data science methods, 6
Data splitting, 261, 277
Data storage
and access, 316
for AI/ML solutions
data pipelines and ETL
process, 377, 378
feature store, 376, 377
model repositories, 378
orchestration tools, 378
vector DB, 377
Data technology solutions, 186
Davies-Bouldin (DB) index, 288
DDM, *see* Drift Detection
Method (DDM)
Decision-making, 148
Decision tree, 277
DeepETA model, 9
Deep learning, 297, 565
architectures, 293, 294
black-box algorithms, 293
vs. classical ML, 306
for CV use cases
image classification, 298, 299
image segmentation, 300, 301
object detection, 299, 300
imagine teaching, 293
multilayer artificial neural
networks, 293
for NLP use cases
machine translation, 305
NER, 305

649

INDEX

Deep learning (*cont.*)
    sentiment analysis, 304
    text summarization, 305
Deep learning architectures, 293, 294
    common loss functions, 296
    for NLP (*see* Natural language processing (NLP))
    vision transformers, 297, 298
DeMoGT framework, 314, 315
Design bias, 499
Designers, 501, 502
Designer's primary responsibility, 470
Design mechanisms, 423
Digital tools, 410
DIKW, *see* Data, information, knowledge, wisdom (DIKW)
Direct instruction learning, 39
Discriminator, 308
DPAs, *see* Data Processing Agreements (DPAs)
DQI, *see* Data quality index (DQI)
Drift Detection Method (DDM), 346
Duplicate detection, 233
Dynamic learning rate, 509, 529

E

E-commerce platform
    ads, 526
    automated retraining, 534, 535
    business metrics, 535
    challenges, 536
    CTR, 527
    data collection and preprocessing, 527, 528
    deployment strategy, 530, 531
    hyperparameter tuning, 529
    limitations, 536
    ML problem definition, 527
    model evaluation, 529, 530
    model monitoring, 532, 533
    model performance, 530
    model performance metrics, 535, 536
    model selection, 529
    model serving, 531, 532
    real-time metrics, 536, 537
Economic principles
    capital investment and opportunity costs, 141, 142
    economic value of money declines over time, 141
    risk-return dynamics in capital allocation, 141
Edge computing, 327
Edge deployment, 327
    automatic number plate recognition system, 329
    AWS services, 328
    CoreML, 328
    devices, 327
    edge computing, 327
    edge ML-powered word generator apps, 329

Edge TPU, 328
face recognition and speech recognition models, 329
industrial robots, 329
NVIDIA Jetson, 328
ONNX Runtime, 328
smart shelves, 329
specialized hardware accelerators, 327
TensorFlow Lite, 328
EDGE framework, 622
Analysis Planning Layer, 163–165
Business Context Layer, 161, 162
Data Collection layer (*see* Data Collection layer)
diagram, 160
enterprise data operating model, 158
functions, 158
key feature, 158
sections and layers, 158–160
Edge TPU, 328
E-learning, 471
Embedding techniques, 259, 343, 624
Emotions, 39, 40
Enterprise architectures, 191, 541, 623, 625
Error handling mechanisms, 423
Ethical AI, 390, 497–502
ETR, *see* Expected time to request (ETR)

Expected time to request (ETR), 10, 11
Explicit learning, 39

F

Facebook, 6, 7
AI-driven initiatives, 23
content, 20
LASER, 23
metric, 21
misinformation, 22
multistep approach, 20
ObjectDNA, 23
responsibilities, 22
Fairlearn, 409
Fairness metrics, 409
AI system, 405
comparison table, 407
demographic characteristic, 406
equal opportunity, 406
loan application system, 406
FastAPI, 510, 531
FastText, 528, 529, 534
Favorability metrics, 416, 417
FCM, *see* Fuzzy C-Means (FCM)
Feature engineering, 259, 260
Feature store, 376, 377
Federated learning (FL)
Apple Watch, 333
benefits, 332
challenges and trade-offs, 332, 333

INDEX

Federated learning (FL) (*cont.*)
    decentralized approach, 330
    Google's Gboard keyboard, 333
    inference attacks, 333
    model initialization, 331
    model poisoning, 333
    sensitive data, 331
    Spotify, 333
    user experience, 331
Feedback Collection, 96, 417
Feedback loops, 470, 475
FICO proprietary algorithms, 393
FICO scores, 392
Financial viability
    discount rate, 144
    evaluation factors, 138
    innovation, 140
    IRR, 144
    long-term transformation, 140
    NPV, 143
    payback period, 144
    practical approach, 145
    prioritization, 139
    resource allocation, 139
    ROI, 142
    stakeholders, 140
    strategic intent, 139
FKM, *see* Fuzzy K-Modes (FKM)
FL, *see* Federated learning (FL)
Forward and reversed gradient, 404
Fraud detection, 11, 606, 625, 627
Fraud detection agent, 625
Fully connected neural network (FCN) layer, 296, 297, 307, 308
Fuzzy C-Means (FCM), 285
Fuzzy K-Modes (FKM), 285

G

GANs, *see* Generative Adversarial Networks (GANs)
Gaussian mixture model (GMM), 290
GDPR, *see* General Data Protection Regulation (GDPR)
GEC, *see* Grammar error correction (GEC)
GenAI, *see* Generative AI (GenAI)
GenAI-powered chatbot, *see* Chatbot
General Data Protection Regulation (GDPR), 332, 349, 428, 431, 448
Generative Adversarial Networks (GANs), 242, 257, 291, 307, 308
Generative AI (GenAI), 59, 306, 543, 566
    BLEU Score, 309
    GANs, 307, 308
    Inception Score (IS), 309
    perplexity (PPL), 309
    transformers-based, 308, 309
    VAEs, 307

Generative Pre-Trained (GPT) language models, 308
GMM, *see* Gaussian mixture model (GMM)
GOFAI, *see* Good Old-Fashioned AI (GOFAI)
Good Old-Fashioned AI (GOFAI), 51
Governance, 557, 563, 608
GPT-4, 53
GPT-4o mini, 513, 514
Gradient reversal layer, 404
Grammar error correction (GEC), 14
Grammarly, 6, 7, 29
  ability, 14
  AI models, 15
  benefits, 19
  communication, 13
  definition, 13
  language rules, 15
  product leaders, 14
  technology-driven innovation, 14
  user experience, 16–18
  user protection, 19
  users services, 13
Group attribution biasness, 398

H

Harvard Business Review paper, 381
Healthcare, 566

HEART Framework, 612
HI, *see* Human intelligence (HI)
Hierarchical clustering, 285, 286
House hunting, 391
Human-AI interaction, 387
Human-centric design, 35, 463
Human intelligence (HI), 37
  cognition, 45
  cognitive and metacognitive capabilities, 58
  context, 44
  DIKW, 64, 65
  learning, 43, 44
  metacognition, 45
  researchers, 57
Human-in-the-loop, 233
Human oversight, 434, 594
Humans train technical algorithms, 395
Human-to-human relationship, 388
Human trust, 383, 387, 490

I, J

IBM Watson, 249–251
Image segmentation, 288, 296, 300, 301
Image validation, 233
Imbalanced datasets, 292
  data resampling, 292
  evaluation metrics, 293
  fraud detection case, 292

Imbalanced datasets (*cont.*)
    product sales data, 292
    SMOTE, 292
Inclusive design methodology, 414, 415
Inclusiveness
    benefits, 412, 413
    characteristics, 413
    design products, 411
    multilingual support, 412
    socio-economic status, 413
Industrial Revolution 4, 5
Industry-leading companies, 401
Industry leading products, 6
Ingestion layer, 173, 174
In-processing techniques
    adversary model, 404
    ensembling, 403
    group re-weighting, 403
    moderation, 403
Instagram, 20–22
Instruction tuning, 309
Integration testing, 352, 372, 373, 376
Intelligence, 42, 43, 68
Internal Rate of Return (IRR), 144
InterpretML, 409
Investment, 558
IRR, *see* Internal Rate of Return (IRR)

K

k-fold technique, 261
Kolmogorov-Smirnov Test, 342
Kullback-Leibler Divergence (KL Divergence), 307, 342

L

Large language models (LLMs), 13, 50, 257, 302, 304, 584
Latent Dirichlet Allocation (LDA), 343
LDA, *see* Latent Dirichlet Allocation (LDA)
Legal Compliance, 428
LightGBM model, 518, 520
Linear regression, 275, 276
LLMs, *see* Large language models (LLMs)
Logging
    for compliance, 349, 350
    contextual information, 350
    create log retention policy, 350
    create separate log categories, 350
    for debugging, 347
        capture performance metrics, 347
        create logs, 348, 349
        track model training process, 347
    role, 347
    use encryption and access control, 351
    use structured log formats, 350
    use tools, 350
Logistic regression, 276

Long short-term memory (LSTM), 47, 294, 302, 303
LSTM, *see* Long short-term memory (LSTM)

M

Machine learning (ML)
    algorithms, 47
    classical, 274
        supervised learning, 275–283
        unsupervised learning (*see* Unsupervised learning)
    deep learning (*see* Deep learning)
    deep learning *vs.* classical ML, 306
    GenAI (*see* Generative AI (GenAI))
    handle OOD (*see* Out-of-distribution (OOD))
    imbalanced datasets, 292, 293
Machine learning (ML) deployments, 316
    benchmarking strategy, 318
        A/B testing, 319
        canary deployment, 319
        shadow deployment, 318
    computational complexity, 317
    create model rollout, 318
    data storage and access, 316
    feature access, 316
    handle traffic and concurrency, 319
    monitor in production, 319
    outcome definition, 317
    outcome delivery method, 317, 318
Machine learning models, 16, 233, 409
Machine translation, 15, 305
MAPE, *see* Mean absolute percentage error (MAPE)
Marketplace model, 11
MDM, *see* Master Data Management (MDM)
Mean absolute percentage error (MAPE), 281, 520
Mean squared error (MSE), 281
Mediapipe, 507, 511
Meta
    AI-powered tools, 20
    algorithms, 21
    applications, 20
    edges, 21
    revenue, 19
    user safety, 22
Metacognition, 41, 42, 57–59
Metadata verification, 233
Metric that Matters (MTM), 614, 615
MHSA, *see* Multihead self-attention (MHSA) technique
Milvus vector database, 531, 532
Mitigation techniques, 401, 415
ML/AI project lifecycle, 272

INDEX

ML-based image segmentation models, 300
MLOps, 312
    challenges, 313
    cross-functional support, 313
    customer experience, 313
    deployments and A/B testing, 313
    model drift, 314
    post-deployment performance, 312
    provide stability, 314
    reduce technical debt, 313
    role, 312
MLOps framework
    DeMoGT, 314, 315
    governance (*see* MLOps governance)
    ML deployments (*see* Machin learning (ML) deployments)
    model deployment infrastructure (*see* Continuous integration and continuous deployment (CI/CD) pipelines)
    model monitoring and maintenance (Mo) (*see* Model monitoring)
    model serving infrastructure strategies (*see* Model serving)
    model testing (*see* A/B testing)
MLOps governance, 354
    security and compliance, 355
    version control, 354, 355
ML-powered search algorithm, 334, 336
Model demand predictions
    alerts and logging, 522
    automated retraining, 523
    business Impact, 526
    data collection and preprocessing, 519
    deployment strategy, 521
    drift detection, 522
    feature engineering, 519
    hyperparameter tuning, 520
    LightGBM model, 518
    metrics
        business, 524
        challenges, 524
        model performance, 524, 525
    ML problem, 519
    model evaluation, 520
    model monitoring, 522
    model performance, 520
    model selection, 520
    model serving, 521
    performance metrics, 522
    underestimation *vs.* overestimation rate, 519
Model drift, 314, 334, 344, 345
Model maintenance, 335, 336
Model monitoring, 334
    benefits, 335

INDEX

challenges, 336, 337
data shift, 335
as early warning
  system, 334
logging (*see* Logging)
and maintenance, 335
metrics, 337
  business metric, 337
  evaluation, 330
  input monitoring, 339–341
  internal factors, 339
  model metric, 337, 338
model retirement and
  replacement, 351–354
model shift/concept shift, 335
Model overfitting, 254
Model retirement/
  replacement, 351
in MLOps, 351
strategies, 351
  graceful retirement, 352, 353
  run old and new models
  simultaneously, 353, 354
Model serving, 320
model deployment, 320
in product management,
  320, 321
strategies, 321
  client-side (*see* Client-side
  model serving technique)
  FL (*see* Federated
  learning (FL))
  server-side
  deployment, 321–324

Modern data architectures, 191,
  192, 197
Modern data systems, 193
Moral and social obligation, 487
MSE, *see* Mean squared
  error (MSE)
MTM, *see* Metric that
  Matters (MTM)
Multi-armed bandit testing, 366
Multihead self-attention (MHSA)
  technique, 297

N

Names Entity Recognition
  (NER), 305
Natural language processing
  (NLP), 230, 301
  artificial intelligence, 301
  ChatGPT, 302
  checks, 233
  FastText, 302
  limitations, 302
  LSTM, 302, 303
  one-hot encoding, 302
  RNN, 302, 303
  TF-IDF vectors, 302
  transformers, 303, 304
  word2vec, 302
NER, *see* Names Entity
  Recognition (NER)
Net Present Value (NPV), 143–145
NLP, *see* Natural language
  processing (NLP)

INDEX

North star metric (NSM), 77, 614, 615
NVIDIA Jetson, 328

O

Observational learning, 39
OCED, *see* Organization for Economic Co-operation and Development (OCED)
Offline metrics, 262, 272
Online metrics, 262
ONNX Runtime, 328
OOD, *see* Out-of-distribution (OOD)
Opaque decision-making, 499
Operant learning, 40
Optimism bias model, 125
Orchestration tools, 365, 378
Organization for Economic Co-operation and Development (OCED), 428
Organizations, 411
Organization's operating model, 157
Out-of-distribution (OOD), 289
    adaptation, 291
    data drift, 289
    detect OOD, 290
        autoencoders, 291
        distance-based methods, 290
        ensemble, deep learning models, 291
        GANs, 291
        GMM model, 290
        single-class SVM, 291
        training data collection, 289, 290
Over-personalization, 476, 480, 482

P, Q

Pattern matching, 233
PCA, *see* Principal Component Analysis (PCA)
Performance metrics, 416, 422
Performance monitoring, 339, 415
Personal data, 421, 619
Personal data protection, 421, 618
Personalization, 466, 476, 482, 484, 485, 501, 502
    AI-powered, 467
    balanced, 482
    balancing, 485
    behavioral, 466
    business goal, 469
    content, 468
    contextual, 466
    demographics, 467
    experience, 477
    feature, 475
    intent based, 466
    preference-based, 467
    privacy and, 482
    responsibilities, 468
    service, 471
    type, 468, 473

Planning fallacy model, 125
Population Stability Index
 (PSI), 341
Post-processing techniques
 binary classification
  models, 404
 threshold values, 404
Precision, 279, 280, 282
Predictive AI, 542
Pre-processing techniques
 dataset, 401
 de-biasing data, 402
Principal Component Analysis
 (PCA), 260, 347, 402
Problem definition stage, 420
Product directors, 30, 31
Production data
 management, 422
Product leaders, 3–5, 7, 14, 25, 26,
 37, 50, 70, 187, 189, 190,
 220, 221, 229, 436
 cross-functional/AI ready
  teams, 79
 investments, 73
 organization, 76
 tasks, 76, 77
 technical fluency, 78
Product leadership roles
 CPO, 31, 32
 PM, 30
 product directors, 30, 31
Product management, 5, 28, 29, 74
Product management best
 practices, 84, 85

Product managers (PM), 5, 6,
 28–30, 35, 38, 217, 219, 222,
 232, 239, 240, 461, 483, 484,
 497, 499, 500
 accessibility, 488
 audit AI-powered features, 488
 and designers, 495
 designing, 488
 design tags and indicators, 489
 personalization, 488
 responsibility, 471
Product metrics, 611, 615
Product recommendation, 351,
 466, 474
Progressive web app (PWA), 326
Prompt engineering, 513, 554
PROMT framework, 210, 244
 data quality metrics, 212, 213
 DQI, 222, 223
 embed data quality, lifecycle,
  216, 217
 foundational tenets, 210, 211
 metrics, 222
 multitiered performance
  indicator layers, 213, 214
 accountability and
  action, 216
 decision-making
  timelines, 216
 executive/strategic
  KPIs, 214–216
 operational metrics, 214
 product/model readiness,
  214, 215

INDEX

PROMT framework (*cont.*)
    ownership, 220
    PM, 219, 222
    policies
        advantages, 218
        components, 218, 219
        elements, 220
    product leader, 221
    SLA/escalation paths, 220, 221
    steps, tenets, 223, 224
    tools, 217
PROMT framework, 210–211, 224, 226, 244, 622
Prototype, 311–379, 501
PSI, *see* Population Stability Index (PSI)
PWA, *see* Progressive web app (PWA)
Python library, 409

R

RADAR, 12
Real-time monitoring, 336, 607
Recall, 279, 280, 282
Recall-Oriented Understudy for Gisting Evaluation (ROUGE), 305
Recurrent neural networks (RNNs), 54, 294, 302, 303
Regular Audit Plan, 432
Reinforcement learning, 54, 59
Relabeling, 402
Representation Metrics, 416

Responsible AI
    behavior, 389
    ethical AI principles, 390
    ethics, 389
    performance and accuracy, 389
    societal values, 389
Return on investment (ROI), 138, 142, 549, 583, 585
Risk assessment, 446
    identification techniques, 446
    likelihood rating, 447
    matrix, 432, 448
    and mitigation, 446
    stakeholders, 449
RMSE, *see* Root mean squared error (RMSE)
RNNs, *see* Recurrent neural networks (RNNs)
Robust AI
    AI system, 452
    characteristics, 452
    configuration driven design, 453
    design principles, 453
    elastic scaling, 455
    human-in-the-loop approach, 454
    overfitting, 456
Robust data operating model, 451
Robust model training techniques, 451
ROI, *see* Return on investment (ROI)
Root mean squared error (RMSE), 281, 520

ROUGE, *see* Recall-Oriented Understudy for Gisting Evaluation (ROUGE)
Routing engine, 9
R squared (R^2), 282
Rule-based method based model, 16

S

Safety principles
    deployment, 425
    integration testing, 424
    performance testing, 424
    Problem definition stage, 420
    regression testing, 424
    robust logging mechanisms, 426
    solution designing and development phase, 421
    system testing, 424
Schema validation, 233
Selection bias
    coverage bias, 397
    nonresponse bias, 397
    sampling bias, 398
Self-attention, 304
Self-reasoning models, 565
Semi-structured data, 229, 257
Sensitive data, 609
Sensitive personal data, 619
Sequence tagging model, 15
Sequence-to-sequence rewriting model, 15
Server-side model serving technique, 322
    batch predictions, 323, 324
    benefits, 322
    streaming applications, 322, 323
Siamese deep learning model, 529
Silhouette coefficient, 287, 288
Silhouette metric, 288
Single-class SVM, 291
Skills and capability, 554, 555
SLA, 224
    components, 224, 225
    designing, 225, 226
SMOTE, *see* Synthetic Minority Over-Sampling Technique (SMOTE)
Social learning, 39
Solution designing and development phase, 421
Source systems, 191
Stakeholders, 72, 123, 140
Statistical checks, 233, 238
Structured data, 243, 257
Supervised learning, 54, 275
    algorithms, 275
        decision tree, 277
        linear regression, 275, 276
        logistic regression, 276
        SVM, 277
    evaluation metrics, 278, 282
        accuracy, 278, 279
        confusion matrix, 280
        F1 score, 280
        FN rate, 282

INDEX

Supervised learning (*cont.*)
    precision, 279
    recall, 279, 280
  regression metrics
    MAPE, 281
    mean squared error (MSE), 281
    RMSE, 281
    R-squared, 282
Support Vector Machines (SVM), 277, 278, 300
SVM, *see* Support Vector Machines (SVM)
Synthetic data, 242–245, 257, 258
Synthetic Minority Over-Sampling Technique (SMOTE), 292
System architecture, 236, 445, 453, 454

T

Technology-centric thinking model, 127
Technology maturity, 93, 95, 97
Technology solutions, 156, 186
Temporal bias, 399
Temporary data storage, 256
Tenserflow.js, 325
TensorFlow framework, 409
TensorFlow Lite, 328
Text summarization, 304, 305, 309
The positive rate (TPR), 406
Top-down strategy, 555
TPR, *see* The positive rate (TPR)

Traceability, 609
Traditional software systems, 440, 441
Training data collection, 289, 290
Transfer of learning, 41
Transformer-based models, 298, 302
Transformers, 54, 303, 304
Transformers-based GenAI, 309, 310
Transparency, 490, 491
  in AI, 491
  designer, 492
  high-stakes decisions, 493
  product managers, 492
Trust
  ability, 384
  of AI, 383
  benevolence, 385
  digital solutions, 382
  elements, 385
  financial angle, 383
  formed and maintained, 386
  human, 383, 387
  human-AI interaction, 387
  human relationships, 381, 384
  integrity, 385
  in machines, 382, 387
  ML model, 433
  navigation app, 387
  in organizational setting, 384
  parents and children, 382
  reliability, 383
  sensitive data, 433

INDEX

SHAP and LIME, 434
survival, 382, 384
technological progress, 382
transparency and
  explainability, 433
trustee, 386
users and stakeholders, 388
Trustworthy AI, 623, 626, 627

U

Uber, 6, 7
  customer support, 12
  customers use cases, 7, 8
  ETA prediction, 9
  fraud detection, 11
  Marketplace
    forecasting, 10, 11
  on-demand transportation/ride
    services, 7
  primary tasks, 8
Unstructured data, 228–231, 257,
  293, 306
Unsupervised learning, 54, 283
  algorithms, 284
    fuzzy clustering, 285
    hierarchical clustering,
      285, 286
    K-means clustering, 284, 285
  common use cases, 283
  customer segmentation, 283
  evaluation metrics, 287
    silhouette coefficient,
      287, 288

Use cases, 121
  bad use cases, 121
    accessible data, 124
    business impact, 123
    high risk/low reward,
      122, 123
    infrequent/one-off
      decisions, 122
    lack of quality, 124
    problems, 122
    stakeholders, 123
  feasibility study
    framework, 130
    data feasibility, 130, 131
    kill switches/non-negotiable
      dimensions, 133
    multidimensional weighted
      scorecard, 134
    operational feasibility, 132
    output decision matrix, 134
    problem statement, 130
    regulatory/ethical feasibility,
      132, 133
    technical feasibility, 131
  identification methods, 152
  mental models, 124
    analysis paralysis/
      perfectionism, 126
    availability bias/over-
      focusing on feasibility, 126
    bandwagon effect/
      confirmation bias, 125
    optimism bias/planning
      fallacy, 125

663

INDEX

Use cases (*cont.*)
  technology-centric thinking/
    neglect of human
    factors, 127
  potential gains, 146, 147
  prioritization framework, 135, 152
  problem statements, 135–137
  selection, 138
  structured approach, 121
  techniques, 127
    business-led discovery,
      127, 128
    cross-functional
      workshops, 129
    data-driven discovery,
      128, 129
    design thinking, 129
    libraries & industry
      benchmarks, 129

User feedback loops, 227

V

VAEs, *see* Variational
  Autoencoders (VAEs)
Variational Autoencoders
  (VAEs), 307
Vector DB, 377
Vector embedding index, 532
Vector embeddings, 531, 532
Vision transformers, 289-297
Visual language models, 309

W, X, Y, Z

WCAG accessibility guidelines, 425
WebAssembly/WebGL, 325, 326
WhatsApp, 20

GPSR Compliance

The European Union's (EU) General Product Safety Regulation (GPSR) is a set of rules that requires consumer products to be safe and our obligations to ensure this.

If you have any concerns about our products, you can contact us on

ProductSafety@springernature.com

In case Publisher is established outside the EU, the EU authorized representative is:

Springer Nature Customer Service Center GmbH
Europaplatz 3
69115 Heidelberg, Germany

www.ingramcontent.com/pod-product-compliance
Lightning Source LLC
LaVergne TN
LVHW021954060526
838201LV00048B/1570